AF520513

SMALL RUMINANT

Emerging Diseases, Containment Under WTO Regime

SMALL RUMINANT

Emerging Diseases, Containment Under WTO Regime

DR. V.S. VIHAN
P.S. and Head, Division of Animal Health

DR. ASHOK KUMAR
S.S. Division of Animal Health

DR. V.K. GUPTA
S.S. Division of Animal Health

DR. N.P. SINGH
Director

Central Institute for Research on Goats
Makhdoom, Mathura (U.P.)

2007

403, Express Tower, Commercial Complex, Azadpur, Delhi-110033 (India)
Phone : 011-27672469, 27672852, Fax : 91-11-27672046
E-mail : info@satishserial.com, hkjain1975@yahoo.com
Website : www.satishserial.com

Published by :

SATISH SERIAL PUBLISHING HOUSE
403, Express Tower, Commercial Complex, Azadpur, Delhi-110033 (India)
Phone : 011-27672469, 27672852, Fax : 91-11-27672046
E-mail : info@satishserial.com, hkjain1975@yahoo.com
Website : www.satishserial.com

First Published 2007

ISBN 81-89304-36-4

Printed at
SALASAR IMAGING SYSTEMS, DELHI-35

PREFACE

Goats and sheep are the two important livestock species in the livestock production systems. They support a large population of the rural landless and marginal farmers and have an important role to play in any programme aimed to alleviate rural poverty and livelihood security particularly in arid, semiarid and hilly regions of the country. They contribute greatly to the agrarian economy in the arid, semiarid and mountainous regions. Their contribution to the economy constituted about 5.40% of the GNP to the Agriculture Sector. Sheep and goats contributed about Rs 10,087 crores in the form of meat, milk, fibre, skin, offals and manure etc. to the Indian economy. Despite the least attention by the planners, goat population in India durmg last two decades has increased at the fastest rate among all the major livestock species.

The incidence of diseases and parasitic infestations in goats is one of the major constraints in the development of this enterprise and is reported to cause huge loss to the poor goat keepers and consequently to the Indian economy. Diseases in goats result into very high mortality which ranges from 5 to 25 %. In addition, diseases also cause morbidity losses in terms of low productivity of animals. Much effort in this country has been made to diagnose diseases, develop vaccines and immunize them. In order to maximize the productivity of the two species, in addition to the components of breeding, nutrition, management, optimum health protection plays an important role. There are heavy losses of good germ plasm due to diseases. Many exotic and emerging diseases are now being introduced in India from other countries for which adequate information for their containment is not fully known. Mortality and morbidity due to diseases are the major constraints in sheep and goat diseases .Some of the emerging disease like PPR, Sheep and Goat Pox and Blue tongue are causing substantial losses. Immunization of important diseases and creation of disease free zones are needed urgent attention. Due to the prevalence of some diseases, the SPS regulation of many OECD countries denies entry of Indian livestock products. There has been a vital neglect for the control of sheep and goat diseases. Thrust for animal disease control in the country should be on the control of List A diseases proposed by the OIE which has serious implication in the international trade of livestock and livestock products which exists in the country. During the past few years, the world has witnessed a global emergence and re-emergence of several infectious sheep and goat diseases that have had, and are still having, significant impact on both animal and human health. Presently, the gross output form sheep and goat sector alone has been estimated to be more than Rs. 100 billion through meat, milk, wool, fiber and other products. It has been worked out that this return could be easily doubled by giving due care to the prevention and control of goat and sheep diseases These infections have not only caused much concern to governments, policy makers and scientists all over the world, but some of them have posed significant threats to the livelihoods of the farming community, especially in rural areas. Among the limitations to stopping the spread of these diseases, is the lack of appropriate reference laboratories in our country for the early detection, notification and control of disease outbreaks. The editors of the book **"Emerging diseases of Small Ruminants and their containment under WTO regime"** have made an attempt to organize and synthesize the information contributed by scientists and researchers working in disease control and food safety areas in the country. Subject matter of the book comprises of Bio- security and SPS measures , Import and export regulatory measures, International regulations for trade of goat meat and their products, bio-containment and quality standards for diagnostics and vaccines, National and international approaches on prevention and control of transboundary diseases, Harmonization of regulatory requrements for veterinary biologicals and diagnostics, advances in bacterial and viral vaccines, molecular approches for the diagnosis of infectius diseases, control of gastro-intestinal infection, coccidiosis and mycotic diseases, quality control of meat and their products in relation to consumer safety, nutritional diseases associated with macro and micro nutrients, genetics of disease resistance, infectious agents transmitted through semen and embryo , quality control of herbal drugs for organic husbandry and containments of important emerging and endemic diseases such as PPR, *Mycobacterium avium subsp. paratuberculosis*, Capri pox, Blue tongue etc. The editors extend special thanks to contributors for their contribution in form of different chapters.

The encouragement and blessings poured by honorable Director General,ICAR and Secretary DARE, Dr Mangala Rai Ji; Dr V.K. Teneja, Ex Deputy Director General (Animal Sciences) ; Dr N.P.Singh, Director CIRG,Makhdoom and Dr V.K. Singh, Director CSWRI, Avikanagar are gratefully recorded. It's pleasure in extending thanks to the Department of Science and Technology and especially Dr G.J. Samathanam and Mr. C.Rajadurai for their financial support to enable us to arrange this National Seminar to discuss different aspects of the subject.

It is pleasure in extending thanks to all our colleagues of different Divisions of the Institute for their help at different stages in preparing this book.

V S Vihan, Ashok Kumar
V K Gupta, N P Singh

LIST OF CONTRIBUTORS

Dr. Lal Krishna
Assistant Director General
(Animal Health)
Indian Council of
Agricultural Research
Krishi Bhawan, New Delhi

Dr. A B Pandey
Senior Scientist and Principal
Investigator
All India Network Programme on
Bluetongue Disease
Centre for Animal Disease
Research and Diagnosis
Indian Veterinary Research
Institute, Izatnagar-243122 (UP)

Dr. S K Ranjhan
Director,
Hind Agro Industries Limited
Phase 1, Okhla Industrial Area
New Delhi

Dr. Nem Singh
Joint Director (Research)
Indian Veterinary Research
Institute, Izatnagar-243122
(UP)

Dr. V P Singh
Principal Scientist
Indian Veterinary Research
Institute,
Izatnagar - 243 122 (UP)

Dr. S Nandi
Senior Scientist
Centre for Animal Disease
Research and Diagnosis
(CADRAD)
Indian Veterinary Research
Insitute, Izatnagar-243122 (UP)

Dr. R S Chauhan
Joint Director
Centre for Animal Disease
Research and Diagnosis
(CADRAD)
Indian Veterinary Research
Insitute, Izatnagar-243122 (UP)

Dr. N P Singh
Director
Central Institute for Research on
Goats, Makhdoom
Mathura (UP) - 281122

Dr. V S Vihan
Principal Scientist and Head
Animal Health Division
Central Institute for Research on
Goats, Makhdoom
Mathura (UP) - 281122

Dr. D K Sharma
Senior Scientist
Animal Health Division
Central Institute for Research on
Goats, Makhdoom
Mathura (UP) - 281122

Dr. V K Gupta
Senior Scientist
Animal Health Division
Central Institute for Research on Goats, Makhdoom
Mathura (UP) – 281122

Dr. J M Kataria
Principal Scientist & Head
Division of Biological Procusts
Indian Veterinary Research Insitute, Izatnagar-243122 (UP)

Dr. S K Gupta
Division of Biological Products
Indian Veterinary Research Insitute, Izatnagar-243122 (UP)

Dr. Rahul S
Division of Biological Procusts
Indian Veterinary Research Insitute, Izatnagar-243122 (UP)

Dr. Ashok Kumar
Senior Scientist
Animal Health Division
Central Institute for Research on Goats, Makhdoom
Mathura (UP) – 281122

Dr. D Singh
Principal Scientist
Animal Health Division
Central Sheep and Wool Research Institute
Avikanagar – 304501 (Raj.)

Dr. C P Swarnkar
Senior Scientist
Animal Health Division
Central Sheep and Wool Research Institute
Avikanagar – 304501 (Raj.)

Dr. M P Yadav
Vice Chancellor,
SVBP University of Agriculture & Technology, Meerut-250110 (UP)

Dr. R K Singh
Head, Division of Virology
Indian Veterinary Research Institute, Mukteswar-263 138
Nainital, District, Uttaranchal

Dr. S C Dubey
Principal Scientist & Head
Animal Health Division
Central Sheep and Wool Research Institute Avikanagar
Via-Jaipur, Rajasthan-304501

Dr. D Mondal
Senior Scientist
Animal Health Division
Central Sheep and Wool Research Institute Avikanagar
Via-Jaipur, Rajasthan-304501

Dr. G K Sharma
Deputy General Manager
Productivity Systems – Animal Health, National Dairy Development Board
Anand- 388 001

Dr. S V Singh
Senior Scientist
Animal Health Division
Central Institute for Research on Goats, Makhdoom
Mathura-281122 (UP)

Dr. P K Rout
Animal Genetics & Breeding Division, Central Institute for Research on Goats, Makhdoom
Mathura -281122 (UP)

Dr. R Roy
Principal Scientist & Head
Animal Genetics & Breeding Division, Central Institute for Research on Goats, Makhdoom
Mathura-281122 (UP)

Dr. G Butchaiah
Dean, Rajiv Gandhi College of Veterinary and Animal Sciences, Kurumbapet, Pondicherry-605009

Dr. P X Antony
Rajiv Gandhi College of Veterinary and Animal Sciences, Kurumbapet, Puducherry–605009

Dr. K P Agrawal
National Coordinator
National Agricultural Innovation Project, Indian Council of Agricultural Research KAB – II, Pusa Campus, New Delhi

Dr. K Prabhudas
Project Director
Project Directorate on Animal disease Monitoring & Surveillance Hebbal, Bangalore- 560 024

Dr. Mahendra Pal
Ex-Professor and Head
Department of Veterinary Public Health, College of Veterinary Science, Anand Agricultural University, Anand-388001 (Guj.)

Dr. Rajneesh Rana
Senior Scientist, Animal Health Division, Central Institute for Research, on Goats, Makhdoom Mathura (UP) – 281122

Dr. S N Singh
Managing Director
Biovet Private Limited
560, C Block, I main. AECS Layout, Kundanahalli
Banglore-560037

Dr. Nagendra Sharma
Vice-Chancellor
Sher-e-Kashmir University of Agricultural Sciences & Technology Jammu

Dr. V Pathak
Division of Livestock Products Technology, Sher-e-Kashmir University of Agricultural Sciences & Technology Jammu

Dr. A K Biswas
Faculty of Veterinary Sciences & Animal Husbandry, SKUAST-J
Sher-e-Kashmir University of Agricultural Sciences & Technology, Jammu

Dr. D D Kulkarni
Principal Scientist
High Security Animal Disease Laboratory, Indian Veterinary Research Institute
Izatnagar -243122

Dr. Mohinder S Oberoi
Regional Coordinator
(South Asia)
Food and Agriculture Organization of the United Nations, New Delhi

Dr. S V S Malik
Senior Scientist
Division of Veterinary Public Health, Indian Veterinary Research Institute
Izatnagar-243122

Dr. Samir Das
Division of Veterinary Public Health, Indian Veterinary Research Institute
Izatnagar -243122

Dr. V M Vaidya
Division of Veterinary Public Health
Indian Veterinary Research Institute, Izatnagar- 243122

CONTENTS

1

Import and Export Regulatory Measures in Context of Small Ruminant Diseases

MOHINDER S. OBEROI

Food and Agriculture Organization of the United Nations
New Delhi

Sheep and goats are an essential part of small ruminant farming systems and are raised for meat, milk, fibre, pelts and skins. The relative importance of each of these products varies from region to region and is largely determined by ecological and economic factors. The small ruminant wealth is in the hands of poor people and its farming carried out as a way of life instead on commercial scale. Highlighting the need for conservation of indigenous breeds of sheep and goat, several States in India have undertaken projects for their identification and conservation, which will further aid in long term improvement of small ruminants.

Development of small ruminants is very important from the point of view of small and marginal farmers and especially in the hilly and backward areas of the country. As per the 2003 Livestock Census there are about 561.47 million sheep and 124.36 million goats in the country producing 234,000 and 473,000 tones meat respectively. Estimated 485 lakh kg wool is also produced. About five million households in the country are engaged in the rearing of small ruminants and other allied activities[1]. The small ruminants in the country have low productivity due to poor exploitation of genetic potential of indigenous animals, low absorption of available technology, inadequate resource of feed and fodder, insufficient health cover, etc.

The Government of India has put sheep and goats for breeding under restricted list of imports so that the applications could be examined prior to the imports to find as to what type of germplasm are being imported. Moreover, animal disease status of the exporting country has to be ascertained before imports in order to prevent introduction of exotic diseases in to the country.

Today trade in livestock and livestock products makes up approximately one sixth, by value, of all agriculture trade worldwide[2]. The steady growth in production, productivity and trade in livestock and livestock products is demand driven in response to growing human populations, economies and consumer preferences for meat and milk products. These growth trends are associated predominantly with developing countries. There is plenty of opportunity for increased production and consumption of food of animal origin in developing countries[2].

Increasing and persistant outbreaks of animal diseases are posing considerable challenges to livestock producers, industries, and policy-makers around the globe in a context of steadily rising demand for locally produced and imported livestock products. The Sanitary and Phytosanitary (SPS) issues would increasingly guide all future world trade in livestock and livestock products. In view of the SPS requirements the regulatory mechanism for the import and export of livestock and livestock products in India has been given appropriate legal support by undertaking the amendment of Livestock Importation Act. However, the absence of a uniform legislation for the control of movement of livestock across the country is a handicap for effective control of diseases. The Infectious and Contagious Diseases Act, which is in final stages of approval, would further strengthen the regulatory mechanism and disease control efforts within country. The imports in the country are mainly for genetic improvement of the indigenous breeds and exports are in terms of meat and wool.

The exports of livestock and livestock products should comply with the requirements of the importing country. The importation requirements depend on both the product and the country/region of origin and are on science based risk assessment. If import risk analysis show that there is a risk of bringing in one or more specific diseases, which are not prevalent in the country and which could adversely affect the health and safety of the human and animal populations of this country the import is refused. Certain restrictions ranging from testing or processing to total import prohibition are placed on both livestock and livestock products if they originate in countries that have a different disease status from importing country. The import restrictions may also be placed on the basis of particular strain(s) of a causative agent being absent in the country. The imported livestock must be accompanied by a health certificate issued by an official veterinarian of the exporting country. Imports of livestock must enter the country through the government controlled quarantine and certification facilities at major sea and airports where necessary inspection and testing could also be done as a further measure before clearing the import consignment. If required in public interest, the quarantine officer can order the destruction of the consignment or its return to the country of origin. It is thus imperative for the countries to develop and further strengthen their capabilities for diagnosis of obscure and exotic diseases to safeguard the livestock wealth in the country and also comply with the SPS provisions under WTO agreement.

The *Office International des Epizooties* (OIE, World Organization for Animal Health) is the international agency laying down standards and regulations in the form of OIE Terrestrial Animal Health Code (updated frequently) for the member countries; developed through various expert committees. The Code provides guidelines to both importing and exporting countries to develop their livestock and livestock products trade policies. The importing countries should invariably conduct science based risk assessment, risk management and risk communication to protect their livestock and human population from any adverse effect of the import but at the same time should not be trade restrictive. Some of the main guidelines are placed in brief below. For more details appropriate documents should be studied in detail.

Import risk assessment

Any country intending to import livestock or livestock products should conduct import risk assessment as per the guidelines laid down in Chapter 1.3.2 of the OIE Terrestrial Animal Health Code (Code)-2006 to safeguard the interest of its livestock and human population.

To quote the Code "An import risk analysis begins with a description of the commodity proposed for import and the likely annual quantity of trade. It must be recognized that whilst an accurate estimate of the anticipated quantity of trade is desirable to incorporate into the risk estimate, it may not be readily available, particularly where such trade is new. Hazard identification is an essential step which must be conducted before the risk assessment."

"The hazard identification involves identifying the pathogenic agents which could potentially produce adverse consequences associated with the importation of a commodity.

The potential hazards identified would be those appropriate to the species being imported, or from which the commodity is derived, and which may be present in the exporting country. It is then necessary to identify whether each potential hazard is already present in the importing country, and whether it is a notifiable disease or is subject to control or eradication in that country and to ensure that import measures are not more trade restrictive than those applied within the country.

Hazard identification is a categorization step, identifying biological agents dichotomously as potential hazards or not. The risk assessment may be concluded if hazard identification fails to identify potential hazards associated with the importation."

The import risk assessment is basically risk assessment, risk management, risk communication and evaluation.

The principles of risk assessment are that it

1. Should be flexible to deal with the complexity of real life situations. No single method is applicable in all cases. Risk assessment must be able to accommodate the variety of animal commodities, the multiple hazards that may be identified with an importation and the specificity of each disease, detection and surveillance systems, exposure scenarios and types and amounts of data and information.

2. Both qualitative risk assessment and quantitative risk assessment methods are valid.

3. The risk assessment should be based on the best available information that is in accord with current scientific thinking. The assessment should be well-documented and supported with references to the scientific literature and other sources, including expert opinion.

As per the Code-2006 the **Principles of risk management** are mainly

1. Risk assessment is the process of deciding upon and implementing measures to achieve the Member Country's appropriate level of protection, whilst at the same time ensuring that negative effects on trade are minimized. The objective is to manage risk appropriately to ensure that a balance is achieved between a country's desire to minimize the likelihood or frequency of disease incursions and their consequences and its desire to import commodities and fulfill its obligations under international trade agreements.

2. The international standards of the OIE are the preferred choice of sanitary measures for risk management. The application of these sanitary measures should be in accordance with the intentions in the standards.

Further the Code-2006 lists the **Principles of risk communication** as,

1. Risk communication is the process by which information and opinions regarding hazards and risks are gathered from potentially affected and interested parties during a risk analysis, and by which the results of the risk assessment and proposed risk management measures are communicated to the decision-makers and interested parties in the importing and exporting countries. It is a multidimensional and iterative process and should ideally begin at the start of the risk analysis process and continue throughout.

2. A risk communication strategy should be put in place at the start of each risk analysis.

3. The communication of the risk should be an open, interactive, iterative and transparent exchange of information that may continue after the decision on importation.

4. The principal participants in risk communication include the authorities in the exporting country and other stakeholders such as domestic and foreign industry groups, domestic livestock producers and consumer groups.

Animal health measures applicable before and at departure (before export)

(Chapter 1.4.1. OIE Terrestrial Animal Health Code-2006)

1. Countries should only authorize the exportation from their territory of animals for breeding, rearing or slaughter which are correctly identified and which meet the requirements of the importing country.
2. Biological tests and/or vaccinations required by the importing country should be carried out in accordance with the recommendations in the Terrestrial Code and Terrestrial Manual, as well as disinfection and disinfestation procedures.
3. Observation of the animals before leaving the country may be carried out either in the establishment where they were reared, or in a quarantine station. When they have been found to be clinically healthy and free from diseases listed by the OIE by an Official Veterinarian during the period of observation, the animals should be transported to the place of shipment in specially constructed vehicles, previously cleansed and disinfected. This must be done without delay and without the animals coming into contact with other susceptible animals, unless these animals have animal health guarantees similar to those of the transported animals.
4. The transportation of the animals for breeding or rearing or animals for slaughter from the establishment of origin to the point of departure from the exporting country shall be carried out in conformity with the conditions agreed between the importing country and exporting country.

Animal health measures applicable on arrival

(Chapter 1.4.4. OIE Terrestrial Animal Health Code-2006)

1. An importing country should only accept into its territory animals which have been subjected to a health examination by an Official Veterinarian of the exporting country and which are accompanied by an international veterinary certificate provided by the Veterinary Authority of the exporting country.
2. An importing country may require adequate advance notice regarding the proposed date of entry into its territory of animals,

stating the species, quantity, means of transport and the name of the border post to be used.

In addition, importing countries shall publish a list of the border posts equipped to conduct control operations related to importation and enabling the importation and transit procedures to be carried out in the quickest and most effective way.

3. An importing country may prohibit the introduction into its territory of animals if it considers that certain diseases exist in the exporting country, or transit countries which precede it in the itinerary, which are capable of being transmitted to its own animals. In the case of transit countries, the prohibition should not apply to bees which are transported in securely closed vehicles or containers.

4. An importing country may prohibit the introduction into its territory of animals if these are found, on examination at the border post by an Official Veterinarian, to be affected by, suspected of being affected by or infected with a disease capable of being transmitted to the animals in its territory.

   Animals which are not accompanied by an international veterinary certificate conforming with the requirements of the importing country may also be refused entry.

   In these circumstances, the Veterinary Administration of the exporting country shall be informed immediately, thereby providing an opportunity for confirming the findings or correcting the certificate.

   However, the importing country may prescribe that the importation be placed immediately in quarantine in order to carry out clinical observation and biological examinations with a view to establishing a diagnosis.

   If the diagnosis of an epizootic disease is confirmed, or if the certificate cannot be corrected, the importing country may take the following measures:

   a. return the animals to the exporting country, if this measure does not involve transit through a third country;

   b. slaughter and destroy in cases where return to the exporting country would be dangerous from the health point of view or impossible from a practical point of view.

5. Animals, accompanied by a valid international veterinary certificate and found to be healthy by the Veterinary Authority at the border post, shall be permitted to be imported and transported in accordance with the requirements of the importing country to the point of destination.

The regulations for individual diseases can be found in the OIE Terrestrial Animal Health Code-2006.

Several of the small ruminant infectious diseases are vaccine preventable. However good quality vaccines produced as per OIE guidelines[3] are always recommended. For example new indigenous vaccines for the control of *Peste des petit ruminants* (PPR) have greatly helped in controlling this scourge of sheep and goats in India. Similarly efforts are on foot to develop effective vaccine against Capri pox (sheep and goat pox) viruses.

The government has put all the OIE notifiable diseases of sheep and goats on import risk but specifically the diseases which have never been recorded in India. These diseases include Rift valley fever, Maedi-visna, Nairobi sheep disease, Border disease, ovine epididymitis (*Brucella ovis)* and salmonellosis (*Salmonella abortusovis)* among others.

References

1. DADF, 2006. Department of Animal Husbandry, Dairying and Fisheries. Website http://www.dahd.nic.in
2. Heath, S E. 2006. Challenges and options for animal and public health services in the next two decades. *Scientific and Technical Review, OIE,* 25(1): 403-419.
3. OIE Manual of Diagnostic Tests and Vaccines for Terrestrial Animals, 2005, http://www.oie.int
4. OIE Terrestrial Animal Health Code-2006, http://www.oie.int

–AAA–

2

Bio-safety, Bio-security, Bio-containment and Quality Standards for Diagnostics and Vaccines under WTO Regime

LAL KRISHNA

Indian Council of Agricultural Research
Krishi Bhawan, New Delhi

Livestock sector plays an important role in national economy and contributing about 27% of total agricultural sector. Livestock keeping is an age-old tradition for millions of rural households and is considered to be fast growing and directly linked with livelihoods of more than 73%of rural households. India has a livestock resource of considerable genetic diversities within species hence it is a challenging task to provide health care to about 16% of world cattle population, 56% of buffaloes, 24% of goats, 4%of sheep and 9% of camel besides more than 200 million poultry.

The global liberalization is likely to bring major changes in trade in livestock products. Such liberalization would provide more opportunities of export of commodities, material, know-how, equipments & machines. Although, the chances of export may increase but greater emphasis will come on quality production and certification. Considering the current changes going on in global livestock trade under the impending World Trade Organization (WTO) agreement, it is important to take a close look on bio-security and IPR issues.

Over the years a number of animal diseases have been introduced through import of livestock and livestock products and many of them have become established and continue to cause economic losses every year. There is always a risk of introducing new disease/strains of pathogen causing serious threat in terms of heavy economic losses. Such pathogens once introduced into a country can cause epidemic due to the absence of diagnostics, vaccine or effective drugs, lack of resistance in the host animals. Hence there is a need to take extra precaution in the import of livestock and their products infected with such pathogens. Trans-boundary animal diseases and emerging, re-emerging and exotic animal diseases are assuming a new dimension to the livestock productions systems in various countries. The key intrinsic factors for making these diseases important at a global scale are mainly the mode of survival and spread of the causal agents, variability and the availability of hosts/ alternate hosts and vectors (incase of viral diseases).

Infectious diseases do not respect national boundaries. Human epidemic of "plague" in 1994 in India, alarming panic created by SARS and bird flue epidemics in 2003-04 in neighboring countries and the recent outbreak of bird flu in India declaring a **Global Alert** for the first time in the history of WHO is unparallel.

A new or reappearing disease or syndrome brought to the attention of a relevant authority unaware of its existence or cause. Some consider that the term "newly recognized diseases" is a better then emerging disease. In India, during the last five decades, trans-boundary transmission of animal diseases has inflicted severe losses due to Bluetongue Peste des petits ruminants (PPR) , swine fever; Equine Influenza & African horse sickness , canine parvo-virus in livestock and Marek's disease, Gumboro disease, egg drop syndrome, chicken infectious anaemia and Lichi heart disease in poultry. Some of the other exotic diseases recognized in the country were, namely: Bovine viral diarrhoea in cattle/buffaloes, Equine coital exanthema in horses, Equine infectious anaemia in horses, Aujeszky's disease in swine and Non-pathogenic avian influenza (strain H9N2) in poultry.

Bio-security

Biosecurity is defined as "Protection from risks posed by organisms to the economy, environment and people's health through exclusion, eradication and control". From this perspective, bio-security is to protect the country's unique biodiversity and facilitate exports by managing risks to animal health and animal welfare. Therefore, it is important that an emphasis need be given to bio-security in order to preserve and enhance the value of our unique biodiversity, and the competitive trading advantages that accrue from our disease free status.

As a trading nation it is impossible to maintain an impervious border due to the economic compulsions of promoting the goals of safe and free trade. Absolute achievement of those goals for all exotic diseases requires mutually exclusive measures. For the most significant animal diseases such as Bovine Spongiform Encephlopathy (BSE), scrapie, etc, we have so far maintained a record of safe trade.

Trans-boundary Animal Diseases

There is also a serious threat in terms of losses due to biotic stresses, which appear in endemic or epidemic proportions but have the potential to be pandemic in nature. There is always risk of introduction of new diseases/pathogenic organisms into a country causing serious disease problems. Such pathogens once introduced in a country can cause epidemics due to the lack of resistance in the host animals and absence of diagnostics, vaccines or effective drugs. Hence there is a need to take extra precaution while importing plant/animal germplasm and their products infected with such pathogens.

In animals, the entry of diseases like Malignant Catarrh Fever (MCF), Rabbit Hemorrhagic Syndrome (RHS), Avian Influenza (AI) has been checked through diagnosis in the imported livestock and poultry at the entry point of quarantining, if not intercepted and controlled, these diseases would have played havoc with our livestock and poultry industry. Efforts need be made to have vigilance at the entry points and diagnostic facilities for exotic diseases strengthened on priority so as to check ingress of exotic diseases.

A globally significant role for the Indian livestock system depends critically on India's ability to eradicate animal diseases. Loss of production on account of diseases alone amounts to abput Rs. 50 billion per year. Moreover, the prevalence of animal diseases precludes India's entry in the lucrative markets for dairy, as well as, meat products. The control of animal diseases would reduce the yield gap i.e., the difference between the actual production and potential production in even the existing low producing livestock of all species, leading to at least 50 per cent more output. Investments for improved feeding, genetic and productivity improvements are all constrained severely by the negative impact of diseases on the animal system, rendering such investments largely in fructuous.

Early detection is the most potent weapon against emerging diseases and to assist bio-security. Biosecurity programmes should encompass an integrated range of activities that include:

- Risk analysis, assessment and bio-security risks.
- Restriction and control of the importation of all risk goods and security to ensure compliance with those import controls
- Post border surveillance to detect new incursions quickly and capability of eradication of new incursions
- Disease control programmes for established diseases
- Awareness and education.

There is a need for continuing rapid progress in science and maintaining maximal transparency in oversight processes while addressing national bio-security requirements and to take a series of actions that collectively enhance bio-security in research.

Resurgence/Emergence of New Diseases

There are new diseases emerging at unprecedented rate over the last decade without clear-out explanation of the reasons for their emergence or recurrence. These diseases now continue to spread over large areas e.g. Contagious agalactiae (Mycoplasmosis), Enzootic abortion (ovine chlamydiosis), Caprine arthritis / encephalitis (CAE), Equine infectious anaemia (EIA), Equine influenza, Equine virual rhino-pneumonitis, Porcine reproductive and respiratory syndrome (PRRS), Infectious bursal disease/Gumboro (IBD/GD), Marek's disease (MD), Avian Mycoplasmosis, Aujeszky's disease (pseudorabies), Bovine viral diarrhoea (BVD), Bovine immunodeficiency virus infection (BIV), Border disease Infectious hydropericardium (Hydropericardium syndrome), Bovine leucosis etc.

Infrastructural Development

A net work of 26,717 Polyclinics/Hospitals/Dispensaries and 28,195 Veterinary aid centers (including stockmen centers/mobile dispensaries), which are supported by about 250 disease diagnostic laboratories, are functioning in the states and Union Territories for quick and reliable diagnosis of diseases. Further for control of major livestock and poultry diseases by way of prophylactic vaccination the required quantity of vaccines are produced in the country at 26 veterinary vaccine production units. Of these, 19 are in the public sector and 7 are in private sector. Import of vaccines/diagnostics by private agencies is also permitted as and when required. Strengthening of these institutions are necessary to meet the international requirements.

Scenario of Veterinary Biologicals and Diagnostics.

"Prevention is better than cure". Hence the development of vaccines and diagnostics has been attracting the attention of the scientists for many years. As a result of concerted efforts in this direction, it has become possible to reduce the production losses substantially and bring down the incidence of several diseases to a minimum level. The uses of vaccines are particularly important in veterinary practice because treatment of individual animal is too expensive that a poor and marginal farmer is not able to afford. In India the production of vaccines and biologicals received attention since long but the development of veterinary biologicals including diagnostics have been rather slow. These production units need to follow the GMP & GLP guidelines in order to maintain the international requirement of Sanitary & Phyto-sanitary conditions and also to maintain a harmonization in the product fomulation and development.

Regional Diagnostic Laboratories

Till 1990, the concept of quality assurance process was not being adopted in India. Recently, the Dept. of Animal Husbandry and Dairying has designated one central and four regional disease diagnostic referral laboratories under the Animal Disease Management and Regulatory Medicine Scheme at Pune, Kolkata, Bangalore and Jalandhar. The Centre for Animal Disease Research & Diagnosis (CADRAD) of the Indian Veterinary Research Institute, Izatnagar has been identified as the referral apex laboratory. There are about 100-district level diagnostic laboratories in country. In addition a large number of State level diagnostic laboratories are also functional in private and public sector to meet out the requirement of the country but these units need to be updated and strengthened to meet the SPS requirements. For exotic animal diseases diagnosis ICAR has established High Security Animal Disease Laboratory (HSADL) at Bhopal with Biosafety Level - 4 (BL-4) containment facilities.

Animal Quarantine Network

In order to regulate the import and export of livestock and livestock products, control of exotic disease and to provide certification as per OIE regulations, Animal Quarantine and Certification Services (AQ&CS) at New Delhi, Calcutta, Chennai and Mumbai under the control of Department of Animal Husbandry & Dairying is existing and the department is contemplating to set up two more Quarantine Stations at Bangalore and Hyderabad, as these are now international airports at which requests for imports are frequently received. Further, there are no facilities are available at the seaports, which are very important imports points. The quarantine facilities at land route border also do not exist. These need to be put in place immediately. The quarantine stations are ill equipped to even undertake basic testing of materials and thus need to be strengthened to handle the quarantine and other sanitary control measures that would be required when the volume of trade goes up on account of removal of quantitative restrictions on imports The Animal Quarantine Certification and Enforcement Authority is being established with necessary autonomy, authority to function as an independent body empowered with legislative authority, so that the movement of animals risking infection/ disease could be adequately controlled through important airport, seaport, international land routes and movement within the states. Establishing Diseases Free Zones for selected diseases by zoning and buffer zoning keeping international view of OIE and WTO, with strict enforcement of phyto-sanitary and zoo sanitary requirements is yet another new initiative.

Accreditation of Laboratories

The well-equipped laboratories with trained manpower are the back bone of any inspection and certification system. The existing laboratories under the Govt., semi Govt. and non-governmental organizations are not so equipped to meet the challenges of today. Therefore, there is an urgent need for establishment of independent laborotries which can cater the diagnostic and analytical services in an efficient way. Once these laboratories will be established and accredited they can undertake quality control and diagnostic services with a more creditability at international level.

Quality Control/ Assurance

The production of veterinary biologicals has been brought under the purview of the Drugs and Cosmetics Act 1940. Drug Controller of

India, Ministry of Health and State Drug Controllers look after the production of veterinary biologicals in addition to their basic function of controlling the production and marketing of Drugs for Human use. At present the Standardization division of the Indian Veterinary Research Institute has been assigned the task of a central vaccine quality control laboratory. The basic function of the Indian Veterinary Research Institute, being research, its Standardization division primarily works on Biologicals Standardization research and additionally it has other multifarious functions. In order to obtain better monitoring of the quality, a new institute Chaudhary Charan Singh Institute of Veterinary Health at Baghpat, Uttar Pradesh is being established.

National Livestock Health Monitoring and Surveillance System/ Networking

Programme planning for execution and research in animal husbandry sector in India has always suffered for want of reliable statistics. In India, there is no organized, effective and efficient system of disease recording, reporting, storage, retrieval, transmission and purposeful use of livestock disease information for developing disease control strategies. The lack of reliable data in India is a fundamental problem to any disease control programme and is usually interpreted as lack of the disease. The task of gathering information, providing health care and promoting diseases control programmes is exceedingly complex. The existing reporting system is passive and do not report about the confirmation of the disease. . In the absence of reliable data on animal disease distribution and population, disease profile, production and marketing patterns, future projections often have an inbuilt draw back of leading strategic plan formulations. A major exercise like National Animal Production and Health Information System (NAPHIS) is required for generating a reliable data basis for profitably utilizing the satellite communication system. It would help in comprehensive planning encompassing all spheres of farm operations and research for multi dimensional growth in the next two decades. Efforts should be made to develop comprehensive data bases on animal and poultry diseases of national and regional importance.

In order to overcome these difficulties ICAR has established Project Directorate with AICRP on Animal Disease Monitoring and Surveillance (ADMAS) which has made a significant contributions by developing simulating animal disease forecasting models and a detailed epidemiology of some of the important conditions. A district and state wise distribution and prevalence of diseases on the basis of agro climatic and eco pathological parameters have been developed. The retrospective simulated disease forecasting study is now in the progress using the various national data basis. The forecasting models need further validation using more field-based data.

Nation-wide communication infrastructure development, strengthening of disease reporting system and diagnostic laboratories in collaboration with State A&H with DICs at district level; and in future informatics Centres at Block level are necessary. Proper collaboration and coordination with state AH department, and more so with their peripheral functionaries, would perhaps secure prompt information in terms of disease outbreaks etc., in livestock and poultry. Promulgation of an Animal Disease control Act, and strict enforcement of quarantine regulations, would also go a long way in localizing disease outbreaks and their speedy control. There is an urgent need to develop a network system of information through

Linkages with codex committees, OIE, FAO, WHO, WTO, ISO, coordination mechanism involving planning and network system and dissemination of information on Codex standards and codes of practices.

Rapid and Sensitive Disease Diagnosis and Control

With the concerted efforts during recent past many of the conventional diseases are on decline but new diseases are emerging due to intensive cross breeding programmes and import of antigens, biologicals, livestock and Livestock products, poultry and poultry products. Therefore, in order to sustain the productivity it is essential to develop research support for diagnostics technologies and immunobiologicals against various diseases of livestock and poultry using newer technological interventions..

Development of protocols and diagnostic kits for existing as well as emerging diseases through newer technologies, up gradation to cost-efficient industrial processes for their mass production are necessary.

Development of potent vaccines against most bacterial and viral diseases with testing for their efficacy and safety under field conditions are the pre-requisites of any diagnostics and vaccines., Application of biotechnological tools like recombinant DNA, hybridoma technology etc. for novel vaccine need to be adopted. The production units are not well equipped with adequate trained personnel and there is no policy for progressive modernization, with the result the technologies used and the skill available are not updated. Most of the Public Sector Units do not follow Good Manufacturing Practices ((GMP) and the private sectors often avoiding expensive quality control procedures which is mandatory as per regulation. Therefore, there is an urgent need for modernization of the existing production units with trained manpower to enable them to cope up the production of biological and diagnostics as per OIE norms.

Control of disease can be achieved only in a phased manner by the application of large-scale immunization of all the susceptible livestock using a potent vaccine. The crux of the problem is therefore, the judicious use of a quality assured vaccine. It has been the experience the work over, that the quality assurances to be adequate, has to be necessarily done at a central laboratory which has a national mandate functioning independently of control of the vaccine manufacturers. For this purpose following need to be adopted.

- A standard national protocol (SOP) for vaccine/diagnostic production has to be used by all the institutions in the country.
- Each institution should have their own built in quality control units for sterility, safety and in vitro evaluation of immunogenecity.
- A centralized facility should be identified for effecting quality assurance, which should function independently of the manufacturing units
- Production potentials of all the institutions should be fully utilized to meet the requirements of India.
- Uniform standard are to be followed for any vaccine/diagnostics.
- Research should be continued on the development of new generation vaccine/diagnostics using recombinant DNA technology with synthetic and sub-unit vaccines and built-in adjuvant

Quality and safety for livestock and livestock products are essential, hence the need for the use of GMP, HACCP approach and risk assessment management is essential in formulating the action points in terms of quality control.

Licensing of Veterinary Biologicals

The Drug Controller of India had been responsible for enforcement of such regulations in consultation with Animal Husbandry Commissioner, Deptt. of A.H. and Dairying, Govt. of India. Before the enforcement of liberalized economic policies of Govt. of India, only killed actual users allowed vaccines for import. The expert opinion of this Department are usually being sought by Drug Controller and on the basis of the recommendations of the technical advisory committees, the import used to be allowed on merit basis for the live vaccines, antigens, sera and diagnostic reagents.

The present export and import policy announced by Govt. of India recently have the philosophy to regulate necessary export and import through the tariff mechanism by abolishing the licensing restriction. The basic objective of policy is to encourage the attainment of high and internationally accepted standards of quality and thereby enhances the picture of India as a whole. The capital goods, raw materials, consumable, spares parts, accessories and other goods are allowed to be imported without any restriction except to the extent such imports are regulated by the negative list of imports. Therefore, in clear terms it can be explained that a free import of any veterinary biologicals except those listed in negative list under Section G can be imported.

Legislative Frameworks

As far as the legislative frameworks are concerned, the present Livestock Importation Act, which dates back to 1989 was very sketchy. The in-effectiveness of its implementation, is evident from the fact that a number of diseases have made ingress into the country over a period of time through importation. These are African Horse Sickness (in 1960), Swine Fever (1962), Avian Leucosis, various chronic respiratory diseases, Infectious Bronchitis, Infectious Laryngo tracheitis, Pullorum Disease and so on. In view of the serious consequences, which the ingress of disease through importation of livestock products could have on the human and animal health and safety in this country, especially after the removal of quantitative restrictions, the livestock Importation Act of 1898 was amended in 2001 to extend the provisions of livestock products. This has enabled the Government to regulate the imports of livestock products after conducting appropriate risk analyses studies with regard to the disease situation prevailing in the exporting country vis-à-vis that prevailing in India.

Risk Assessment and Risk Analysis

- The process of risk analysis can be made transparent so that the interested personnel/parties can be provided with the documented basis on which a risk is accepted or declined.

- Risk management is the process of identifying and implementing measures which can be applied to reduce risk to an acceptable level and documenting the decisions. While risk communication is the process by which the results risk assessment and risk management are being communicated to the decision makers. Therefore, this activity is an important to be undertaken while considering implications of GATT.

Bio-containment Facility

The disease causing agents of both humans and animals are broadly classified into risk groups I, II, III, IV and V. The risk group I and II are of low individual and community risk, and moderate individual risk and limited community risk respectively. These disease-causing agents (risk group I and II) can be handled in conventional laboratories existing all over the country. Risk group III organisms having high individual and low community risk should be handled in Biosafety level-III laboratories. Risk group IV organisms pose high individual risk and high community risk, while Risk group V organisms are exotic (non-native) to a country. These organisms (risk group IV and V) should be handled in maximum containment laboratories (Biosafety level-IV). In India the disease causing agents are handled in conventional non-containment laboratories irrespective of their groups and because of this we are not able to control the diseases effectively inspite of having huge manpower in veterinary and human health services. Although data are not available, many of the organisms handled might be escaping from the work places endangering the environment and surrounding human and animal population.. The recently established High Security Animal Disease Laboratory (HSADL), IVRI, Bhopal in 1998 has BL-4 facility to safely handle high-risk pathogens.

-∆∆∆-

3

Safety Assurance of Exported Indian Meat Meeting W.T.O. Guidelines

S. K. RANJHAN

Hind Agro Industries Limited, India
A-1, Phase-1, Okhla Industrial Area
New Delhi-110020

1. INTRODUCTION

India has the largest livestock population in the world comprising of 189 million cattle, 96 million buffaloes, 60 million sheep, 120 million goats, 185 million layers and 1,100 million broilers. The animal food production through these sources is about 92 million tonnes of milk (first position), 5.9 million tonnes meat (fifth position), 44 billion eggs (number 4) and 1.8 million tonnes of poultry meat (5th position).

2. MEAT INDUSTRY IN INDIA

Meat production is estimated at 5.9 million tonnes, standing fifth in rank in the world's meat production. Buffalo in India contributes about 30% of total meat production. The contribution by cattle, sheep, goats, pigs and poultry is 31%, 5%, 10%, 10% and 13%, respectively.

In spite of big potential because of large livestock population, the meat industry in India has not been given its due share mostly due to socio-religious concern. Although India has acquired number one status (projected production 92 million tonnes 2005–2006) in the world contributing 13% of world milk production, the meat production, which jibes well with dairying, is very low. During the last 25 years, it has only increased from 764,000 tonnes in 1970-71 to 5.9 million tonnes valued at Rs. 34,000 crores (US$ 7,000 million), (GOI - 2004). The compound average growth rate during the last two decades works out to 4.6% as against 10% during the last few years, it is now the fastest growing segment of livestock sector.

The trends in livestock population, slaughter rate (number slaughtered as percentage of population), carcass weight and meat production in India in 2003 are shown in Table 1. It is noticed that about 8% cattle, 10% buffaloes, 19% sheep, 38% goats and 89% pigs and 73% poultry are slaughtered each year. The share of bovine meat in the total meat production in India is about 61% as against small ruminants (15%), pigs (10%) and poultry (13%).

Table 1. Trends in Livestock Production and meat Production in India -2004.

Livestock Spices	Population in (Million)	Animal Slaughtered (Million)	Percent Slaughtered (%)	Carcass weight (kg)	Meat production (million Tonnes)	Share in total meat production (%)
Cattle	189.1	14.2	7.9	103	1.46	31.1
Buffalo	96.0	10.3	10.0	138	1.43	30.5
Sheep	40.1	19.2	47.9	12	0.23	4.9
Goats	124.0	47.0	37.9	10	0.47	10.0
Pigs	18.0	16.0	88.9	31	0.47	10.0
Poultry	820.0	604.0	73.6	0.8	0.63	13.4
Total					4.69	100

Source: Department of Animal Husbandry & Dairying, GOI, FAO-2004

The contribution of buffalo meat towards the total meat production in India is about 25% (Table 2) which accounts for 1.2% of the total meat production.

3. EXPORT OF MEAT

India's international trade in livestock and livestock products is mainly in live animals (17%), meat and meat products (82%), dairy products and eggs (1%). At the global level, India's export and import accounts for only 0.17% of each. Meat and meat products have dominated the exports from livestock.

The major export of meat is buffaloes, which is shown in Table 2. The export of sheep and goat meat is less and is shown in Table 3. It may be seen that export of buffalo meat has increased significantly in the last three years. The export of buffalo meat in 1997-98 was only 176,328 MT, which had increased to 243,356 MT in 2001-2002 accounting for an increase of 43 %. In 2002-2003, the export was almost 300,000 MT valued at 13054 million Indian rupees equivalent to about US $ 300 million. In 2004-2005, the export was valued at Rs 1700 crores (US $ 330 million). The export of small ruminant meat (sheep and goat) has started picking up during the last three years. The export of sheep and goat meat in 1997-1998 was 7546 MT, which came down to 3915 MT in 2001-2002, thus showing a fall of 35%. During 2002-03 the export was 4973 tonnes'valued at 399.5 million Indian rupees equivalent to 9 million US $. The fall in export of small ruminant meat has been due to the reduction in the import by Saudi Arabia. Now that the market has opened for Indian meat, the export of meat from bovine, sheep and goat will increase substantially. The top importing countries of buffalo meat are shown in Table 4.

Table 2. Buffalo Meat Exports from India

Year	Total Meat Produced	Buffalo Meat Produced	Total Meat Export	Export Percentage
2001-02	8425000	1421000	240989	17%
2002-03	9622000	1428000	295456	21%
2003-04	5898000	1443000	338940	20%
2004-05	5922000	1471000	302280	20%
2005-06	6212000	1582000	449013	29%

Source: DGFT (2006)

Table 3. Export of Sheep/Goat Meat from India (MT)

Country	2002-2003	2003-2004	2004-2005
Bahrain	78.8	127	120
Jordan	37.5	1124	39
Kuwait	13.8	126	126
Oman	672.4	1541	392
Qatar	464.2	936	1278
Saudi Arabia	27.7	1587	4002
UAE	3340.0	4351	2739
Others	-	-	-
TOTAL	4,973-00	16,030	8,767

Source:
?Average Price Rs. 89/kg
?Average Growth - 45%

Table 4. Top Importing Countries of Indian Buffalo Meat Quantities and Values

	2003-04		2004-05		2005-06	
	Qty. Tons	Rs. million	Qty. Tons	Rs. million	Qty. Tons	Rs. million
Malaysia	80,442	3,723	65,529	3,559	82,337	4,994
Philippines	49,202	2,085	35,625	1,789	48,918	2,693
Angola	30,494	1,156	24,686	1,197	39,570	2,185
Jordan	36,759	1,642	20,092	1,098	39,044	2,198
Saudi Arabia	997	89	23,941	1,336	36,757	2,281
Kuwait	11,099	619	14,227	821	34,108	2,162
U.A.E..	20,694	1,094	19,099	1,068	33,098	2,028
I ran	7,878	334	7,877	389	16,986	750
Georgia	21,726	853	9,349	489	13,948	722
Vietnam	577	40	2,216	130	11,917	732
Congo	4,993	188	6,609	296	10,994	570
Oman	8,810	455	7,534	461	10,819	689
Grand Total including other countries	**343,817**	**15,368**	**306,971**	**16,156**	**459,937**	**26,296**

Source: DGFT (2006)

4. MEAT QUALITY AND SAFETY MEASURES

Most of the export-oriented meat processing plants in India follow world class sanitary and phytosanitary (SPS) measures given in the OIE guidelines, a referral institution of WTO. These plants are certified for HACCP (Hazard Analysis Critical Control Points), ISO-9001 and SGS, meeting the OIE norms. These measures are for meat safety, which starts right at the primary production level either with the farmers raising between 5 - 20 animals or in the feedlot. The identification and traceability of the animals from production source to the abattoir is fully maintained. It is ensured that animals have been raised under disease free conditions given in List 'A' of OIE. Except for Foot and Mouth Disease (FMD) which is endemic in a few pockets in India and has an insignificant incidence (0.0007%). Other trade-related diseases like the Rinderpest, Contagious Bovine Pleuropneumonia (CBPP), BSE etc. have not been reported. India has now launched a massive FMD control programme comprising of 54 districts in three zones of the country (North, Central and South India) around the major export oriented slaughter houses where there is maximum concentration of the buffaloes. The vaccination is at present limited to large animals.

5. RISK ANALYSIS

Hazard Analysis Critical Control Point (HACCP) is a risk management tool which provides a scientific approach to the control of manufacturing and processing products rather by traditional inspection and quality control of the end product. The Codex Alimentarius Commission has promulgated the concept of HACCP by adopting Guidelines for application of the Hazard Analysis Critical Control Points (HACCP) system during its 20th Session. The Critical Control Points in processing safe and hygienic meat followed at Hind Agro Industries Limited (HAIL), which are by and large a model code, have been given in Annexure 1.

6. TRACEABILITY - BACKWARD INTEGRATION

In the identification of risk in food trade, traceability has become very important. Indian livestock production is by "masses rather than mass production" observed in some developed countries. Thus, the production of animal products and their safety problems are unique.

In Hind Agro Industries Limited (HAIL), this problem was overcome by initiating an Animal Rearing Project (ARP) with the farmer cooperator which was initiated in 1995. At present, there are 7,200 villages in 8 districts around the plant in District Aligarh (UP-India) where more than 750,000 animals are provided animal health and production services at the farmers' door with individual animal health card. Although the State of Uttar Pradesh where the plant is located is free from FMD since January 2001, the 100 km area around the plant where ARP is operating since 1995, is free from FMD for the last 7 years. It has its own commercial farm where 3000 to 5000 animals are reared at any one time. These animals are procured at the age of 9 to 12 months and are fed on high protein, high energy diet to produce lean quality meat. The animals have the individual identification before they are sent for slaughter.

7. VETERINARY SET UP IN INDIA

Veterinarians play an important role in monitoring sanitary and phytosanitary measures and provide support services to the safety of the product for export.

There are 7,872 Veterinary Hospitals/Polyclinics, 17,820 Veterinary dispensaries and 25,433 Veterinary aid centers/mobile dispensaries to look after animal health and assist in disease monitoring and surveillance. There are about 250 disease diagnostic laboratories for quick and reliable diagnosis of animal diseases. About 26 Veterinary biological centers are established to produce wide range of vaccines and diagnostics using modern technologies. There are 35 Veterinary Colleges and five Veterinary Universities in India to produce veterinary professionals.

8. DISEASE MANAGEMENT

Prevention, control and eradication of diseases are important in safe animal food production and trade.

9. STATUS OF O.I.E. LIST 'A' AND LIST 'B' DISEASES IN INDIA

Among the List 'A' diseases of cattle and buffaloes, except for Foot and Mouth Disease, India is free from all other diseases namely, Rinderpest, Vesicular Stomatitis, Contagious Bovine Pleuropneumonia, Lumpy Skin Disease and Rift Valley Fever.

In the case of FMD, effective control measures are in place actively supported by Central and State Governments. The entire State of Uttar Pradesh has not reported any FMD outbreak since January 2001 (OIE Reports 2002-2006). For control of FMD, 54 districts have been identified where 100% vaccination programme has been in operation since August 2003. The animals are vaccinated every six months under the programme.

In addition, for meat exports, the practices followed in India are as per OIE guidelines given in Chapter 2 on FMD of Animal Health Code for Terrestrial Animals which completely eliminates the FMD risk from Indian buffalo meat, and has been demonstrated in the past 40 years in the export. In a study sponsored by APEDA to assess the risk free status of Indian buffalo meat in respect of FMD, it was reportred that acidic pH below 6 in the meat significantly eliminates FMD Virus in the meat. Data of more than 3,000 samples of buffalo meat showed an average of pH less than 6, suggesting that Indian deboned and deglanded lean buffalo meat is risk free which was confirmed by absence of FMD virus using ELISA and cell culture isolation tests[2].

In the list 'B' diseases, India is free from BSE (Bovine spongiform Encephalapathy), Anjeszky's Disease, Heart Water, Trichanellosis and Enzootic Bovine Leubkosis.

Vaccination and other control measures, including ante mortem and post mortem inspection, are followed in modern meat processing plants to ensure safety of meat.

10. REGULATION OF MEAT SAFETY

Indian meat export is regulated as per Export (Quality Control and Inspection) Act, 1963 and the Export (Quality Control and Inspection) Rules, 1964. Under the Act, Export of Raw Meat (Chilled/ Frozen) Quality Control and Inspection Rules 1992 (Amended Rules 1993) and Processed Meat (Quality Control and Inspection) Rules, 1995 have been made for regulation of fresh, frozen and processed meats.

Registration of abattoirs and meat processing plants with APEDA, a Government of India Organization, is mandatory for export of meat from India. APEDA is assisted by Meat Plant Inspection Committee with members drawn from the Ministry of Agriculture, Ministry of Food Processing Industries, Export Inspection Council, Veterinary Experts from Central and State Veterinary Departments etc. The Committee inspects the infrastructural facilities, operation and sanitary aspects as per the meat export rules. A Plant Registration Certificate is issued by APEDA which is renewed every year. Thus, the quality aspects of meat for export are monitored and assured effectively at every step.

11. ANIMAL WELFARE AND MEAT QUALITY

Animal welfare issues are important in the production of quality meat and international trade. India follows the Act on the Prevention of Cruelty to Animals Act, 1960.

12. ORGANIC LIVESTOCK FARMING

Livestock sector by and large in India remained untouched with feeding of Meat-cum-Bone Meal (MBM) and growth hormones. Animals are raised on natural feeds and fodder. The pesticide and drug residues are monitored regularly.

13. FOOD SAFETY REGULTIONS IN INDIA

- The Food Safety and Standards Act , 2006.
- Milk and Milk Product Order, 1992,
- Meat and Meat Product Order, 1973
- Water (Prevention and Control of Pollution) Act 1974
- Air (Prevention and Control of Pollution) Act, 1981
- Prevention of Food Adultration Act, 1954
- The Prevention of Cruelty to Animal Act, 1960
- Export (Quality) Control and Inspection) Act, 1963
- Export (Quality Control and Inspection) Rules 1964
- Export of Raw Meat (Chilled/Frozen) Rules 1992
- Export of Processed Meat (Quality Control and Inspection) Rules 1995

14. WTO, OIE AND CODEX STANDARDS

Harmonization of food standards are the pre-requisite to the protection of human as well as animal health in the International Trade Agreement as the application of Sanitary and Phytosanitary measures (SPS Agreement) and Technical Barriers to Trade (TBT) Agreeement of WTO recognize international standards, guidelines and recommendations, including the Codex Alaimentarius, as refrerence points for facilitating international trade and resolving trade disputes in international law.

The Codex Alimentarius Commission (CAC) is responsible for making proposals to FAO and WHO on all matters pertaining to the implementation of the Joint FAO/WHO Food Standards Programme.

To adopt Codex standards, the member countries require adoption of adequate Food Laws as well as Technical and Administrative Infrastructure with the capacity to implement them and ensure compliance. WTO under the SPS agreement recognizes food safety, developed under the auspices of OIE.

India is scrupulously following these standards which have been provided in the above Acts. They have now been brought under a compact legislation, namely, Food Safety and Standards Bill 2005 which has been approved by the Parliament in 2006 and has received

Presidential assent, and has become an Act namely, Food Safety and Standards Act 2006. However, the Prime Minister is yet to decide whether it will be administered by the Ministry of Health or the Ministry of Food Processing Industries.

15. CODE OF HYGIENIC PRACTICES FOR CARABEEF IN INDIA

- Microbiological Criteria for Foods
- Control of the use of Veterinary Drugs
- Contamination of Food with Chemicals
- Good Animal Feeding Practices

16. CONCLUSIONS

Indian meat exports are safe and are of good quality. It addresses all the issues of safety, namely microbiological, chemical, pesticides and drugs residues, feeds and pathogens. The EOUs have HACCP, ISO:9001:2000 and SGS Certification which ensure good manufacturing practices and independent monitoring of the meat plants.

References

1. FAO of the United Nations 2004 Animal Production Year Book.
2. Kondaiah, N.K. 2005 Safe Animal Food Production. Asian Buffalo Magazine – Page 30.

-ΔΔΔ-

4

Harmonization of Regulatory Requirements for Veterinary Biologicals and Diagnostics

J. M. KATARIA, S. K. GUPTA AND RAHUL S.

Division of Biological Products, Indian Veterinary Research Institute, Izatnagar, Bareilly - 243 122, Uttar Pradesh

A reliable supply of pure, safe, potent, and effective vaccines is essential for maintenance of animal health and the successful operation of animal health programmes. Immunization of animals with high quality vaccines is the primary means of control for many animal diseases. In other cases, vaccines are used in conjunction with national disease control or eradication programmes. The approach to ensuring the purity, safety, potency, and efficacy of veterinary vaccines may vary from country to country depending on local needs. However, proper standards and production controls are essential to ensure the availability of consistent, high quality products for use in animal health programmes.

As the pathogenesis and epidemiology of each disease varies, the role and efficacy of vaccination as a means of control also varies from one disease to another. Some vaccines may be highly efficacious, inducing an immunity that not only prevents clinical signs of the disease, but also may prevent infection and reduce multiplication and shedding of the disease-causing agent. Other vaccines may prevent clinical disease, but not prevent infection and/or the development of the carrier state. In other cases, immunization may be completely ineffective or only able to reduce the severity of the disease. Thus the decision whether to recommend vaccination as part of an animal disease control strategy requires a thorough knowledge of the characteristics of the disease agent and its epidemiology, as well as the characteristics and capabilities of the various available vaccines. There is also growing public interest in the beneficial implications for animal welfare of the use of veterinary vaccines as a means of disease control. In any case, if vaccines are used, successful performance requires that they be produced in a manner that ensures a uniform and consistent product of high quality.

Nomenclature

The nomenclature for veterinary biological products varies from country to country. For example, in the United States of America (USA) the term 'vaccine' is used for products containing live or inactivated viruses or protozoa, live bacteria, or nucleic acids. Products containing killed bacteria and other microorganisms are identified as bacterins, bacterial extracts, conventional or recombinant subunits, bacterintoxoids, or toxoids, depending on the type of antigen they contain.

Vaccine Types or Forms

Vaccines may be prepared as live or inactivated (killed) products. Some live vaccines are prepared from low virulence, mild, field isolates of a disease-causing agent that have been found to be safe and effective when administered by an unnatural route or under other conditions where exposure to the microorganism will immunize rather than cause disease. Other live vaccines are prepared from isolates of disease-causing agents that have been modified by passage through laboratory animals, culture media, cell cultures, or avian embryos to select a variant of reduced virulence. The development of recombinant DNA (rDNA) procedures has provided some unique opportunities for vaccine production. Modified live vaccines may now be specifically produced by deletion of virulence-related genes from a microorganism. Others are produced by the insertion of genes that code for specific immunizing antigens from a disease-causing microorganism into a non-virulent vector microorganism. Nucleic-acid-mediated vaccines containing plasmid DNA are being developed. The DNA is usually in plasmid form and codes for immunizing antigens from disease-causing microorganisms.

Killed products may contain:

1) Cultures of microorganisms that have been inactivated by chemical or other means;
2) Inactivated toxins; or
3) Subunits (antigenic parts of microorganisms) that have been extracted from cultures or that have been produced through rDNA procedures.

Both live and inactivated vaccines may be formulated with adjuvants designed to enhance their efficacy.

Quality Assurance

The consistent production of pure, safe, potent, and efficacious vaccines requires quality assurance procedures to ensure the uniformity and consistency of the production process. As production processes for vaccines provide a great opportunity for variability, care must be taken to control variability to the greatest extent possible, preferably using validated procedures, and to protect the product from contamination through all stages of production. Vaccine purity, safety, potency, and efficacy must be ensured by consistency in the production process. Consistent product quality (batch-to-batch uniformity) must be built in at each stage. Final product testing is used as a check to verify that the controls on the production procedures have remained intact and that the released product meets the specification previously agreed with the licensing authority.

Regulatory authorities in different countries have developed various approaches to ensuring the quality of vaccines. Although alike in their ultimate goal, these systems may vary in the emphasis given to control of the production process (process standards) in comparison with control through testing of the final product (performance standards). The control procedures selected should be those that best fit the conditions under which vaccines are being produced and, where possible, comply with good manufacturing practice. The control standards and procedures established for a product define the risk or possibility of producing and releasing a product that is worthless, contaminated, dangerous, or harmful. The acceptable degree of risk may depend on the benefits to be gained by having the product available to prevent disease losses. Thus standards may justifiably vary from country to country or product to product, depending on local animal health conditions. However, control authorities should strive to establish control standards and procedures that ensure a finished product of the highest purity, safety, potency and efficacy possible.

The optimal quality assurance system should address both production procedures and final product testing in proper balance. An absolutely fail-safe system that would result in no risk of releasing an unsatisfactory product would probably be too expensive with regard to cost of production as well as control. Thus regulatory officials and manufacturers of vaccines must select control procedures that are capable of ensuring an acceptable low level of risk in relation to hazard. Such procedures, however, must not be burdensome to the extent that they inhibit the development and availability of the products needed to provide proper preventive medical care at a cost that is acceptable to the consumer.

Production Facilities

Facilities used for the production of vaccines should be designed to protect the purity of the product throughout the production process and to safeguard the health of the personnel. They must be constructed so that: 1) they can be readily and thoroughly cleaned; 2) they provide adequate separation of preparation rooms 3) they have adequate ventilation 4) they have ample clean hot and cold water and efficient drainage and plumbing; and 5) they have dressing rooms and other facilities for personnel that are accessible without passing through biological product preparation areas. Facilities must be adequate to provide for all applicable production functions, such as: storage of master seeds, ingredients, and other production materials; preparation of growth media and cell cultures; preparation of glassware and production equipment; inoculation, incubation, and harvest of cultures; storage of in-process materials; inactivation, centrifugation, addition of adjuvant, and formulation of product; filling, desiccation, sealing of containers, labeling and storage of final product; quality control testing of in-process materials and final product; and research and development.

Separate areas are generally required for different activities. All rooms and air-handling systems must be constructed so as to prevent cross-contamination from other products and to prevent contamination by people or equipment. Virulent or dangerous microorganisms must be prepared and stored in rooms separate from the rest of the establishment. In particular, challenge organisms must be completely separated from vaccine strains. All equipment that comes into contact with product must be sterilized using validated procedures.

Production facilities have to be designed in such a way that contamination of the external environment is prevented. Any material used during production has to be made safe before leaving the facility. If highly contagious microorganisms are propagated, the exhaust air must be treated to prevent escape of infectious agents. Personnel must follow safety procedures such as showering, and avoid contact with susceptible animals after leaving the production facilities.

Although the quality and design of production facilities may vary significantly, they must always meet standards considered to be appropriate for the vaccines that are to be produced. For example, the requirements for facilities for the production of chicken embryo vaccines administered by oral, intranasal or intraocular routes in chickens may not need to be quite as demanding as those for the production of cell culture vaccines administered subcutaneously or intramuscularly.

Documentation of The Manufacturing Process

A detailed Outline of Production, a series of SOPs, or other documents should also be prepared to describe the protocol for the manufacture and testing of each product produced in an establishment. Criteria and standards for source materials should be clearly and accurately documented. Documentation should also address such things as: the source, isolation, and passage (sub culturing) history of each strain of microorganism; the source and sequence of nucleic acid elements, or peptides included in products derived from biotechnology, including plasmids or other vectors used in the construction of genetically modified microorganisms for use as master seeds; methods for identifying the microorganisms and determining their virulence and purity; the medium or cell culture system used for seed and production cultures, including the methods used to demonstrate that media are free from contamination; the source of ingredients of animal origin; methods of media sterilization; storage conditions of cell lines and seed cultures; size and types of containers used for growth of cultures; methods for preparing seed cultures and inoculating production cultures; time and conditions for incubation; observations during growth; criteria and specifications for satisfactory harvest material; and harvest techniques. There should be documentation on measures implemented by the firm to minimize the risk of transmissible spongiform encephalopathies (TSE) agent contamination in ingredients of animal origin and procedures to ensure that fetal bovine serum is free of pestiviruses. It should also include: a description of all tests conducted to assess the purity and quality of the product as it proceeds through the production process; each step in the formulation of the final product; the tests used for assessing the purity, safety, potency, and other requirements of each batch/serial of completed product; the specifications for finishing, including packaging and labeling with complete indications and recommendations for use; and the expiry date established for the product.

Competent control authorities publish guidelines for the preparation of such documents for veterinary vaccines. This documentation is intended to define the product and to establish its specifications and standards. It should serve along with the blueprints and blueprint legends (or production plan and SOPs) as a uniform and consistent method of producing the product that should be followed in the preparation of each batch/serial.

Record Keeping

The producer should establish a detailed record-keeping system capable of tracking the performance of successive steps in the preparation of each biological product. Records kept should indicate the date that each essential step was taken, the name of the person who carried out the task, the identity and quantity of ingredients added or removed at each step, and any loss or gain in quantity in the course of the preparation. Records should be maintained of all tests conducted on each batch/serial. All records relevant to a batch/serial of product should be retained for at least 2 years after the expiry date on the label, or in line with the requirements of the competent control authority. In addition, a record should be maintained of all labels used on all products, with each label identified as to its name, product number, product license number, package size, and label identification number. All labels printed should be accounted for. Records must be kept concerning sterilization and pasteurization procedures. These are usually made by means of automatic recording devices. The manufacturer must also keep complete records for all animals at the establishment, including health prior to being used for any tests, results of tests performed, treatment administered, maintenance, necropsy, and disposal.

Master Seed

The objective of testing the master seed is to ensure vaccine safety, quality and efficacy. Safety should be tested in an early stage. A master seed should be established for each microorganism used in the production of a product to serve as the source of seed for inoculation of all production cultures. Working seeds and production seeds may be prepared from the master seed by sub culturing; generally the final production cultures should not be more than five (sometimes ten) passages from the master seed. The number of passages should be determined by data and designated in each case. Using a master seed and limiting the number of passages of seed microorganism in this manner assists in maintaining uniformity and consistency in production. Records of the source of the master seed should be maintained. For genetically modified microorganisms, the source of the gene(s) for the immunogenic antigens and the vector microorganism should be identified. Furthermore, the gene sequences introduced into the seed microorganism genome during construction of the modified seed should be provided. The master seed should consist of a single uniform batch/serial of seed that has been mixed and filled into containers as one batch/serial. Master seed should be frozen or desiccated and stored at low temperatures such as -40°C or -70°C, or under other conditions found to be optimal for maintaining viability. Each master seed should be tested to ensure its identity, safety and efficacy. Genetically modified seeds should also be tested to ensure stability and safety of the inserted gene sequences. Purity should also be established by testing to ensure freedom from extraneous bacteria, fungi, mycoplasma, and viruses.

Master Cell Stocks

When cell cultures are used to prepare a product, a master cell stock (MCS) should be established for each type of cell to be used. Records of the source of the master cell stock should be maintained. For each product, the highest and lowest passage levels of cells that may be used for production should be established and specified in the Outline of Production or SOP. Some control authorities do not permit more than 20-40 sub cultivations. Each MCS should be characterized to ensure its identity, and its genetic stability should be demonstrated when sub cultured from the lowest to the highest passage used for production. The karyotype of the MCS should be shown to be stable with a low level of polyploidy. Freedom from oncogenicity or tumorogenicity should be demonstrated by *in- vivo* studies in appropriate species using the highest cell passage that may be used for production. Purity of MCSs should be established by testing to ensure freedom from extraneous bacteria, fungi, mycoplasma, and viruses.

Ingredients

The specifications and source of all product ingredients should be defined in the Outline of Production, SOP, or other appropriate documents. The Outline of Production must be approved by the National licensing agency. All ingredients of animal origin that are not subject to a validated sterilization procedure should also be tested to ensure freedom from extraneous bacteria, fungi, mycoplasma, and viruses. Their country of origin should be known. Measures should be implemented by the firm to avoid the risk of TSE agent contamination by ingredients of animal origin. Some control authorities discourage the use of preservatives or (more importantly) antibiotics as a means of controlling adventitious contamination during production and prefer the use of strict aseptic techniques to ensure purity. However, they sometimes allow the use of preservatives in multidose containers to protect the product during use. These control authorities usually limit any addition of antibiotics in the manufacture of the product to cell culture fluids and other media, egg inocula, and material harvested from skin or possibly other tissues. They normally permit the use of no more than three antibiotics in the same product. Some control authorities prohibit the use of penicillin or streptomycin in vaccines administered by aerosol or parenterally. If the antibiotics used are not recommended for use in the target species, they should be shown to have no harmful effects in the vaccinated animals and not result in the contamination of food derived from vaccinated animals.

Safety Tests

The intrinsic safety of vaccines should be demonstrated early in the development stage and documented as part of the licensing dossier. Safety studies during development and licensing for all products should include the safety of a single dose, of an overdose and of repeated single doses. Additional data are derived for live vaccines from the increase in virulence tests and by assessing risk to the environment and in-contact animals. Safety should be demonstrated in each species for which the product is indicated. As a general rule, overdose studies are required for all vaccines: ×10 for live and ×2 for inactivated vaccines (if this is not practical, an indication of safety may be obtained from the results of the potency tests). For inactivated virus or bacterial products, where host animals are used for potency testing, safety may be determined by measuring local and systemic responses following vaccination and before challenge in the potency tests. Further evidence concerning the safety of products is derived from field safety trials. Vaccines derived through biotechnology should be evaluated as discussed in the classification of biotechnology-derived vaccines and release of live rDNA vaccines below.

Increase In Virulence Tests

With live vaccines, there is concern that the organism might be shed from the host and transmitted to contact animals, causing disease if it retains residual virulence or reverts to virulence. Therefore, all live vaccines should be tested for virulence by means of passage studies. Vaccine organisms are propagated *in vivo* by inoculating a group of target animals with master seed, usually using the natural route of infection for that organism. The vaccine organism is recovered from tissues or excretions and is used directly to inoculate a further group of animals, and so on. After not less than five passages (more for poultry products), the isolate must be fully characterized, using the same procedures used to characterize the master seed. Regulatory authorities opinion varies in whether or not it is acceptable to propagate *in vitro* between passages organisms that otherwise cannot be passaged five times because of their degree of attenuation. The vaccine organism must retain an acceptable level of attenuation after propagation in this way.

Assessing Risk to the Environment

The ability of each live vaccine to shed, to spread to contact target and non-target animals, and to persist in the environment must be evaluated to provide information for assessing the risk of the vaccine to the environment, taking into account human health. In some cases this may be done in conjunction with the increase in virulence tests. These and additional considerations are especially important in the case of products based on biotechnology or recombinant DNA techniques.

Efficacy Tests

The efficacy of veterinary vaccines should be demonstrated by statistically valid vaccination–challenge studies in the host animal, using the most sensitive, usually the youngest animals for which the product is to be recommended. Data should support the efficacy of the vaccine in each animal species by each vaccination regimen that is described in the product label recommendation, including studies on the onset of protection when claims for onset are made in the product labeling and for the duration of immunity. The tests should be performed under controlled conditions starting, wherever possible, with seronegative animals. Where validated potency tests are available, target species vaccination–challenge studies may not be required if predictive serological test results are available. The application of procedures to replace, reduce, and refine animal tests (the 'three Rs rule') should be encouraged whenever possible.

Efficacy studies should be conducted with final product vaccine that has been produced at the highest passage level from the master seed that is permitted in the Outline of Production, or other documentation of the manufacturing process. This will have specified the minimum amount of antigen per dose that must be in the final product throughout the entire authorized shelf life. Where a range of antigen level per dose is permitted, the antigen level per dose in the vaccine tested for efficacy must be at or below the minimum permitted amount. The precise challenge method and the criteria for determining protection vary with the immunizing agent and should be standardized whenever possible.

Field efficacy studies may be used to confirm the results of laboratory studies or to demonstrate efficacy when meaningful vaccination-challenge studies are not feasible. However, it is generally more difficult to obtain statistically significant data to demonstrate efficacy under field conditions. Protocols for field studies are more complex, and care must be given to establish proper controls to ensure the validity of the data. Even when properly designed, field efficacy studies may be inconclusive because of uncontrollable outside influences. Some problems include: a highly variable level of challenge; a low incidence of disease in non-vaccinated controls; and exposure to other organisms causing a similar disease. Therefore, efficacy data from both laboratory and field studies may be required to establish the efficacy of some products, as well as 'posteriory' field trials linked to vaccinovigilance.

Interference Tests

For products with two or more antigenic components, tests must confirm that there is no interference between individual components, that is, one component causing a decrease in the protective immunological response to another component. Interference testing should be conducted for each combination product prior to approval.

A loss of potency may also result when residual inactivating agent in a killed liquid product used as a diluent for a desiccated live fraction reduces the viability of the live organisms because of viricidal or bactericidal activity. Each batch/serial of liquid killed vaccine that is to be used, as a diluent for live vaccines must, therefore, be tested for viricidal or bactericidal activity prior to release. Consideration must also be given to possible interference between two different vaccines from the same manufacturer recommended to be given to the same animal within a 2-week period.

Consistency of Production

Prior to marketing approval of any new product, each establishment should produce in its facilities three consecutive production batches/serials of completed product to evaluate the consistency of production. These batches/serials should be prepared according to the procedures described in the Outline of Production and blueprints and legends, SOPs or other documentation of the manufacturing process. The size of each of the three batches/serials should be at least one-third the size of the average batch/serial that will be produced once the product is in production.

Stability Tests

Stability studies (based on an acceptable potency test) are required to establish the validity of the expiry date that appears on the product package. Some authorities allow the use of accelerated stability tests to determine a provisional expiry date for new products, e.g. incubating at 37°C for 1 week for each year of dating. Such estimates must be confirmed by periodic real-time potency tests on at least three different batches/serials through the period of time indicated by the expiry date, and 3–6 months beyond. For products containing viable organisms, testing should be done at release and at the approximate expiry date until a statistically valid record has been established. For non-viable products, each batch/serial presented for licensing is tested at release and at periodic intervals through, or past, the requested expiry date.

If at the end of the dating period (shelf life) specified, the product is tested and found still to be above the release quality, consideration can be given to extending the designated shelf life, by request to the control authority. Stability testing also provides the opportunity to test for residual moisture and for other important parameters, such as the stability of adjuvant emulsions.

Batch/Serial Release for Distribution

Prior to release, the manufacturer must test each batch/serial for purity, safety, and potency, as well as perform any other tests described in the firm's Outline of Production or other documentation of the manufacturing process for that product. In countries that have national regulatory programmes that include official control authority re-testing (check testing) of final products, samples of each batch/serial should also be submitted for testing in government laboratories by competent authorities. If unsatisfactory results are obtained for tests conducted either by the manufacturer or by competent authorities, the batch/serial should not be released. In such cases, subsequent batches/serials of the product should be given priority for check testing by competent authorities.

Other Tests

Depending on the form of vaccine being produced, certain tests may be indicated and should be provided as appropriate in the Outline of Production or other documentation of the manufacturing process. These tests may concern: the level of moisture contained in desiccated products, the level of residual inactivant in killed products, the complete inactivation of killed products, pH, the level of preservatives and permitted antibiotics, physical stability of adjuvants, retention of vacuum in desiccated products, and a general physical examination of the final vaccine.

Sampling

Samples should be selected from each batch/serial of product. The selector should pick representative final containers from each batch/serial and store these samples at the storage temperature recommended on the label. The producer should keep these reserve samples at the recommended storage temperature for a minimum of 6 months after the expiry date shown on the label, so that they are available to assist in evaluating the cause of any field problems reported from the use of the vaccine. The samples should be stored in a secure storage area and be tamper-evident.

Labeling

Standards for labeling products will vary from country to country; however, the label indications and all claims that are made on the label should be supported by appropriate data that have been reviewed and approved by competent authorities. It is recommended that all labels for veterinary vaccines be water-proof and contain the relevant information, although for very small containers, the label may instead refer to the carton label or to an enclosed package insert for some of the less prominent information. Labels may also include other factual statements that are not false or misleading. Special restrictions concerning the use or handling of the product, when applicable, should also be indicated. Similar information should also be given in a Product Data Sheet that is provided as a package insert. This will also contain much more detail about method of use and possible adverse reactions.

Field Tests (Safety and Efficacy)

All veterinary biological products administered to animals should be tested for safety and, if possible, for efficacy in the field, using good clinical practice, before being authorized for general use. Field studies are designed to demonstrate efficacy under working conditions and to detect unexpected reactions, including mortality that may not have been observed during the development of the product. Under field conditions there are many uncontrollable variables that make it difficult to obtain good efficacy data, but demonstration of safety is more reliable. The tests should be done on the host animal, at a variety of geographical locations, using appropriate numbers of susceptible animals. The test animals should represent all the ages and husbandry practices for which the product is indicated; unvaccinated controls must be included. The product tested should be one or more production batches/serials. A protocol should be developed indicating the observation methods and the recording methods.

Inspection of Production Facilities

Establishments that are approved to produce veterinary biologicals should be subject to in-depth inspections of the entire premises by national competent authorities to ensure compliance with the Outline of Production and blueprints and legends, SOPs, or other documentation of the manufacturing process. These inspections may include such items as: personnel qualifications; record keeping; general sanitation and laboratory standards; research activities on products being developed; production procedures; operation of sterilizers, pasteurizes, incubators, and refrigerators; filling, desiccating, and finishing procedures; care and control of animals; testing procedures; distribution and marketing; and product destruction. It is desirable to have good manufacturing practice (for manufacturing) and good laboratory practice (for quality assurance testing). The inspectors should prepare a comprehensive report documenting the findings of the inspection and stating the actions that the establishment must take to improve its production processes. The establishment should receive a copy of the report. When necessary a follow-up inspection should be conducted to determine whether appropriate action has been taken to correct deficiencies. Continued reassessment in this manner is needed to ensure that production facilities continue to be operated in an acceptable manner.

Updating the Outline of Production

Before production procedures are changed, the corresponding Outline of Production or other documentation of the manufacturing process should be changed. Establishments should have internal review procedures to evaluate all changes in production before they are initiated. Changes should also be reviewed and approved by competent authorities prior to their implementation. In cases where a significant production step is altered, revisions may require additional data to support the purity, safety, potency, and/or efficacy of the product. If the regulatory programmes include check testing the final product at national laboratories, revisions should entail testing of the new product by competent authorities.

Performance Monitoring

Manufacturers are required to maintain an adverse reaction notification system and an effective mechanism for rapid product recall. These should both be subject to audit by regulatory bodies. In many countries, the manufacturer must notify all adverse reactions immediately to the regulatory authority, along with any remedial action taken. An alternative used in some countries is that if at any time, there are indications that raise questions regarding the purity, safety potency, or efficacy of a product, or if it appears that there may be a problem regarding the preparation, testing or distribution of a product, the manufacturer must immediately notify the regulatory authorities concerning the circumstances and the action taken.

After release of a product, its performance under field conditions should continue to be monitored by competent authorities. Consumer complaints may serve as one source of information; however, such information needs to be investigated to determine whether or not the reported observations are related to the use of the product. Users of veterinary vaccines should be informed of the proper procedures for making their complaints. The manufacturer of the product should be informed of all complaints received by competent authorities. Competent authorities should also ascertain whether they have received other similar complaints for this product and, if so, whether the manufacturer has taken appropriate action. Control laboratories may test samples of the batch/serial of product involved, if necessary.

When the investigation is complete, a final report should be prepared and a summary of the findings sent to the complainant and to the manufacturer. When it is determined that a product is causing serious problems, immediate action should be taken to remove the product from the market and to notify animal health authorities.

Enforcement

National programmes established to ensure the purity, safety, potency, and efficacy of veterinary vaccines must have adequate legal authority to ensure compliance with product registration conditions and other programme requirements. The goal should be to obtain voluntary compliance with established regulatory requirements. However, when violations occur, competent authorities must have adequate legal authority to protect animal and human health. Authority for detention, seizure, and condemnation of products found to be worthless, contaminated, dangerous, or harmful may be valuable for this purpose. Under such authority, product may be detained for a period of time, and if during that time compliance cannot be achieved, competent authorities may seek a court order or decree for seizure and condemnation. The authority to remove or suspend establishment and/or product licenses, obtain injunctions, and stop the sale of product is also needed. Civil penalties or criminal prosecution may also be necessary for serious or deliberate violations.

Licensing of Products Derived Through Biotechnology

Recent advances in biotechnology have made possible the development and commercialization of new biological products with useful antigenic and diagnostic properties. Many such products have now been licensed or approved, and more are being developed. Products of rDNA technology do not differ fundamentally from conventional products. Therefore, existing laws and regulations are fully applicable to these new products.

Classification of Biotechnology-Derived Vaccines

Each competent authority with power to regulate organisms and products derived from recombinant techniques should ensure that the public health and the environment are protected from any potentially harmful effect. For the purpose of evaluating license applications, veterinary vaccines derived through rDNA technology may be divided into three broad categories. The division is based on the products' biological properties and on the safety concerns they present.

1) Category I products consists of nonviable or killed products that poses no risk to the environment and present no new or unusual safety concerns. Such products include inactivated microorganisms, either whole or as subunits, created by using rDNA techniques.

2) Category II products contain live microorganisms modified by adding or deleting one or more gene(s). Added genes may code for marker antigens, enzymes, or other biochemical by-products. Deleted genes may code for virulence, oncogenicity, marker antigens, enzymes, or other biochemical by-products. The license application must include a characterization of the DNA segments added or deleted, as well as a phenotypic characterization of the altered organism. The genetic modifications must not result in any increase in virulence, pathogenicity, or survivability of the altered organism in comparison with the wild-type form. It is important that the genetic modification does not cause deterioration in the safety characteristics of the organism.

3) Category III products make use of live vectors to carry recombinant-derived foreign genes that code for immunizing antigens. Live vectors may carry one or more foreign gene(s) that have been shown to be effective for immunizing target host animals. The use of DNA vaccines containing recombinant-

derived foreign genes that code for immunizing antigens (plasmid DNA vaccines) constitutes a new approach to vaccine development. The proper categorization of this type of rDNA-derived product will be established as biological properties and safety characteristics are determined.

These new vaccines may find application in a wide variety of situations much as conventional products have. Guidelines for the development, production, characterization, and control of these new products are still preliminary and subject to change as new data and knowledge are developed.

Release of Live rDNA Products

The release of live rDNA microorganisms (Categories II and III) for field testing or general distributions as an approved or licensed product may have a significant effect on the quality of the human and animal environment. Before release is authorized, the manufacturers of the vaccine should conduct a risk assessment to evaluate the impact on the human and animal environment. In the USA, for example, a procedure is adopted that could be used as a model system in other countries. The European Union has adopted a similar system. It is performed as follows:

A risk assessment is carried out that should contain the following information: the purpose and need for the proposed action; the alternatives considered; a list of the government agencies, organizations, and persons consulted; and the affected environment and the potential environmental consequences. The topics discussed should include: the characteristics of the vaccine organism, human health risks, animal health risks for both target and non-target animals, persistence in the environment, and increase in virulence.

If the risk assessment results in a finding by competent authorities that the proposed release of the recombinant vaccine into the environment for field trials or general distribution would not have a significant impact on the environment, a notice should be published and distributed to the public announcing this and that the risk assessment and findings are available for public review and comment. If no substantive comments are received to refute the findings, competent authorities may authorize the field-testing or grant the license or approval for general distribution.

The preparation of a risk assessment and the findings made from the assessment may also include the scheduling of one or more public meetings if a proposed action has ecological or public health significance. Such meetings should be announced through a public notice. Interested persons should be invited to make presentations, along with presentations by the producer of the product, and government personnel. The transcripts of such meetings should become part of the public record.

If, in the course of preparing a risk assessment, competent authorities conclude that the proposed action may have a significant effect on the human environment, an Environmental Impact Assessment Statement (EIS) should be prepared. The EIS provides a full and fair discussion of the significant environmental impacts, and informs decision-makers and the public of any reasonable alternatives that would avoid or minimize the adverse impacts.

The Role of Official Bodies in the International Regulation of Veterinary Biologicals

The official control of veterinary biologicals is vested in various national and regional organizations that differ in their approach to ensuring the quality, safety and efficacy of the products. International harmonization of regulations concerning biological products did not begin until well after those concerning chemically defined products. The first biological products for veterinary use were not manufactured and distributed until the end of the nineteenth century. They were often produced under unsophisticated conditions, and distributed or sold without any control other than those of their manufacturers. Later, each manufacturer developed its own standards. In Europe, these were subject to State controls as early as 1895 for certain diagnostic products (e.g. mallein, tuberculin) or vaccines. Gradually the conditions for international harmonization of standards evolved, beginning with the comparative testing of products being issued by different European laboratories. It was only in the second half of the twentieth century that national laws covering veterinary biological were imposed. These demanded that precisely defined techniques be followed before biological products for veterinary use could be licensed. This was followed by considerable efforts to harmonize these national regulations, first at the regional level (notably in Europe and the Americas) then at the global level, notably by the Office International des Épizooties (OIE) with the publication of the first edition of the OIE Manual of Standards for Diagnostic Tests and Vaccines in 1989.

World-wide harmonization of standards for veterinary biologicals will be of help to National controlling authorities who must follow the instructions given in the OIE International Animal Health Code, as they apply to all biological products for use in international trade. It will also be of assistance to vaccine producers, who have expressed their wish for world-wide harmonization of registration rules so as to simplify and facilitate marketing of their products. Evidently, it will also be of interest to farmers and to consumers, who would benefit from the fact that the safety and efficacy of the products that they use would have been assured to a uniformly high level.

The Role of International Organizations

Most nations have a range of official acts that regulate the sale and use of veterinary biologicals. Almost all of these acts stipulate 'minimum requirements' for quality, safety and efficacy of veterinary biologicals (mostly vaccines), to be tested at independent laboratories, usually under State supervision. These acts and tests may differ from one country to another, and they involve costs and restrictions for producers, users and testers.

The idea of harmonizing this testing to simplify and reduce costs on a regional, or even world scale is not new, and much has been accomplished during the past 20 years. The purpose of this section is to review the current situation by describing the role of international organizations in the regulation of veterinary vaccines. In this section the term 'international organization' refers to those concerned with animal health on a world-wide scale (OIE, the Food and Agriculture Organization of the United Nations [FAO] and the WHO).

The role of the OIE

The OIE was founded in Paris in 1924 as the world organization for animal health, and comprised of 167 Member Countries in the year 2005. The principal aims of the OIE are: to ensure transparency in the global animal disease and zoonosis situation, to collect, analyze and disseminate scientific veterinary information, to provide expertise and encourage international solidarity in the control of animal diseases, within its mandate under the Agreement on Sanitary and Phytosanitary Measures (SPS Agreement) of the World Trade Organization (WTO), to safeguard world trade by publishing health standards for international trade in animals and animal products, to improve the legal framework and resources of national Veterinary Services and to provide a better guarantee of the safety of food of animal origin and to promote animal welfare through a science-based approach.

Within the OIE there are four Specialist Commissions: the Terrestrial Animal Health Standards Commission, which deals with the Terrestrial Animal Health Code, the Biological Standards Commission, the Scientific Commission for Animal Diseases and the Aquatic Animal Health Standards Commission (including diseases of molluscs and crustaceans). In addition, there are three Working Groups: the Working Group on Wildlife Diseases, the Working Group on Animal Production Food Safety and the Working Group on Animal Welfare

Among the Specialist Commissions, the one most closely connected with standardization is the Biological Standards Commission. This Commission establishes standards for diagnostic methods (including diagnostic preparations) and for vaccines. Its terms of reference reflect the Commission's obligation to participate in the standardization of biological products, including vaccines used for prophylactic purposes. The Biological Standards Commission is responsible for the preparation of the OIE Manual of Diagnostic Tests and Vaccines for Terrestrial Animals, and the organization of Reference Laboratories for many of the diseases on the OIE List.

However, full standardization of vaccine testing can be achieved only when the necessary standards have been devised. It is hoped to reach the goal of standardization and wide availability of standards through the participation of OIE Reference Laboratories. The functions and responsibilities of experts at the over 150 OIE Reference Laboratories include the provision of a centre of excellence in a designated activity; standardization of methods; preparation, storage and distribution of standard antisera, antigens and other reagents.

Among the 15 OIE Collaborating Centers, three may be involved at some stage in veterinary vaccine control and/or harmonization: the Collaborating Centre for Veterinary Medicinal Products in Fougères (France), the Collaborating Centre for ELISA (enzyme-linked immunosorbent assay) and Molecular Techniques in Animal Disease Diagnosis in Vienna (Austria), and the Collaborating Centre for the Diagnosis of Animal Diseases and Vaccine Evaluation in the Americas in Ames (USA).

In 1994, following discussions with the International Technical Consultation on Veterinary Drug Registration (ITCVDR), the OIE set up an Ad hoc Group on the Harmonization of Veterinary Medicines, which was the first step towards the creation of the VICH (International Cooperation on Harmonization of Technical Requirements for Registration of Veterinary Products). In May 2003, the OIE International Committee adopted a resolution entitled OIE Procedure for Validation and Certification of Diagnostic Assays (Test Methods) for Infectious Animal Diseases. This resolution requires the OIE Director General to make provisions to establish an OIE registry for assays with levels of validation specified. Fitness for purpose should be used as a criterion for validation.

The role of the Food and Agriculture Organization (FAO) of the United Nations

FAO, established in 1945, is responsible for agricultural development and food production. The Animal Production and Health Division ('AGA') within the Agriculture Department is concerned with livestock development, and it includes the Animal Health Service ('AGAH'), the main role of which is to assist member countries in the control of animal diseases, with the aim of improving livestock production as an integral component of general social, economic and agricultural development. FAO's involvement in testing veterinary biologicals is primarily through its technical assistance system to Member Countries to set up and even execute independent quality control of vaccines and other biologicals. One example is FAO's assistance to the AU (African Union) in setting up a system for continental testing of veterinary vaccines, especially against rinderpest and contagious bovine pleuropneumonia, by the Pan African Veterinary Vaccine Center (PANVAC). FAO also commissions, at the request of member countries, initiatives for either quality assurance of vaccines and other biologicals or expert consultations on this subject, or publication of manuals on the production and quality control of vaccines. Furthermore, two auxiliary services can be asked to intervene on matters concerning these products, namely Codex Alimentarius and the Division of Nuclear Techniques in Food and Agriculture. The latter is operated jointly by FAO and the International Atomic Energy Agency (IAEA) based in Vienna (Austria). It has an Animal Production and Health Section, which assists veterinary services and research institutes in developing countries to establish radio-immunoassay (RIA) and ELISA techniques. Linked to this activity is a quality assurance scheme under which laboratories in receipt of FAO/IAEA ELISA kits are required to routinely monitor internal quality controls and to periodically (once or twice a year) test a batch of unknown samples, and to forward the results to IAEA. The overall aim is to provide assurance to all concerned that the results being generated through the use of such internationally standardized and validated kits can be relied upon as correct.

The role of the World Health Organization

Currently WHO is not directly involved in preparing international reference preparations (i.e. antigens or antibodies) for purely veterinary use, but has developed and still retains in the National Institute for Biological Standards and Control, Potter's Bar (UK) some materials related to purely veterinary diseases (e.g. Newcastle disease live vaccine, anti-swine fever serum). WHO wishes to retain a role in this area in instances where the veterinary reference preparations and guidance documents have a direct relevance to human health. This involves zoonotic and potentially zoonotic agents and other infectious agents of animal origin that are potential contaminants of biological products, whether these are vaccines produced in cell cultures or organs for xenotransplantation. At the meeting of the Expert Committee on Biological Standardization in October 1998, a review of currently retained international standards and reference preparations for veterinary medicine was carried out and a list of candidates for discontinuation, replacement and revision was suggested. The Expert Committee however decided to defer taking action on preparation of veterinary reference materials pending an evaluation by WHO with its partners in the veterinary field of the need for these various biological products. In addition, the present day topicality of certain preparations, especially veterinary vaccines against known zoonoses (e.g. anthrax, brucellosis) adopted and/or revised in the 1960s and 1970s, also needs to be evaluated.

The Role of VICH

VICH is a trilateral (EU-Japan-USA) programme aimed at harmonizing technical requirements for veterinary product registration. Its full title is the International Cooperation on Harmonization of Technical Requirements for Registration of Veterinary Medicinal Products. VICH was officially launched in April 1996.

The initiative to begin the harmonization process came in 1983 when the first International Technical Consultation on Veterinary Drug Registration (ITCVDR) was held. Since then a series of government and industry initiatives have been developed, culminating in the formation of the VICH. The Codex Alimentarius formed a Committee on Residues of Veterinary Drugs in Foods in 1985. Standard requirements for veterinary product registration were adopted in Europe in 1981. The US Food and Drug Administration and the European Commission have held regular bilateral meetings for the last decade to discuss common areas of interest. This has involved mutual exchange of guidelines for consultation. The first International Conference on Harmonization of Technical Requirements for Registration of Pharmaceuticals for Human Use (ICH) was held in Brussels in November 1991. The meeting brought together regulators and industry representatives from the USA, the EU and Japan to address quality, safety and efficacy requirements in the three regions. In January 1993 the GHOST (Global harmonization of standards) discussion document was published by FEDESA (Fédération Européenne de la Santé Animale). In 2002, FEDESA became the International Federation for Animal Health [IFAH]). It set out a programme for the international harmonization of registration requirements for veterinary pharmaceuticals and biologicals. Following discussions at ITCVDR and the OIE conferences, the OIE set up an ad hoc group on harmonization of veterinary medicinal products in 1994.

Preparatory work for the establishment of VICH was carried out by this OIE ad hoc group. During 1994 and 1995, two meetings were held at which the scope of veterinary harmonization was discussed and the membership and objectives of the VICH proposed. On the subject of food safety standards, it was decided that the VICH should complement the work of Codex and JECFA (Joint FAO/WHO Expert Committee on Food Additives). Issues related to GLP and GMP that are already the subject of mutual agreements will not normally come within the remit of the VICH. Issues related to biologicals were considered appropriate to fall within the scope of VICH.

With all the ground-breaking work completed, the Steering Committee of the VICH held its first meeting in April 1996, at which the membership and the working procedures were agreed and a work programme established.

The objectives of the VICH are along the same lines as those of the ICH. They are:

1) Establish and implement harmonized regulatory requirements for veterinary medicinal products in the VICH regions, which meet high quality, safety and efficacy standards and minimize the use of test animals and costs of product development.
2) Provide a basis for wider international harmonization of registration requirements.
3) Monitor and maintain existing VICH guidelines, taking particular note of the International Conference on Harmonization of Technical Requirements for the Registration of Pharmaceuticals for Human Use (ICH) work programme and, where necessary, update these VICH Guidelines.
4) Ensure efficient processes for maintaining and monitoring consistent interpretation of data requirements following the implementation of VICH guidelines.
5) By means of a constructive dialogue between regulatory authorities and industry provide technical guidance enabling response to significant emerging global issues and science that impact on regulatory requirements within the VICH regions.

Progress towards achieving the VICH objectives

For veterinary immunologicals there is an ongoing programme of harmonization in a number of areas including target animal safety studies, reversion to virulence and tests for the presence of *Mycoplasma*. To date only a relatively small number of VICH guidelines has been developed for veterinary biologicals and it is worth emphasizing the difficulties in reaching agreement on veterinary biologicals between the three regions. The only two adopted VICH Guidelines for veterinary biologicals are testing of residual formaldehyde adopted May 2003 and testing of residual moisture also adopted in May 2003.

The Indian Scenario

In India, under the Drugs and Cosmetics Act 1940, the Indian Pharmacopoeia is the book of standards for drugs and the standards as included in the Indian Pharmacopoeia (IP) are official. In order to fulfill the growing needs of medicines for the enormous livestock and poultry wealth of the country, the Government of India published a Veterinary Supplement to the Indian Pharmacopoeia 1996 containing information relating to substances intended for use in veterinary practice.

The Indian Pharmacopoeia 1996 contains monographs in respect of drug substances and their dosage forms used mainly in human beings. Monographs of certain excipients commonly used in preparation of official dosage forms are also included therein. Under the Drugs and Cosmetics Act 1940, a drug includes all medicines for internal or external use in human beings or animals and all substances intended to be used for or in the diagnosis, mitigation or prevention of any disease or disorder in human beings or animals. The Drugs and Cosmetics Rules 1945, formulated under the Drugs and Cosmetics Act, include Schedule F (I) wherein provisions especially applicable to the production and quality control of bacterial vaccines, antisera and diagnostic antigens for veterinary use are included. A number of new entities got introduced into veterinary practice over the years. The requirements included in Schedule F (I) were no longer adequate. It was therefore considered prudent to review the standards in respect of various veterinary drugs, pharmaceuticals and biologicals used in India so as to harmonize them with those of similar products used elsewhere. The preparation of a Veterinary Supplement to the Indian Pharmacopoeia 1996 thus became inevitable. The Veterinary Supplement to the Indian Pharmacopoeia 1996 also addresses the need for harmonizing and integrating to the extent possible the pharmacopoeial standards in India with those adopted by other pharmacopoeias. The supplement is divided into two parts; Part I deals with non-biological preparations whereas biological products including antisera, diagnostics and vaccines are grouped together in Part II. It provides a collection of minimum standards with which a manufacturer must comply before release of a product for sale or distribution. It is, therefore, the responsibility of the manufacturer to ensure that the product is manufactured in accordance with Good Manufacturing Practices and that sufficiently stringent limits are applied at the time of release of a batch of a product or material so that the pharmacopoeial standards are met until its expiry date under the storage conditions specified. The general principles applicable to the manufacture and quality assurance of drugs and preparations meant for human use would thus apply equally to veterinary products as well.

Conclusion

Presently, there is a clear intention to achieve greater international harmonization of regulatory requirements for veterinary biologicals.

Progress has already been achieved through international organizations to allow fair competition in the marketing of veterinary products. Although past efforts by international organizations have not resulted in a level of harmonization sufficient to facilitate international trade, they have laid the groundwork for current efforts. National authorities recognize the advantages of harmonization and are now committed to working towards this goal. The efforts of international organizations have made the goal of harmonization possible and have resulted in an organization and process for proceeding towards this goal. Success in achieving this goal will depend on the willingness of participating national authorities to work together and accept the compromises that will be necessary to resolve the difficult scientific and policy issues.

References

1. Indian Pharmacopoeia-Veterinary Supplement, 2000. Published by The controller of Publications, Government of India, Delhi.

2. O. I. E. Manual of Diagnostic Tests and Vaccines for Terrestrial Animals., Chapter I.1.9. The role of Official bodies in the international regulation of Veterinary Biologicals, May, 2005., Published by The Office International des Épizooties (OIE), Paris.

3. O. I. E. Manual of Diagnostic Tests and Vaccines for Terrestrial Animals., Chapter I.1.7. Principles of Veterinary Vaccine Production, May, 2006., Published by The Office International des Épizooties (OIE), Paris.

-ΔΔΔ-

5

National and International Approaches on Prevention and Control of Transboundary Diseases and their Global Economic Impact

M. P. YADAV AND R. K. SINGH

SVBP University of Agriculture and Technology
Meerut - 250 110, Uttar Pradesh
Division of Virology, Indian Veterinary Research Institute
Mukteswar - 263 138, Nainital, District, Uttaranchal, India

Abstract

All animal diseases have the potential to adversely affect human populations by reducing the quantity and quality of food, other livestock products (hides, skins, fibers) and animal power (traction, transport) that can be obtained from a given quantity of resources and by reducing people's assets. Of these, Transboundary Animal Diseases (TADs) tend to have the most serious consequences. The TADs may be defined as "those epidemic diseases which are highly contagious or transmissible and have the potential for very rapid spread, irrespective of national borders, causing serious socio-economic and possibly public health consequences". The recent major TAD epidemics or epizootics (e.g. SARS, avian influenza, Nipah Virus, West Nile virus) demonstrate the global impacts of emerging and re-emerging diseases of animal origin on the health and well-being of the public, as well as the tremendous socio-economic and trade damages that this group of diseases can cause to both developed and developing countries. These sanitary crises have also demonstrated the important role of official veterinary services in disease prevention and control (early detection, diagnosis, surveillance, rapid response and research activities), as well as underlined the importance of strengthening the capacity of these services in compliance with OIE international standards, and the need for creating new and more effective partnerships with public health services and the private sector (a global sanitary governance). Thus, the potential socio-economic and cultural consequences of TADs are of such a magnitude that their occurrence may also have a significant detrimental effect not only at national level but at regional and global level which warrants internationally coordinated effective and foolproof strategies to control the TADs.

1. INTRODUCTION

Transboundary Animal Diseases (TADs) are defined as those epidemic diseases which are highly contagious or transmissible and have the potential for very rapid spread, irrespective of national borders, causing serious socio-economic and possibly public health consequences. Almost all the diseases affecting animals have the potential to adversely affect human population as these diseases might result in reduction in the: (i) quantity and quality of food and other products (like hides, bone, skin and fibres) of livestock origin, (ii) draught animal power (traction, tilling and transport) obtainable from resources available, (iii) people's assets. Of all the animal diseases, TADs resulting in high morbidity and mortality in susceptible animal populations have the potential for most serious adverse consequences. The consequences due to TADs are of very high magnitude and as such occurrence of any TAD could have a significant detrimental effect on economy and public health not only of the affected nation(s) but a serious threat to whole of the world.

Because of their great diversity and capacity for rapid evolution, and because therapeutic interventions are more limited, viruses have received the greatest attention. Some important TADs of significance to India include African Horse sickness, African swine fever, Avian influenza, BSE (not really a TAD), CWD of deer, Morbillivirus infections of dolphin/porpoise/seals, feline immuno-deficiency, Simian immuno-deficiency, foot-and-mouth disease (FMD, new serotypes/strains not existing in India), Malignant Catarrhal Fever (MCF), Rinderpest virus (??) including in endangered free-living ruminants, LCM in mice, Eastern Equine Encephalitis (EEE) in crane, Feline infectious peritonitis in cheetah, Rabies in endangered animals, Myxoma virus of rabbits, Contagious bovine pleuropneumonia (CBPP), Rabbit haemorrhagic disease, Hyper-virulent strain of IBD, American strain of NDV in pigeon, Equine influenza, Vesicular stomatitis, Swine vesicular disease,

Lumpy skin disease (LSD), Rift valley fever (RVF), West Nile virus (WNV), Nipah & Hendra viruses, Akabane virus, Border Disease etc.

2. SOURCE OF EIDS/TADS

It seems plausible that the infectious agents causing TADs are always present in environment and they get activated under certain circumstances. The transmission of the infectious agent could be between (a) animal and human (Animal Û human transmission), (b) wildlife-human and domestic animals (Wildlife Û human Û domestic animal) and wildlife, domesticated animal and human (Wildlife Û domesticated animal Û human). The zoonotic pool, spill-over and spill-back, host/parasite translocation, pathogen pollution (or biological) pollution constitute the main sources for maintenance and transmission of infectious agents in nature.

3. POTENTIAL CONSEQUENCES OF TADS

The TADs have the potential to: (i) threaten food/feed safety and nutritional security through loss of animal protein as well as draught animal power for agriculture; (ii) increase poverty of nations which are highly dependent on livestock farming for sustenance; (iii) significant production losses for livestock products reducing farm incomes, (iv) restrict opportunities for augmenting the animal productivity by local livestock industries as they may not be able to utilize exotic high producing breeds since they may be more susceptible to TADs in a new niche; (v) add significantly to the cost of livestock production through the necessity to apply expensive disease control measures; (vi) seriously disrupt or inhibit trade in livestock and livestock products either within a country or internationally; (vii) adversely affect the public health system when TADs are of zoonotic nature and can be transmitted to humans; (viii) burden the environment with undesirable gasses and elements in the atmosphere due to large die-offs of domestic animals and their disposal by either burial or incineration and die offs in wildlife populations disturbing the ecosystem; (ix) pain and suffering for affected animals; and ultimately emotional distress to human population as a consequence of TADs.

Concern about TADs and their implications for national security revolves around their zoonotic nature, economic significance and ever-growing threat of more TADs in future. Many TADs are zoonotic in nature as they affect both human and animals. The bovine spongiform encephalopathy (BSE, the Mad Cow Disease), Rift Valley Fever, West Nile Virus (WNV) outbreak, SARS Corona outbreak, and Highly Pathogenic Avian Influenza (HPAI) are all the diseases which have been in the news in recent past that adversely affected animal as well as human health. The fear that HPAI if becomes easily transmissible to and among humans is the major cause of concern not only to medical professionals but also to policy makers and general public alike as it raises the possibility of a major influenza pandemic. TADs can also have dramatic economic impacts which are obviously a concern for national security. Further, the risk of animal disease outbreaks will only increase in the future as the higher incomes in developing countries continue to stimulate more demand for meat and animal products. As a result livestock industries are growing larger, production is more intensive, and international trade is increasing. The number of animals raised for meat is growing very rapidly. During the 1990s, poultry production in East Asia rose nearly 12% per year—a doubling every 5 to 6 years. Evidence suggests that threats from TADs are increasing. We are familiar with the relatively new human diseases like AIDS, Lassa Fever, SARS, and Ebola which seem to have developed over the past 30 to 40 years without any warning. At the same time, new animal diseases have also been appearing at an alarming rate - HPAI, Nipah, BSE, hendra virus as well as new biotypes or antigenic forms of existing infectious diseases, such as a hyper-virulent strains of infectious bursal disease virus in poultry in Europe and Newcastle disease virus in USA.

3.1 Loss of Biological diversity

Rinderpest, a fatal morbillivirus infection of cattle, has been known to decimate the African bovid population. Similarly, canine distemper virus infection decimated African wild dog population and avian influenza threatens endangered game breeds of chicken especially fighting cocks. Infectious diseases also threaten endangered species of wildlife like Trypanosomosis resulted in death of many lions in India in many. Of late, Ebola virus infection has been reported to decimate 95% of the western lowland Gorilla (*Gorilla gorilla gorilla*) population in Odzala-Kokoua National Park, Republic of Congo in 2004 (Nature 442: 13 July, 2006).

3.2 Global economic loss and compromised developmental goals

Infectious diseases of crops and livestock cost the global economy uncounted billions of euros every year, besides sudden epidemics of infectious diseases also deliver humanitarian and economic shocks on a scale that is difficult to absorb. According to World Bank, the 2003 SARS epidemic, which killed fewer than 1000 people, was responsible for an estimated 2% fall in GDP across East Asia, and an influenza pandemic could kill millions of people and cost about US$ 900 billion globally in a single year (Brahmbhatt, M. Avian and Human Pandemic influenza-economic and social impacts, See: http://www.worldbank.org.). The direct and indirect costs of the foot-and-mouth disease (FMD) outbreak in the United Kingdom in 2001 were estimated to be over $9 billion. Classical Swine Fever in the Netherlands in 1997-98 had economic impacts of $2.5 billion. Over 150 million chickens died or were culled in an attempt to control HPAI in SE Asia in 2004. At one point in the same year; with HPAI, BSE and other animal diseases, the Food and Agriculture Organization (FAO) of the United Nations estimated that nearly one-third of the world meat trade was facing import bans due to animal diseases.

Further, infectious diseases also adversely impact global developmental plans. The United Nation's Millennium Development Goals have explicit targets of (i) reducing the burden of human diseases (particularly the "Big Three" tropical diseases: HIV/AIDS, tuberculosis, malaria), and (ii) reducing the hunger and poverty. However, these targets of Millennium Development Goals are compromised by crop and livestock diseases. In most developing regions, where the impacts of infectious diseases are greatest, there is now little hope of meeting any of the Millennium Development Goals by 2015.

3.3 Impact of infectious diseases on culture

Some infectious diseases which get established also become part of the culture. In the nations in which smallpox was endemic, it became part of life, with evidence of its impact in various cultural practices still intact although smallpox has been eradicated worldwide for 26 years now. One of the famous citations has been that Priestess Omolajaya Oloroke of Nigeria presides over the Obulawaye Shrine of Obulawaye, God of Smallpox, formerly the most dreaded disease in Yorubaland. The mural decorations on the walls behind the Priestess Omolajaya Oloroke suggest the marks of smallpox when Obulawaye strikes (http://www.stanford.edu/group/virus/1999/reynafue/history.html). Similarly, temples of "Shitla Devi" exist all over India. Other regional names of the devi are Pochamma and Mariamma. Full description of devi exists in Shabdakalpadruma (c. 1820). The devi is shown seated next on a donkey. In one hand, she carries a broom and with the other she pours water over an indigenous winnower (supa) covering her head. This description is more appropriate for a female demon than for a goddess. Skanda-purana (c.100 AD) describes the mode of worship. The festival, Shitalashtami" is celebrated in Hindu month, eighth day of the dark fortnight of Chaitra (March-April) (Joshi, 1976).

4. WHY INCREASING THREAT FROM TADS?

Besides many TADs discussed earlier, vector-borne pathogens like West Nile Virus, bluetongue and Rift Valley Fever are also spreading to a much wider area. The increasing threat from TADs stems from : (i) Globalization and trade, (ii) Increased conflict and civil unrest, (iii) Rising meat demand vis-à-vis intensive animal production and trade, (iv) Translocation due to change in forest ecology, (v) Climate change and Global warming, (vi) Bioterror threat

5. WHY TADS CAN POSE A SERIOUS NATIONAL SECURITY THREAT?

The increase in risk factors, as described above, coupled with the fact that the response is slower fro animal diseases as compared to human diseases, clearly demonstrates why TADs can pose serious national and/or international security threat. A combination of deficient technical capacity and perverse economic incentives slows down the response to TADs. Veterinary Public Health Services in developing countries are usually much weaker than Medical Public Health Services. At the same time, unlike human disease reporting, animal disease reporting systems are usually based on passive reporting rather than active disease surveillance. Some of the important points to consider include:

(i) lack of farmer awareness in developing countries,

(ii) lack of diagnostic capabilities in developing countries,

(iii) disease reporting,

(iv) poor compensation schemes deter disease reporting,

(v) failure of disease reporting hastens spread of TADs,

(vi) shifting of focus of international agencies and donor

6. STEPS TO CONTROL TADS

TADs will be an ever-growing threat to animal/human health, economic sustenance of the world and global environmental wellbeing. It is difficult to predict the number of these diseases that could rapidly escalate in a national or region threatening the human and animal life as well as economy of that nation or region. The rising meat demand has put humans and animals together in numbers never seen before in the world, particularly in East and South Asia. This is compounded by the fact that the biosecurity on farms in these countries where meat production is growing most rapidly is often poor. The intense animal-human interaction coupled with poor biosecurity measures creates great potential for animal diseases to jump species to create human health problems. Some of the issues and solutions to these issues were thoroughly described in IOM Report, 2003 and discussed and debated by Charles H. Riemenschneider (2005) during his presentation at "Health in Foreign Policy Forum 2005" in Washington, DC on February 4th, 2005 and are detailed below:

6.1 Early detection and early response

Early detection and early response to TADs is the key issue in meeting this threat. Huge investment in veterinary public health services as well as human public health in developing countries will go a long way to accomplish this task. Even the best public health strategies in a nation will be jeopardized in thee absence of an up-to-date veterinary public health system having better trained personnel and sophisticated laboratories.

6.2 Preparedness plans

Preparedness to detect and respond to emerging infections particularly that appear suddenly and are capable of spreading over large areas is a must. Preparedness plans for human diseases are weak in many developing countries—yet they are even weaker for animal diseases. There is also a need for improving incentives to report animal diseases with better compensation to affected farmers. Import bans in response to an animal disease outbreak must be based on sound scientific evidence so as to assure that countries also have incentive to report internationally. The strategic Plan for preparedness developed by WHO and designated as **"Epidemic and Pandemic Alert and Response (EPR): Preparedness for Deliberate Epidemics"** as accessed on the World Health Organization site (http://

www.who.int/csr/delibepidemics/en/ accessed 10-10-2006) is detailed as below:

In May 2002, the World Health Assembly passed a resolution entitled "*Global public health response to natural occurrence, accidental release or deliberate use of biological and chemical agents or radionuclear material that affect health*". The diseases covered under WHO's EPR include Anthrax, Avian influenza, Crimean-Congo haemorrhagic fever (CCHF), Dengue/dengue haemorrhagic fever, Ebola haemorrhagic fever, Hepatitis, Influenz, Lassa fever, Marburg haemorrhagic fever, Meningococcal disease, Plague, Rift Valley fever, Severe Acute Respiratory Syndrome (SARS), Smallpox, Tularaemia, and Yellow fever (http://www.who.int/csr/disease/en/; accessed 10-10-2006).

For details, Resolution can be accessed. The Secretariat provided a background paper entitled "Deliberate use of biological and chemical agents to cause harm" for consideration by the World Health Assembly. The strategy developed by WHO to respond to this resolution includes 4 main areas as described below:

(i) International preparedness

The objective is to respond to the increased number of requests by Member States for technical assistance on national CBW preparedness and response programmes and training. The *Public health response to biological and chemical weapons: WHO guidance* has been revised and published in 2004. This second edition of WHO's 1970 publication *Health aspects of biological and chemical weapons* includes information designed to guide preparedness for and response to the deliberate use of biological and chemical agents that affect health. The *Chemical and Biological Weapons (CBW) Scientific Advisory Group* is being established as a permanent resource for the WHO Secretariat and its members.

(ii) Global alert and response

The objective of *The Global Outbreak Alert and Response Network* provides an operational framework to link the expertise and skills needed to keep the international community constantly alert to the threat of outbreaks and ready to respond. In the event of the intentional release of a biological agent, WHO's global alert and response activities and operational framework together with the technical resources of the Global Outbreak Alert and Response Network would be vital for effective international containment efforts in responding to potential use of biological agents.

(iii) National preparedness

The objective is to respond to the increased number of requests by Member States for technical assistance on national CBW preparedness and response programmes and training.

Guidelines for the assessment of national CBW health preparedness and response plans are being developed according to the recommendations provided by a group of experts that met in Rome, Italy, in 2002.

Work is in progress with the *United Nations Disaster Management and Training Programme* to develop a training module on the *management of preparedness and response programmes on chemical, biological and radionuclear incidents.* The laboratory and epidemiology country capacities are being strengthened through a programme targeting microbiologists and epidemiologists from several countries of the African, Eastern Mediterranean and European regions. This includes an in-depth review of surveillance systems leading to national plans of action for strengthening surveillance and early warning systems for epidemic-prone diseases, including those associated with deliberate use. The *WHO Biosafety Programme* contributes to these activities by providing information, training, and advocacy for laboratory biosafety procedures and practices.

(iv) Preparedness for selected diseases and intoxications:

The objective is to contribute to international preparedness on specific diseases associated with biological weapons by: (a) establishing global networks of experts and laboratories; (b) establishing standards and procedures, and disseminating information; and, (c) setting up and implementing training. Eleven biological agents have been identified for biological weapons by a risk assessment carried out by WHO. In addition, a *WHO Working Group on Chemical and Biological Weapons (CBW)* been established to exchange information and coordinate activities within the relevant departments and WHO Regional Offices. For further information on these activities one can contact access the WHO site (deliberate@who.int).

Preparedness needs foresight

Governments and international agencies need a vision of how threats such as infectious diseases are likely to evolve in the future so that they can identify effective science and technology strategies to chalk out the preparedness plans. Many Foresight programmes were recently initiated, largely originating from Japan and the USA. The latest Foresight Project report assessed the projected risks from infectious diseases of humans, animals, and plants over 10- and 25-year horizons, focusing on detection, identification, and monitoring of diseases, the aspects perceived as neglected and where the development of new technologies and systems could have major impacts. The project identified eight categories of infectious diseases for which improved detection systems would make a difference over the next 10 to 25 years. These categories are not mutually exclusive and are not exhaustive, but the list captures the priority concerns identified by the project. The project, after thorough review, (i) selected appropriate technologies (novel information technologies for capture/ analysis/modeling of data; Genomics and postgenomics approaches for characterization of pathogens, non-invasive detection systems, and portable devices for diagnosis), (ii) highlighted the importance of fostering multi-disciplinary approaches, (iii) key choices for policy makers like international coordination, and (iv) setting up of "Regional Reference and Coordination Centres (RRCCs) which would form a network of high-quality laboratories linked to one another, to existing facilities, and to appropriate partners in developed countries. Such projects like Foresight, the OFFLU, and the Global Initiative on Sharing Avian Influenza Data (GISAID, http://gisaid.org) would really provide sufficient feedback to scientists, policy makers and funding agencies including governments to in their efforts towards preparedness to ward off and/or contain dangerous disease incursions. Thus, the individual countries, Unions or regions need to implement Foresight projects so as to assess the risk and chalk out strategy for preparedness which will be more relevant to their own country, region or union.

6.3 Decentralization of government structures

Public health and national security issues are the perspective of national governments. The decentralized government structures and improved international coordination are essential to address the growing threat of TADs as they do not respect either local or national boundaries. Government support at the lowest administrative level is essential to assure early detection which also requires adequate coordination at national and regional level to assure that sufficient response is given to avoid spread of disease through livestock movement controls, closing of live markets, sharing of diagnostic services, etc.

6.4 International coordination

Finally, help of international agencies like OIE-the World Animal Health Organization, WHO and FAO is also crucial. In 1994, FAO established the EMPRES program for the emergency prevention of transboundary animal and plant pests and diseases to improve FAO's early warning and early reaction to disease threats. This effort has greatly increased our ability to respond to emerging animal disease risks.

6.5 International efforts at Data sharing

The perception that infectious disease threat is diverse and dynamic has lead to emergence of a consensus that rapid and full sharing of disease data, samples and genetic information is in the best collective interest of the research, surveillance and control. The pledge by leaders of the rich G8 nations at the recently concluded summit in Russia in July 2006 to boost the meager quality of international surveillance of infectious diseases correctly identifies many key needs viz. (i) better coordination between the animal and human health communities, (ii) building laboratory capacities, and (iii) full transparency by all nations in sharing, on a timely basis, virus samples (Nature News, August 31 2006). Several countries and international agencies have recently taken steps to improve sharing of influenza data (Nature 441) following the initiative of leading veterinary virologists. Realizing the inadequacy of the current level of collection and sharing of data, given the magnitude of the threat, it was proposed to expand and complement existing efforts with the creation of a global consortium ¾ the Global Initiative on Sharing Avian Influenza Data (GISAID, http://gisaid.org) ¾ that would foster international sharing of avian influenza isolates and data. The GISAID policies for rapid and complete data release are modeled on guidelines for sharing data in large-scale biological research (see www.wellcome.ac.uk/doc_wtd003208.html) or those established for community resource projects. Such policies have proven effective as in International HapMap Project (www.hapmap.org) sequencing of SARS coronavirus. Success of such initiatives will largely depend on openness of scientists and governments in sharing the data. A similar initiative, the OFFLU ¾ started in April 2006 as a network on Avian Influenza by OIE and FAO with the commitment to make public the material on outbreaks in animals ¾ had mixed success as it relies on the cooperation of affected countries.

6.6 Understanding of ecology, microbial evolution and viral traffic

An integral approach to ecology of disease including consideration of social factors, viral and microbial traffic, ecological changes, major demographic changes (human and animal population migration) and trade practices is necessary as the above mentioned factors work in tandem and result in precipitation of emerging infections. These signals for viral and microbial traffic should be seen as warning signs. Consideration of biodiversity should include viruses and microbes, and environmental impact assessments in development planning should take health effects into account. Microbial (more specifically viral evolution) evolution is a continuous process and, therefore, basic research on emerging infectious diseases both of viral and other microbial origin and the factors responsible for emergence of new microbes should be encouraged to provide important insights. Understanding of viral traffic more systematically and strengthening predictive capabilities by effectively using molecular epidemiology, geographic methods and modeling systems to track disease and better anticipate its spread is most immediately warranted.

6.7 Expanded surveillance systems

Effective global disease surveillance (with national, regional and referral centres) to give early warning of emerging infections is the essential first step in the chain of medical and veterinary public health action. National Animal Health Monitoring System being maintained by USDA should be common feature with other countries as well and data from such national studies can be used to identify risk factors or trends associated with zoonotic diseases. World surveillance capabilities are critically deficient as a result of financial constraint and diminished interest. Adequate financial assistance, political will, interdisciplinary contributions from biomedical, veterinary, agriculture, ecologists, social scientists, administrators and planners etc is necessary. It is also important that international world bodies concerning animal health, human health, education, meteorology, environment, agriculture, financial institutions, NGOs work in tandem to solve this onerous problem which concerns not only a nation and a region but the whole world. Maintaining a network of more than 100 laboratories world over by WHO for constant survey of influenza viruses is one of the best examples of networking laboratories for emerging infections including TADs.

To tackle the emerging infectious disease problem, the usual answer, "increased surveillance," is correct but insufficient. We also need "tripwires" that will launch active investigations. These investigations need to be multidisciplinary, involving physicians, veterinarians, entomologists, mammalogists, ecologists, molecular biologists, vaccinologists, epidemiologists, and a large number of other specialties (including pathogenesis and immunology). Formal designation of the specialties is unimportant, but having the right expertise present on the team is crucial. Most of the current surveillance systems include active surveillance to detect particular known pathogen and passive systems for more generalized monitoring. The sensitivity of passive surveillance systems is one area of concern. Most of the surveillance systems are capable of tracing major outbreaks but the real challenge is finding and assessing smaller outbreaks, scattered cases of disease, or finding large outbreaks at incipient stage. Enhancing surveillance systems by establishing Laboratory Response Network at National, regional and international level is particularly important and finances should not be a constraint. By providing a link between public and private laboratories, such networks will definitely enhance the capabilities, at all levels, to detect and prevent the spread of emerging infections whether they are transmitted naturally or intentionally.

6.8 Disease intelligence

Better disease intelligence is required where other factors may help anticipate when and where disease may be likely to occur for more targeted surveillance. The production of actionable intelligence may derive from analyses of changes in such factors as climatic conditions, vegetation, wild-life demographics, trade pattern, or vector demographics and distribution.

6.9 Collaboration and cooperation among government agencies

Collaboration and cooperation among government agencies at all levels (local, state, national) as well as between medical public health, veterinary public health, Trade and commercial organization/ authorities, farmers and general public is most fundamental in preventing the emerging infections. To improve interdisciplinary collaboration, one step might be for federal governments to develop tripartite cooperative programme to address infectious diseases in humans, in domestic animals and in wildlife. Such collaboration would serve as focus for regular communication through working groups: (a) to address information transfer, (b) to improve response to disease emergencies, (c) to establish priorities for collaborative focused investigations, and (d) to pursue other areas of mutual interest. Collaborative programmes can also be developed to integrate emergency response capabilities within the public health, domestic animal, and wildlife conservation communities. Response to emerging infections of wildlife need to be hastened as needed by combined capabilities of the different programmes to minimize the potential for establishment and spread of wildlife diseases capable of infecting other species including humans. Collaboration at international level has also to be improved as a part of preparedness and early emergency response. The decisive actions of Hong Kong public health authorities and involvement of international collaborators were exemplary which helped in containing the spread of influenza in 1997. Post-1997 avian influenza outbreaks in many Asian countries including the Indian outbreak in 2006 could effectively be contained following the Hong Kong model which is a standard package of practices recommended by WHO and OIE for tackling avian flu outbreaks.

6.10 Changing Disease Patterns Demand Multi-disciplinary Collaboration (i.e. Cross-Field Partnerships)

The subject of climate change and global warming and its possible adverse effects on medical and veterinary public health has been debated hotly for more than last three decades now by a large number of world organizations through international conferences and seminars. A number of new world bodies have been established exclusively to address the issue of global change and its effects thereon. Large-scale variations in weather patterns are very likely to cause shifts in regions suitable for farming or alter the distribution, frequency, and magnitude of diseases such as arboviruses, plague, or hantavirus. The public health community has neither the power nor authority to implement the social changes necessary to reverse the trend of global warming, but public health professionals do have a role to play in preparing for the possible results of such large-scale change. Today few scientists doubt the atmosphere is warming. However, there is considerable debate in both the popular press and scientific literature concerning the extent to which this might occur - the viability of the projected increases, and the consequences of this temperature change. The majority of the articles focus on the adverse effects of a temperature increase of between 2°C (35.6°F) and 5°C (41°F). Long-term temperature shifts of this magnitude will likely have a major effect on a wide variety of social and ecological factors, including zoonotic diseases. As changing climates widen the range of vector-borne diseases, human and animal health workers need to start talking to each other. This places emphasis on cross-field partnership where experts from all the related areas are drawn in a group to tackle such complex issues.

Wyoming Department of Health's West Nile Virus Surveillance and Epidemiologic Project is one of the leading examples of such partnership leading to success. Wyoming's West Nile project is a partnership among the University of Wyoming State Veterinary Laboratory, USDA Agricultural Research Service, Wyoming Game and Fish Department, and Wyoming Department of Health. Terry Creekmore, the West Nile Project coordinator, is an employee of the health department, but is stationed at the Veterinary Laboratory. He coordinates both human and veterinary surveillance as well as leading Wyoming's West Nile and Vector-borne Disease Task Force. This collaborative project has also strengthened ties between veterinary and human health in Wyoming in other areas, including bioterrorism, emerging diseases, and zoonoses.

7. LESSONS LEARNT FROM TADS

7.1 Biosecurity

Biosecurity can be defined as "those efforts designed to prevent the introduction and spread of disease in a population, herd, or group of animals. These efforts can further be categorized into (a) External measures (external biosecurity) those directed at prevention of entry of new diseases into a group, and (b) Internal measures (internal biosecurity) those directed at prevention of spread of disease within a group. Further, although the term "biosecurity" is not specifically defined in the legislation, it has been referred to as "the cost effective protection of any natural resources from organisms capable of causing unwanted harm. In the Act *natural [and physical] resources* means organisms of all kinds; air, soil, water in or on which any organism lives or may live; landscape or landform; geological features; and systems of interacting organisms and their environment".

The concept of biosecurity is not new but there has been a renewed global interest in biosecurity which has been driven by many factors like recent international disease outbreaks, potential risk of agri-terrorism, and increasing commercialization of livestock production. Some most significant recent international disease outbreaks include WNV in USA, Hendra virus outbreaks in Australia, Nipah virus outbreaks in Malaysia, FMD in UK, BSE in Europe and USA, Avian Flu in Asia, Equine influenza in dogs in USA and Undefined Morbillivirus in cattle in USA. Some of these diseases were controlled but some of them have spread and covered more areas. The devastating effect of these TADs made the policy makers and general public to realize the importance of biosecurity. Now the "Minimum Biosecurity Plan" has been defined and to be valid it should address the issues viz., (i) the means of isolating new animals introduced into an existing herd or group of animals, (ii) the regulation of animal, worker and equipment trafficking (movement), and (iii) the design and implementation of cleaning and disinfection procedures directed at reduction of pathogen load within herd or group of animals.

7.2 Multi-disciplinary Collaboration

Multi-disciplinary Collaboration is Essential in research, disease diagnosis, implementation of biosecurity measures from farm till enforcement of legislation, maintaining vigil after effectively controlling the disease, rapid & effective control of outbreaks of TADs. Benefits of multi-disciplinary approach in controlling diseases like West Nile, Hendra, Nipah, Influenza, BSE and FMD has already been demonstrated. Multi-disciplinary collaboration should involve (i) inter-departmental cooperation between Medical Public Health, Veterinary Public Health, Forest and Environment, wild-life; (ii) Integration of research activities between various labs as "Network mode" of research delivers fastest as has been demonstrated by SARS Coronavirus episode - sequencing of the complete virus genome in least possible time (seven days) with involvement of 16 labs world over. Another example of "Network Mode" is "European Med-Vet-Net" which is made up of 16 European partners. The aim is to provide a 'network of excellence' for integrated research on the prevention and control of zoonoses. Similar networks are required world over for tackling TADs.

Most Successful Models for Cross-Field Partnership had been for tackling West Nile Virus (WNV) outbreaks in USA, Avian Influenza outbreaks and Bovine Spongiform Encephalopathy (BSE). The WNV, Avian influenza and BSE are not the only zoonotic infections to illustrate the importance of integrated working but have again highlighted the need to understand and develop multi-disciplinary approach.

7.3 Most Infectious Diseases are continuously changing

Infectious diseases have shaped the evolution & history of animal kingdom especially human and animals. Infectious diseases are dynamic and will always continue to influence the overall activities of human being in all the spheres of their lifestyle. The pattern of infectious disease has been changing continuously which has been responsible for sustained microbial evolution. These changes include the description of diseases caused by (i) pathogens long present but not previously identified, (ii) recognition of apparently new microbes, (iii) changes in epidemiology (changes in distribution, disease incidence, virulence of microbial strains, microbial resistance to drugs) of old pathogens, (iv) new disease — disease interactions, and (iv) spread (to human/animals) of organisms never previously known to be human/animal pathogens.

7.4 TADs are always worst disasters

Since the first outbreak of Nipah virus in Malaysia in 1997, the Nipah virus outbreak was first detected in Bangladesh in 2001, followed by second outbreak in 2003, and widespread outbreak in 2004 with a CFR of 75% where man-to-man transmission was also recorded along with detection of anti-Nipah virus antibody in flying foxes. Now the virus has been isolated in Bangladesh and man-to man-transmission has also been demonstrated. Presence of a disease like this is a threat looming large over not only to nations neighboring Bangladesh but to whole region and the world.

8. CRITICAL ISSUES IN MANAGEMENT OF TADS

8.1 International funding is essential for management of TADs

Since a TADs rapidly beyond national borders and may involve substantial part of a region of the world, stopping or controlling TADs is therefore a true global public good, wherein all the countries will have interest. Thus, while country level leadership and engagement is essential for success, it must be backed by global resources. Even though the benefits of containing an epidemic or a pandemic are overwhelming, individual governments may still be daunted by the social, political and economic costs of various policy measures, especially when these measures are in the nature of global public goods that benefit many more than just the people of that nation (http://web.worldbank.org/WBSITE/EXTERNAL/COUNTRIES/EASTASIAPACIFICEXT/ EXTEAPREGTOPHEANUT 0,,content MDK:20713527~pagePK:34004173~piPK:34003707~theSitePK:503048,00.html).

8.2 Global strategy is essential for management of TADs

The WHO, FAO and OIE have, in close coordination, prepared a global strategy for the progressive control of HPAI which is a "master coordination plan ... defining ... short, medium and long term priority activities, to be endorsed and supported by individual countries and regional organizations ..." which is to be complimented by more detailed country-specific HPAI control plans, the preparation of which is being technically supported by the FAO. This global strategy, in effect, addresses the WHO's two strategic actions in the pre-pandemic phase. The immediate and short-term objective is to prevent further spread of HPAI in those countries that are currently infected; specific measures are tailored to the circumstances of individual countries. The medium- and long-term objectives are to eradicate the disease progressively from the remaining "compartments" of infection and prevent its spread to those countries currently free of HPAI. This global strategy emphasizes a number of key points where action is required and these key issues are (i) inadequate capacity is the main factor limiting the management of TADs; capacity-building in terms of trained manpower and infrastructure for disease surveillance and control, policy development and socio-economic analysis needs to be taken care, the dynamics of the rapid spread and persistence of many TADs like HPAI remain unclear; research into the ecology of the diseases(s), evaluation of the vaccines, the development of improved vaccines and rapid diagnostic tests are all essential. In addition, analysis of production and marketing systems and the risks associated with them is required in order to target effective disease control, and (iii) country-specific projects need to be formulated, under-pinned by sub-regional TAD support units developing harmonized technical standards and regional policies related to live-animal movement, compensation plans, capacity building, and disease reporting (http://web.worldbank.org/WBSITE/EXTERNAL/ COUNTRIES/EASTASIAPACIFICEXT/EXTEAPREGTOPHEANUT/0,,contentMDK:20713527~pagePK:34004173~piPK:34003707~theSitePK:503048,00.html).

8.3 Economic analysis

Economic analysis of the country-specific projects must take into account the unique nature of the problems to be solved. The three categories of economic impacts associated with outbreaks of TADs and potential epidemic or pandemic include: (i) the economic consequences and costs of sickness or death resulting from the disease outbreaks; (ii) the economic consequences and costs associated with public and private efforts to prevent the emergence or spread of the disease and to treat its effects, and (iii) the economic impacts are those associated with the policy efforts of the government to prevent the start of an epidemic, to contain the epidemic once it has begun, and to mitigate its harmful effects on the health of the population. The first two categories se two are clearly related; for example, a greater effort at prevention and/or treatment for a given severity of epidemic would be expected to reduce the spread of sickness and/or the percentage of mortality and thus reduce the economic impact and costs (http://web.worldbank.org/WBSITE/EXTERNAL/COUNTRIES/EASTASIAPACIFICEXT/EXTEAPREGTOPHEANUT/0,,contentMDK:20713527~pagePK:34004173~piPK:34003707~theSitePK:503048,00.html). The calculation of economic costs needed to be calculated include direct and indirect loss, which is discussed below in succeeding text.

8.3.1 Primary or Direct costs

The direct costs include losses due to mortality of large number of affected of animals and/or humans due to the disease and expenditure incurred in control measures such as culling animals and birds, with impacts extending not only to farmers but also to upstream and downstream sectors such as traders, feed mills, breeding farms etc. Additional losses also occurred because of lowered production and reduced activity in distribution channels. On the other hand there will also be important substitution effects, especially towards other production affiliates.

8.3.2 Secondary or indirect economic costs

Secondary or indirect economic costs also arise due to fall in international tourism because of the disease fear and travel restrictions. This may not be much affected where TADs only affect animals but if TADs are of public health significance, the tourism gets adversely affected as, of late, the global media have started prominent and aggressive reporting on TADs as has been witnessed after SARS and avian influenza epidemics. Finally, the costs incurred on prevention and control, costs to the government of purchase of poultry vaccines, medications and other inputs, hiring workers for culling and cleanup, surveillance and diagnosis as well as hire of transportation, is also taken into account. Governments also face the need to pay compensation to poultry owners, which is important in inducing owners not to conceal that the disease outbreak. While such payment is in the nature of a transfer payment for the economy as a whole, it can impose a significant fiscal burden on the government. (http://web.worldbank.org/WBSITE/EXTERNAL/COUNTRIES/EASTASIAPACIFICEXT/EXTEAPHALFYEARLYUPDATE/0,contentMDK:20708543~menuPK:550232~pagePK:64168445~piPK:64168309~theSitePK:550226,00.html).

8.3.3 Prioritization of handling of TADs

All these public policy measures entail economic costs. Even though the human and economic benefits of preventing or containing an a TAD like influenza pandemic are overwhelming, governments may still be daunted by the cost of various policy measures, especially when these measures are in the nature of global public goods that benefit many more than just the citizens of that nation. The investment cost for scaling up the research for understanding the disease ecology, developing vaccines, antiviral medicines would be a case in point. Thus careful economic analysis of different aspects in management of TADs in an efficient and cost-effective manner will go along way in optimizing the resource utilization making the resource allocation an easy job in the effort to prevent or contain TADs. Similarly, the use of blanket measures to curb movement within and between countries could cause major economic disruption due to the increasingly globalized nature of modern production processes. Again, careful analysis of critical logistical chains in the world economy could allow consideration of targeted prophylactic use of antiviral medicines to protect transport and other key groups of workers (http://web.worldbank.org/WBSITE/EXTERNAL/COUNTRIES/EASTASIAPACIFICEXT/ EXTEAPREGTOPHEANUT/0,,contentMDK:20713527~pagePK:34004173~piPK:34003707~theSitePK:503048,00.html).

8.4 Future Research Thrusts

Future thrusts will include basic and molecular virology researches to develop immunodiagnostics and prophylactics, infrastructure development including information database, preparedness for diagnosis and control of exotic diseases, emerging and re-emerging diseases etc. various points which warrant immediate attention are (i) researches on molecular biology of pathogenic microbes to develop highly sensitive and cost-effective immunodiagnostics, improved prophylactics and safe and effective therapeutic regimens, (ii) identify strain variations of viruses isolated from vaccine outbreaks through nucleotide sequencing, (iii) identify sequences responsible for virus persistence and virulence, (iv) molecular basis of disease resistance, (v) improving mechanisms of disease forecasting and monitoring to ensure strategic control and eradication, (vi) creation of Animal Disease Database and Information System (ADDIS) infrastructure and interfacing with human disease databases, (vii) the Emergency Preparedness and Emergency Control Measures for emerging and re-emerging diseases, (viii) create Data Bank for exotic disease situations in India and its neighboring countries and network for dissemination of the information, (ix) develop diagnostic test capabilities for exotic diseases and be in readiness for providing diagnosis of exotic viral diseases at a short notice, (x) development of recombinant antigens (diagnostic reagents) for disease pathogens which have either been eradicated like Rinderpest or for exotic disease like avian influenza, SARS corona virus, BIV, BVDV etc. Critical reagents like monoclonal antibodies, if not available, may be procured commercially or from Reference Labs, (xi) provide referral and consultancy services and advise on procedures of diagnosis and containment of exotic disease problems, (xii) train Field Veterinarians, medical staff and other staff who could be of help in diagnosis and control of exotic/emerging diseases, and, (xiii) initiation of disease control and eradication programmes when a new disease emerges so that the disease can not establish itself.

8.5 Concept of "one-medicine" for Human and Veterinary Disease is relevant for TADs

The concept of "One-Medicine" for human and veterinary disease has been best quoted by Professor E.P.J. Gibbs while delivering Wooldridge Memorial Lecture at BVA Congress in 2005, held at Royal Society of Medicine. At the start of his lecture, Professor Gibbs drew attention to a portrait of Edward Jenner that hangs on the walls of the Royal Society of Medicine and he pointed out that "the Jenner's observation that the zoonotic character of cowpox could be used to protect against smallpox", was 'arguably the most important observation in the history of infectious disease'. In honour of Sir Edward Jenner's scientific approach and the role of the cow in protecting mankind from smallpox, Louis Pasteur named the process of inoculation 'vaccination' (from vacca, latin for cow) - appropriate recognition for the 'one medicine' for human and veterinary disease, as practiced by Jenner and Pasteur (Gibbs, 2005). A particular zoonotic viral infection in animals and human results due to the same virus which, in general, will not have any significant genetic change which might adversely affect the immune response after giving a vaccine which has been developed using the same organism isolated either from animal host or human. For example, H5N1 virus caused disease in poultry and humans and can remain in many species of water fowls without causing any disease. A vaccine developed using virus isolated from either bird flu affected poultry, human or water fowl should be able to protect both birds as well as human. The Avian flu, Hendra, Nipah, SARS Corona, are WNV are some of the best examples of TADs of public health significance where if a vaccine is developed using one of the virus isolates either from human or animals/birds can definitely be used as "One-vaccine" for human and veterinary disease. Similarly, many other vector-borne viral zoonoses can be controlled just by developing vaccine(s) using one of the viruses isolated either from human, animal host or vector.

9. THE BOTTOM LINE

It is likely that viruses of the future (as well as the ancestors of others that may yet evolve) are here now, in other species and in previously isolated human populations, awaiting their opportunity to emerge. This historical process has been going on for centuries, but the conditions of modern life provide rich new opportunities. The factor responsible for emerging infections are continuing and increasing, and the rapidity and extent of movement worldwide render all places increasingly vulnerable to new introductions. If we wish to prevent tragedies such as the flu pandemic from occurring, we must be prepared to find these infections and control them when they first appear. Thus, the bottom line in control of TADs is that nations should always be vigilant and be in preparedness to tackle the disease emergency by rapidly identifying the infected animals and eliminating these infected and in contact animals in the affected population following the OIE pathway for exotic animal viral disease control programme. Stamping has proved to be one of the best method to get rid of a TAD. Vaccination alone will not be enough to contain a newly emerged disease. Once the disease gets established, it may take decades or even centuries to eradicate the disease entailing huge sum of money from developmental grants of any nation or the International donor agencies.

Interesting readings

1. Avian Influenza Control and Eradication - FAO's Proposal for a Global Programme http://www.fao.org/ag/againfo/subjects/documents/ai/HPAIGlobalStrategy31Oct05.pdf
2. Delgado, C. L., Rosegrant, M. W. And Meijer, S (2001) Livestock to 2020: The revolution continues. Paper presented at the annual meeting of the International Agricultural Trade Research Institute. Auckland, New Zealand. IFPRI, Washington, DC, 38 pp.
3. FAO Statistical Databases 2004, FAOSTAT http://faostat.external.fao.org/
4. FAO - EMPRES (Emergency Prevention System against Transboundary Animal and Plants pests and Diseases) Avian Influenza website: http://www.fao.org/ag/againfo/subjects/en/health/diseases-cards/special_avian.html
5. OIE/FAO Expertise on Avian Influenza http://www.offlu.net/
6. OIE Regional Office for the Americas http://www.rr-americas.oie.int/es/proyectos/es_gf_tads/es_gftads_bsas_f06_recursos.htm

-ΔΔΔ-

6

Quality of Meat and their Products in Relation to Consumer Safety

NAGENDRA SHARMA
V. PATHAK AND A.K.BISWAS

Sher-e-Kashmir University of Agricultural Sciences and Technology, Jammu

Food safety is the key issue in the new pattern of meat trades, in the recent era. With the highly publicized World Trade Agreement (WTA) and Sanitary and Phyto-Sanitary (SPS) measures, consumers are more health conscious even past. Indeed, this demand a holistic approach in the reorganizations and strict implementation of quality control systems considering current agricultural and animal husbandry practices. Quality of meat and their products are affected by various factors during their production, processing, handling, packaging and then transportation. However the fact, among all quality attributes, hazards associated with physical, chemical and biological origin comes under first priority in the quality assurance schemes for consumer's safety and promotes export trade.

The office of the United States Trade Representative has estimated that international trade has increased five fold since signing of General Agreement on Tariff and Trade (GATT) in 1947. Following this agreement, the creation of World Trade Organization (WTO) in 1995 resulted in further significant increased trade, in food of animal origin and live animals among different countries. Globalization of food supplies means that people are exposed through foods purchased locally, to hazards native to remote part of world. The residual contaminants, emergence and re-emergence of diseases are the key issues in the new pattern of food trades. The monitoring and strict implementation of these emerging hazards continues to be a major demand for both importing and exporting countries. Emergence and reemergence of diseases are reflected by the evidence of Enterohaemorrhagic *Escherichia coli,* once confined to North America until mid 1990s, are now found throughout the world. Likewise, multidrug resistant *Salmonella typhimurium* DT-104 has spread widely since their first detection in United Kingdom [1]. Other food borne emerging diseases include *Listeriosis,* which spread throughout the France and also in Canada, where meat and meat products were implicated as a source of *Listeria monocytogenes*[2]. *Staphylococcal* food poisoning or food intoxication syndrome first reported in 1894 is now a global problem in meat industry.

The Codex Alimentarius Commission (CAC) is the reference point of World Trade Organization for disputes involving food which has achieved a great deal of international consensus on food export or import inspection and certification system. Realizing the situation and facts now it is essential to work together in order to develop a comprehensive concept for ensuring the quality and origin of meat.

Status of meat production, slaughtering conditions and processing

India the second most populous country in the world and third largest economy in Asia is yet to take its due share in meat production, in spite of the largest livestock resource in world. The present meat production of India is over 6 million tones as compared to world's 272 million tones i.e. 2.2% of world production. Global meat production has increased more than five fold since 1950 and more than doubled since the 1970s. Total meat production in developing countries grew by 5.4 percent per year between the early 1980s and mid-1990s, more than five times the developed-world rate. In India also out of last 25 years, the compound average growth rate during the previous two decades remained 4.6% but reached 21% during the last 5 years indicating it as fastest growing segment of livestock sector

At present, there are 3600 licensed slaughterhouses in the country. Most of the slaughterhouses are under civic bodies and are old, out-dated, sub-standard with poor hygienic conditions. These slaughterhouses operate as service abattoirs where butchers are paid a fee for slaughter either in form of cash or edible and inedible part of carcass. However, the recent trend in India is to establish latest state of art abattoirs cum meat processing plants. India has already established 10 most modern state-of-art merchandized abattoirs cum meat processing plants in various states and a few are under construction. These plants follow all the sanitary and phyto-sanitary (SPS) measures required by the International animal heath code of O.I.E. There are also four integrated poultry meat processing plants which follow all the sanitary and phyto-sanitary (SPS) measures.

Most meat in India is sold in butcher retailer shop as hot meat and cooked in domestic kitchen. Less than 3% of total meat is sold as processed meat, whereas in developed countries it is 65-80%. Instead of export of live animals and fresh meat, processing and export of processed meat products has become inevitable to fetch more profit and employment generation. There are 170 licensed meat processing units and 6 bacon factories in the country. Production of Western meat products like cured ham, bacon, sausages, frankfurters, hot dog, luncheon meat etc have also started with good pace. Presently among the processed meat, cured products comprises (50%), sausages (20%), canned products (20%). Situation is changing in favour of processed meat products especially in big cities and competition at global level demands a strict quality monitoring system to assure the consumer safety.

Cause of Concern and Hazards Associated With them

The quality of meat and meat products are at high risk because of existence of various contaminants either in form of microbes or pesticide, drug and chemical residue. All these after making their way to product for consumption by either means, produces a serious threat to wholesomeness of product and therefore consumer safety. The important cause of concern has been broadly classified under four heads here.

A) Microbial Contaminants

Microbial contamination and spoilage pose the most dangerous threat to meat and meat products. The number and types of microorganisms found on meats are determined by the environment under which the animals are raised and processed, properties and microbiological quality of any added ingredients, packaging, handling, storage and distribution conditions. The natural surface floras of meat animals usually of least importance as the contaminating microorganisms are from their intestinal or respiratory tracts. However, hides, hooves and hair contain not only large numbers of microorganisms from soil, manure, feed and water but also important kinds of spoilage organisms. The skin of many meat animals may contain *Micrococci, Staphylococci* and beta-haemolytic *Streptococci.* Staphylococci on the skin or from the respiratory tract may find their way onto the carcass and then to the final raw product. The faeces and faecal contaminated products of animals may contain many enteric organisms including *Salmonella, Escherichia coli* O157:H7, *Campylobacter jejuni or Campylobacter coli.* Personnel engaged in meat processing operations also can act as vector of many foodborne pathogenic bacteria, but this represents only little importance [3].

Meat can acquire a large variety of pathogenic and spoilage microorganisms during primary and further processing. Pathogens can include *Clostridium perfringens, Staphylococcus aureus, Salmonella* spp, pathogenic *E. coli, Campylobacter* spp, *Yersinia enterocolitica, Listeria monocytogenes, and Aromonas hydrophila.* Spoilage of meat is largely dependant upon initial microbiological quality, handling and subsequent storage conditions [4]. Yeast and molds are not a major concern for fresh muscle foods, because they develop slowly and predominate only on carcasses after prolonged storage and aging which reduces growth of bacteria due to surface drying. The flora includes *Torulopsis, Candida, Rhodotorula, Cryptococcus, Tricosporon, Sporotrichum, Cladosporium, Thamnedium, Mucor, Chaeotostylum, Penicillium, Rhizopus, Aspergillus, Monilia* and *Alternaria.*

Growths of these microorganisms not only hamper the physico-chemical and sensory attributes of the products but also render them unsafe for consumption because of their toxic metabolite produced during their growth. There are considerable evidences of foodborne pathogens responsible for major health hazard effects. A number of outbreaks of meat borne infections and intoxications are being reported in human beings every year.

B) Pesticide Residues

These are either synthetic or natural compound which can leave residues in edible food products from animal exposed to that compound. A number of chemicals are used in modern agriculture for producing food commodities as cheaply and efficiently as possible. Pesticides the agro-chemicals first came to the attention of an environmentally conscious public in early 1960s. Most pesticides are acutely toxic to human and animals however; ingestion of low levels over a long time can also have serious adverse effects. In recent year, violative levels of pesticides are relatively uncommon. But despite of this, pesticide remains an important public health consideration because of their widespread use in meat and poultry production, their persistence in the environment, and their varying toxicity. It has been estimated that nearly 80% of all contaminants in food animals are estimated to occur through the feed. The major contaminations have resulted from application of pesticides in crop fields; however, feed may be contaminated by improper use of certain pesticides during storage, processing or transportation or by storing grain where other chemicals have previously been stored. Other sources of exposure include drinking water contaminated through crop field run off pollution, the mixture of misapplication of pesticide to animals or their housing also can result in toxicosis [5].

Pesticides via food are the source of several mysterious diseases unheard in the past. The very horror of these would drag the patient fast to the graveyard. The catalogue of such diseases and disorders include cancer, epilepsy, liver and kidney dysfunction, somatic cell growth, depression and neuritis [6]. The United Nations has estimated that about 2 million poisoning and 10,000 deaths occur each year from pesticides, with about three-fourths of this occurring in developing countries [7].

C) Antimicrobial Drugs Residue

In contrast to pesticides, residues of veterinary medicinal products are most common in food of animal origin as are directly exposed to the animals. The range of veterinary medicinal products used in regular animal husbandry practices is extremely wide, ranging from teat dips to hormones. Approximately 42% of all veterinary pharmaceuticals used world-wide are used as feed additives, 19% are used as anti-infectives, 13% as parasiticides, 11% are used as biologicals and 15% represent other pharmaceuticals [8]. The most commonly used antimicrobials in food animals include â lactams (benzyl-penicillin/ procaine benzylpenicillin), tetracyclines (oxytetracycline, chlortetracycline and tetracycline), aminoglycosides (streptomycin and gentamicin), macrolides (erythromycin) and sulfonamides (sulfadimidine) administered to animals either by injections (intramuscularly, intravenous, subcutaneous) or orally in the feed and water, topically on the skin, and by intra-mammary and intra-uterine infusions [9]. A study reported that injectables were responsible for

46% of the violative residues in meat followed by oral administration at 20% (feed, water and bolus) and intra-mammary infusions at 7%. The National Research Council and Institute of Medicine have noted a link between the use of antibiotics in food animals and the development of bacterial resistance to these drugs causing human diseases [10].

D) Others

Several industrial chemicals posing threat to human health have also become matter of concern as they are hormone disruptors. Polychlorinated Biphenyls (PCBs) and Polychlorinated Dibenzo-p-dioxins (PCDDs) compounds are environmental contaminants that are fat soluble and chemically stable. PCBs were used in electrical transformers, in electronic parts, and flame reductants, whereas PCDDs originate from combustion of chlorine containing organic compounds. These compounds get entry into the animal models and food chain thus causing several diseases like liver disorders, cancer, reproductive and immunological disorders in animals and human beings.

Besides these there are some substances that get their way in meat or meat products during treatment, processing or preservation. As regards the desirable substances, they must be present at certain level in the product to achieve the desired effect than too high a level that leaves undesirable residues behind. This group include nitrate, nitrite that form nitrosamines in cured meat products, benzopyrenes from smokehouse processes in smoked products are carcinogenic in nature.

The physical hazards like glass, hair, metals, grease paints, rubbers, insects, plastics etc. may also get entry into meat and their products through incoming raw material, poor personnel hygiene & practices, faulty processing, facilities, and equipment. Physical hazards can cause varying degree of injuries, however these hazards, when compared to chemical and biological hazards, may not be distributed as uniformly through out the meat products.

Control Measures

In most of the countries, the meat safety and quality assurance have been regulated by ante-mortem and post-mortem inspection. In case of quality of meat products, the emphasis has been laid on microbiological and sensory profile of the products. However with advancement in awareness, science, methodologies etc, safety from other hazards like physical, pesticide & drug residues have also gained importance and regulations are being developed globally to monitor the quality of meat and meat products, but still a lot of efforts are required for establishing perfect limits and monitoring systems for newly recognized hazards.

The meat wholesomeness is attributable to many factors starting from farm to consumer's table. Strict quality control is needed for each and every processing step to ensure consumer safety of the product. Animal welfare and ethical considerations should be compulsory as it is not only favors raising of quality livestock crops sustainable but also helps in the improvement of quality and safety of meat and their products substantially. The important factors identified are

1. Production of healthy, disease free animals
2. Hygiene & animal welfare considerations during transportation of animals from farm to production units
3. Cleanliness and sanitation of all parts abattoirs
4. Pre-slaughter care, ante-mortem and post-mortem inspections
5. Proper handling, packaging, storage and distribution.

Similarly in meat products hygienic conditions of plant/premises, personnel, equipments, post-processing handling are very critical factors along with meeting packaging and storage requirements. In the present scenario, there is tremendous increase in demand for processed products resulting in flood of such commercial products in the market. The processing, storage and distribution of processed meat products needs to be more strictly monitored because unlike raw meat there is destruction of natural spoilage bacteria by partial or regular cooking and food poisoning bacteria may grow without any evidence of spoilage.

The most effective way to control microbial problems is through prevention. The Hazard Analysis Critical Control Point (HACCP) system is a preventive scientific approach to quality control. Both the concept and practice of the HACCP system began with the Pillsbury Company's projects in food production and research for the space program. The HACCP system was first made public during the 1971 National Conference on Food Protection. Since then, it has been adopted by many countries around the world.

There are seven steps of the HAACP program, beginning with hazard analysis and ending with verification. Principles of Hazards are:

1. Conduct Hazard Analysis
2. Identification of Critical Control points (CCPs)
3. Establish critical limits (CLs)
4. Establish CCPs monitoring procedures
5. Establish corrective actions
6. Establish verification procedures
7. Establish documentation concerning all procedures appropriate to these principles and their applications.

These principles must be used to develop a HACCP plan. It is designed to prevent problems by making sure that controls are applied at any point in a food production system where hazardous or critical situations could occur. Implementation of HACCP also ensures Good Manufacturing Practices, Standard Operating Procedures and Sanitation Standard Operating Procedures HACCP certification is mandatory for export of meat and meat products in many countries of the world.

Regulations of contaminants in India are governed under different Acts/Rules: Insecticides Acts, 1968; Hazardous (Management and Handling) Rules, 1989; Prevention of Food Adulteration Acts, 1954; The Factories acts, 1948; Bureau of Indian Standards Act, 1986 and Drug and Cosmetic Act, 1940.The Ministry of Agriculture (Govt. of India) regulates the manufacture, sale, import, export and use of pesticides through the Insecticides Act. The monitoring of residue level in food comes under the purview of Ministry of Health and Family Welfare (Govt.of India) .The Central Committee on Food Standards (CCFS), a statutory committee under the PFA Act, 1954. Ministry of Health and Family Welfare is responsible for laying down standards for various food items [11]. National drug committee is responsible for approval of new drugs intended for use. The various controlling and monitoring agencies need to device set standards and effective and acceptable methodologies for regulating different hazards throughout the food chain from primary production to final consumption.

Conclusion

During the past several decades, several scientists all over the world have been performing a variety of fruitful and valuable research activities to promote meat safety. However the milestone of ensuring meat safety is still far ahead. Practices are needed at the live animal level that can be implemented realistically which will reduce the hazards in the final products. Application of HACCP system as a part of quality control measures is inevitable for ensuring the safety and wholesomeness of meat and meat products. In our countries various acts/ rules exist regarding the regulation of contaminates in food articles however, regulatory authorities have to be prompt to play their role in enforcing relevant standards and laws.

References

1. JECFA. 2002a. Future trends in veterinary public health. Fifty-eight report of the Joint FAO/WHO Expert Committee on Food Additives. Report of a WHO study group. WHO technical Report Series. No. 907.
2. Anon. 2002a. Environics International. Food Issues Monitor. Available: http:/ www.environisinternational.com. /default.asp? sp-fim.asp.
3. Frazier W C and Westhoff D C. 1999. Contamination of foods. In: Food Microbiology. Frazier, W.C. and Westhoff, D.C., eds.4th edn. Tata McGraw-Hill Publishing Limited, New Delhi, pp59-68.
4. Gill C O. 1998. Microbiological contamination of meat during slaughter and butchering of cattle, sheep and pigs. In: Davies A. and Board R (eds.). *The Microbiology of Meat and Poultry.* Blackie Academic & Professional, New York. pp 118-157.
5. Heise S. 1992. Pesticide residues in food as a potential health hazards in the third world. In: Proceeding of 3rd Congress on foodborne infection and intoxication held at Berlin, June16, 1992,Vol.II.pp706-707.
6. Daniels J L, Olsan AF and Savitz D A. 1997. Pesticides and childhood cancers. *Environmental Health Perspectives,* 105(10): 1068-1077.
7. Quiyano R, Panganiban L. and Cortex-Maraba N. 1993. Time to blow the whistle, danger of toxic chemicals. World Health, 46 (5): 26-27.
8. Miller D J S. 1993. Present state and trends in the use of veterinary drugs. *Proceeding of the Euro Residue II Conference*, Veldhoven, The Netherlands, 3-5 May.
9. Mitchell J M, Griffith M W, McEwen S A, MeNabb W B. and Yee A J. 1998. Antimicrobial drug residues in milk and meat: causes, concerns, prevalence, regulations, test and test performance. *Journal of Food Protection* 61:742-756.
10. NRCIM. 1999. The use of drugs in food animals: Benefits and risks. Natural Research Council and Institute of Medicine, National Academy Press, Washington, DC.
11. Dikshit A K and Parmar B S. 2005. Pesticide residue and food safety. In: National Seminar on Food Quality and Safety Standards of agricultural Raw Produce.apri26-27.New Delhi, pp74-77.

–ΔΔΔ–

7

Development and Quality Control of Herbal Veterinary Formulations for Effective Health Management in India

V S VIHAN AND ASHOK KUMAR

Division of Animal Health
Central Institute for research on Goats
Makhdoom, Farah , Mathura (UP) - 281122

Medicinal plants have been a major source of cure of human diseases since time immemorial. Today, one fourth of the world population depends on traditional medicines. Despite the introduction of antibiotics since the 1940's, even 80 per cent of the population today relies on indigenous medicinal plants and the drugs. It is estimated that the global traditional medicine market is growing at the rate of 7 - 15 per cent annually. The medicinal plant value is about Rs.5000 crores in India and it is estimated that the country exports about Rs.550 crore worth of herbal drugs. But with the rich and diverse botanical resources in our country, this is not an impressive export performance considering the worldwide herbal market worth US 60 billion dollar[6].

There is urgent need that Indian pharmaceuticals have to stress on various standardization processes to establish their products in the global markets. Standardization starting from production of quality materials, analysis of raw materials for authentication, foreign matter, organoleptic evaluation, microscopic examination, extractive values, chromatographic profiles, pesticides residue, heavy metal detection etc. are necessary for standardization of drugs. Similarly, the standardization methods of medicinal plants and its extracts have great importance in the fields of cosmetics and neutraceuticals, which are emerging as two most important segments in the global markets for human being. Due to the need and emphasis on quality control of drugs, chemical assay, stability, safety assessment, pharmacokinetic studies etc. have become necessary to establish the products and its efficacy in the global markets. Setting up of high profile and modern R&D quality control labs have become an unavoidable tool for herbal manufacturers to sell their products worldwide. By referring global standards and international pharmacopoeia like Herbal B.P., China, Japanese Herbal, Ayurvedic Formulary of India, WHO Guidelines on Herbal Medicines, this could be met with. It seems to very necessary that Indian herbal companies have to take steps to establish a good quality control mechanism to survive in the domestic and international markets, for which the government should consider assisting the industry in setting up a well-established unit which shall work towards excellence in standardization of drugs to meet international requirements in the coming years.In order to withstand competition in the global market, it is necessary to create a brand image, especially in cosmeceuticals and natural products. It is also necessary to integrate modern knowledge with traditional knowledge.

Side effects of conventional medicine, efficiency of plant-derived drugs and growing interest in natural products has increased scientific interest in medicinal plants. A look at the research and investments that are going into medicinal plants looks like; they will continue to play an important role in human and animal health. Today, there are at least 120 distinct chemical substances derived from plants that are considered important herbal drugs. Statistical data available in USA shows that 41 percent of pharmaceutical development has originated from herbal drugs. As per WHO estimates, 80 percent of about 4000 million inhabitants one this planet rely on plant products.

Herbal medicines have been used since the dawn of civilization to maintain health and to treat diseases. The WHO estimates that about three-quarters of the world's population currently use herbs and other forms of traditional medicines to treat their diseases. Even as we commence the new century with its exciting prospect of gene therapy, herbal medicines remain one of the common forms of therapy available to the world population.

Most of time, herbal products are thought safe, however, herbal products can be as toxic as or even more toxic than prescription medicine. They can also have unwanted side effects, cause drug interactions and possibly create surgical problems. Another problem is that people tend to rely more on testimonials on the benefits of herbal products than solid, scientific evidence. According to a study in the Journal of the American Medical Association, roughly 15 million adults are at risk of possible adverse interactions between prescription medicines and herbs or high dose vitamins. There is an example of Kava, the popular herbal relaxant, causing severe liver damage, reported in German, Britain , France and Canada.

Herbal formulations are popular among the rural and urban community in India. The reasons for the popularity and acceptability of natural products are due to its safe nature. In the literature, therapeutic efficacy or toxicity is some times attributed to the herb without describing the brand or its content of bioactive or quantitatively significant components. Pharmaceutical products used as medicines are usually single chemical entities with specific actions at receptors, enzymes and other cellular sites. These drugs or preparations are marketed after vigorous clinical trials to support rational pharmacotherapy. The most important question regarding any drug to be used is how safe it is for clinical use. Unfortunately this aspect is not addressed or transparent to the consumers who want to use the herbal formulation.

The herbal formulations, which are sold as over, the counter (OTC) have a different protocol regarding preparation, acquiring license and marketing. The active principles of herbal preparations are not often well defined. Scientists or consumers do not know regulations regarding safety and efficacy.

The herbal medicines are poorly regulated. Assurance of safety, quality and efficacy of medicinal plants and herbal products is a key issue, which needs to be addressed.

A number of countries including Germany, France, Canada, USA, China, etc. are registering standardized plant extracts of proven clinical efficacy and safety obtained from natural sources as herbal drugs or dietary supplements. Inspite of the fact that India has a vast resource of drugs of natural origin, we are unable to exploit the vast world market because we have an unsatisfactory system of their quality control and registration[6]. Therefore, there is need to strengthen the quality control system in India. Moreover, a separate directorate at National levels for approval, regulation and functioning of veterinary drugs and pharmaceuticals include Ayurverdic and homeopathic medicines.

1. Status of medicinal Plants Wealth of India

India is rich in medicinal plant diversity. All known types of agro climatic and ecologic conditions are met within India. The bio geographic position of India is so unique that all known types of ecosystems. India is rich in all the three levels of biodiversity-such as species diversity, genetic diversity and habitat diversity. There are about 426 biomes representing different habitat diversity that gave rise to one of the richest centers in the world for plant genetic resources. Out of 17,000 plants, the classic systems of medicines like Ayurveda, Siddha and Unani make use of only about 2000 plants in various formulations. The rural people who constitute 70 to 75% of the Indian populations live in about 5,76,000 villages located in different agro climatic conditions. The village people have their own diverse systems of health management. While most of the common ailments were managed in the house by home remedies, which included many plant species. There is however no systematic, inventory and documentation about the folk remedies of India. There is urgent need to document this fast disappearing precious knowledge system.

Indian medicine is not reflected merely in the treatment of human beings but other important dimensions like veterinary medicine are addressed in detail through these systems. This represents a whole new spectrum of knowledge and opportunity. This area has not been exploited at all. The definition of Ayurvedic medicines under the Drugs & Cosmetics Act includes veterinary medicines. There are several authentic books of Ayurveda dealing with veterinary medicines such as Nakul Samhita, Palkapya Shastra, Go Ayurveda, Hasti Ayurveda, Baj Nama etc. Central Government will encourage two institutions in the 10th Plan for introducing courses and undertaking documentation of the classical books in various languages. Homoeopathy also has effective treatment for care of animals and a similar approach would be followed.

ICAR New Delhi, under a Net work project (NATP) on collection and documentation of Indigenous technical knowledge in the field of treatment of animals and subsequently their scientific validation to find out potential ITKs for their use by farmers and also drug development by pharmaceutical Industry . As a result, Inventory have been prepared and published for scientist of the country. The oral traditions of the villagers use about 5000 plants for medicinal purposes. India is also inhabited by a large number of tribal communities who also posses a precious and unique knowledge about the use of wild plants for treating human ailments. A survey conducted by the All India Coordinated Research Project on Ethnobiology (AICRPE) during the last decade recorded over 8000 species of wild plants used by the tribals and other traditional communities in India for treating various health problems. Some interesting observations made in the study is the use of the same species found in different regions for the same ailments while some other species are used differentially.

IUCN (International Union for Conservation of nature and Natural Resources) categorized red list plants, which are threatened[4]. Species (or sub specific taxa) facing any threat of extinction should be assigned to one of the following categories on the basis of detailed and specific criteria, given by the IUCN[6,9,1].

1.1 Threat status at the global and Indian levels

The following is a comparative statement of the threat status of vascular plant species in India and at the global level as per the 1997 IU CN Red List of Threatened Plants:

Nature	Global	India
Extinct (Ex)	380	19
Extinct/Endangered (Ex/E)	371	41
Endangered (E)	6522	152
Vulnerable (V)	7951	102
Rare (R)	14505	251
Indeterminate (I)	4070	690
Total under threat	33418	1236
Total number of species	242013	16000
Per cent under threat	13.8	7.7

Cited from "*Red list of threatened vascular plant species in India*", 2003 Ministry of environment and forest , Government of India

Acanthaceae, Annonaceae, Asclepiadaceae, Compositae, Euphorbiaceae, Labiatae, Leguminosae, Rubiaceae, Umbelliferae, Araceae, Gramineae, Orchidaceae and Palmae appear to be under greater threat than others. In these families both the genera and species under threat are in double digits, the highest being in Orchidaceae with 46 genera and 105 species. If the actual number of genera and species in different families are accurately known, meaningful percentages can be calculated. Though the sizes of the families and the actual numbers of taxa under threat are the parameters, the actual numbers may be misleading. If a family with one genus and one species is under threat, it will be an alarming 100 per cent threat. There should be a rational assessment of the degree of threat for the families, genera and species, which requires a lot more of dependable data than now available[8].

1.2 Strategies of conservation in India

A holistic and systematic approach envisaging interaction between social, economic and ecological systems will be a more desirable one(IUCN, UNEP & WWF). The most widely accepted scientific technologies of biodiversity conservation are the *in situ* and *ex-situ* methods.

Establishment of biosphere reserves, national parks, wild life sanctuaries, sacred groves and other protected areas are some examples. In India 4.5% of its total geographical area constitute protected area network, comprising 8 designated biospheres, 87, national parks, 447 wild life sanctuaries. Conservation of medicinal plants can be accomplished by the *ex-situ* i.e. outside natural habitat by cultivating and maintaining plants in botanic gardens parks, other suitable sites, and through long term preservation of plant propagates in gene banks (seed bank, pollen bank, DNA libraries, etc.) and in plant tissue culture repositories and by cryopreservation). Tropical Botanical Gardens & Research Institute (TGBRI), located in a degraded forest region of Western Ghat Mountains in Kerala has an excellent example in ex-situ conservation of plant diversity in India. The National Parks and Wildlife Sanctuaries are protected areas encompassing the Biological Diversity in its pristine condition. The Ministry of Environment and Forests funded an All India Coordinated Research Project of Ethnobiology (AICRPE). The programme was executed by Tropical Botanical Garden Research Institute, Thiruvananthapuram and in association with other institutions such as Central Drug Research Institute (CDRI), Lucknow, Ethnobiology and Plant Systematic Laboratory, Garhwal, Botanical Survey of India, International Institute of Ayurveda, Coimbatore, etc.

1.3 Research and Development in Medicinal plants

Urgently main thrust should be on evolving and optimizing the most appropriate technologies for conservation, especially for endangered or endemic specie and molecular methods for characterization; detailed studies on life cycle and breeding behaviour, taxonomy, seed biology ; Population and habitat viability studies ;optimizing appropriate methods for post-harvest handling, processing and storage and Investigation on quality control standardization and shelf life of raw materials and finished products. A multidisciplinary approach to drug development from medicinal plants used in traditional medicine was tried out in several technical assistance programmes by UNIDO beginning in the mid-1970s. The important points are, economic mapping of the spontaneous flora; Selection and authentication of plant species; Collation of ethnomedical and ethnobotanical data; trial propagation to develop high yielding varieties; medium scale plant propagation, chemical studies on plant constituents; analytical studies on quantitative assessment of major constituents; pharmacological and toxicological studies; Pilot plant scale processing of plant extracts, standardization of extracts; formulation studies on extracts into dosage forms; toxicity studies on formulations and analytical studies on formulated products

Additionally and importantly, pharmaceutical importance/ therapeutic value of medicinal plants is due to specific constituents/ combination of secondary metabolites present in them. Changes in the proportion of secondary metabolites are often required for the improvement of therapeutic values of medicinal plants. Genetic manipulations can help increase/decrease the contents of specific compounds in medicinal plants. These genetic techniques will also help develop designer medicinal plants. For example it will be possible to produce medicinal plants that will help raise the immunity towards infectious diseases in the use of such plants.

1.4 Priorities for drug development

It is necessary to priotise the thrust areas to obtain the output of research efforts and other resources. Several factors help in determining the priorities. These include the distribution of flora, national or regional disease pattern, availability of modern health care etc. In addition we have to keep in mind the global priorities in developing new drugs so as to get a good financial return. Primary health care usually requires comparatively milder medication and the acceptability of herbal medicines for such conditions is also much more. The main considerations should be adequate availability or possible cultivation on required scale, lack of toxicity and ease of formulation. In a Seminar, organized by DST, New Delhi at Hyderabad on "Veterinary drugs and pharmaceuticals" on 23-24, March 2002. in which veterinary scientists prioritized the diseases and condition for the research in the field of haemato-protozoan diseases, ectoprastic infesation,, helminthic infections, bacterial diseases, wound healing, mastitis, reproductive disorders, inflammatory diseases, immuno-stimulation, haematinics and coccidiosis.

2.0. Regulatory Situation of Herbal Medicines (The WHO Traditional Medicine Programme)

The World Health Assembly (WHA) has adopted a number of resolutions drawing attention to the fact that a arge section of the population in many developing countries still relies on traditional medicine. In 1978, the Declaration of Alma-Ata recommended, *inter alia*, the inclusion of proven traditional remedies into national drug policies and regulatory measures. The policy of the World Health Organization regarding traditional medicine was presented in the Director-General's report on Traditional Medicine and Modern Health Care to the Forty-fourth World Health Assembly 1991. Major objectives of the Traditional Medicine Programme are: to facilitate the integration of traditional medicine into national health care systems; to promote the rational use of traditional medicine through the development of technical guidelines and international standards in the field of herbal medicine and acupuncture; and to act as a clearing house for the dissemination of information on various forms of traditional medicine. In resolution WHA42.43 (1989), the Health Assembly urged Member States: to make a comprehensive evaluation of their traditional systems of medicine; to make a systematic inventory and assessment (preclinical and clinical) of the medicinal plants used by traditional practitioners and by the population; to introduce measures for the regulation and control of medicinal plant products and for the establishment and maintenance of suitable standards; and to identify those medicinal plants, or remedies derived from them, which have a satisfactory efficacy/side-effect ratio and which should be included in national formularies or pharmacopoeias.

In recent years, many developed countries have shown growing interest in alternative or complementary systems of medicine, with a resulting increase in international trade in herbal medicines and other types of traditional remedies[5]. A WHO consultation in Munich, Germany, June 1991, drafted Guidelines for the Assessment of Herbal Medicines, which was adopted for general use by the Sixth ICDRA in Ottawa, October 1991. These guidelines (WHO/TRM/91.4) define basic criteria for the evaluation of quality, safety and efficacy of herbal medicines to assist national regulatory authorities, scientific organizations, and manufacturers to undertake an assessment of the documentation, of submissions and/or the dossiers in respect of such products.

These guidelines contain basic criteria for the assessment of quality, safety, and efficacy and important requirements for labeling and the package insert for consumers' information. The WHO Guidelines are intended to facilitate the work of regulatory authorities, scientific bodies and industry in the development, assessment and registration of herbal medicines, reflecting scientific results which could be the basis for future classification of herbal medicines and would also accommodate cross-cultural transfer of traditional herbal medicinal knowledge between different parts of the world. The aim of such national policies would be to develop regulatory and legal reforms to ensure good practice, and to extend primary health care coverage, while ensuring the authenticity, safety and efficacy of these medicines. Main objectives include the recognition of traditional medicine as an integral part of national health care systems, co-operation between modern and traditional medicine, promotion of the rational use of products, the introduction of quality assurance systems, the guarantee of regular supplies, the promotion of research and development of regulatory measures. To develop criteria and general principles to guide research work on evaluating herbal medicines, the WHO Regional Office for the Western Pacific, in 1992, organized a meeting of experts to develop guidelines for research on herbal medicines.

Research approaches should differentiate between herbal medicines with a long documented experience and those the "traditional" use of which has not yet been established. In accordance with the WHO Guidelines for the Assessment of Herbal Medicines (WHO/TRM/91.4), traditional experience with the respective preparation, which includes long-term use as well as the medical-historical and ethnological background, should be taken into consideration as a general rule in conducting research .

Herbal medicines have two special characteristics which distinguish them from chemical drugs; use of crude herbs and prolonged usage. A single herb may contain a great many natural constituents and a combination of herbs even more. Experience has shown that there are real benefits in the long-term use of whole medicinal plants and their extracts, since the constituents in them work in conjunction with each other. However, there is very little research on whole plants because the drug approval process does not accommodate undifferentiated mixtures of natural chemicals, the collective function of which is uncertain. To isolate each active ingredient from each herb would be immensely time-consuming at unsupportable cost, and is almost impossible in the case of preparations. The summary and recommendations of the Sixth ICDRA prompted WHO to continue to develop pharmacopoeial monographs on herbal medicines on the basis of the Guidelines for the Assessment of Herbal Medicines. In response to the request from Member States, WHO's Traditional Medicine Programme decided to prepare a technical document entitled "WHO Monographs on Selected Medicinal Plants" for primary health care. The information in the monographs includes two parts: Part I consists of summaries of the botanical characteristics, major active chemical constituents and quality control of each plant; Part II consists of summaries of clinical applications, pharmacology, posology, possible contraindications and precautions, and potential adverse reactions.

2.1. Regulatory situation for market and use of herbal medicines in India

In India, a great deal of folk knowledge exists among ordinary people about the traditional use of herbal medicines. It is difficult to quantify the market size of the traditional Indian systems since most practitioners formulate and dispense their own recipes. The present annual turnover of products manufactured by large companies is estimated at approximately US $ 300 million, compared to a turnover of approximately US $ 2.5 billion for modern drugs. According to a study on the attitude of modern medicine practitioners towards Ayurvedic products, general practitioners are relatively unfamiliar with Ayurvedic products even though some are prescribed. They are willing to try an Ayurvedic product if its efficacy is scientifically proven, and would try Ayurvedic products if no modern medicinal remedies were available. People use self-medication for minor ailments such as cough, cold, diarrhoea and stomach problems. Patent and proprietary Ayurvedic medicines are sold over the counter in pharmacies. These products appear to represent a major share of branded traditional products in India. Nevertheless, systems like Ayurveda still need to gain an empirical support of modern medical science to make them credible and acceptable for all. An innovative research effort to define the advantages of traditional systems of medicine with respect to their safety and efficacy could result in a better utilization of these complementary systems of medicine[7].

Legal Status

In India, traditional medicines are governed by the Drugs and Cosmetics Act of 1940 and the Drugs and Cosmetics Rules of 1945. They regulate the import, manufacture, distribution and sale of drugs and cosmetics. In 1959, the Government of India recognized the traditional Indian systems of medicine and amended the Drugs and Cosmetics Act to include drugs, which are derived from traditional Indian medicine.

GMP Notification on herbal drugs (Ayurvedic drugs)

It was published in Gazette of India Part II Section 3, sub-section iii] by MINISTRY OF HEALTH AND FAMILY WELFARE (Department of Indian Systems of Medicine and Homoeopathy) NOTIFICATION New Delhi, the 2003.G.S.R. Whereas a draft of certain rules further to amend the Drugs and Cosmetics Rules, 1945. It ensure that (i) raw materials used in the manufacture of drugs are authentic, of prescribed quality and are free from contamination;(ii) the manufacturing process is as has been prescribed to maintain the standards;(iii) adequate quality control measures are adopted;(iv) the manufactured drug which is released for sale is acceptable quality;(v) to achieve the objectives listed above, each licensee shall evolve methodology and procedures for following the prescribed process of manufacture of drugs which should be documented as a manual and kept for reference and inspection. General Requirements like Location and surroundings (Buildings, Water Supply, Disposal of Waste, Container's Cleaning, Storage). All raw materials procured for manufacturing will be stored in the raw materials store. The manufacture based on the experience and the characteristics of the particular raw material used in Ayurveda, Siddha and Unani system shall decide the use of appropriate containers, which would protect quality of the raw material as well as prevent it from damage. All workers employed in the factory shall be free from contagious diseases.

For carrying out manufacturing depending on the size of operation and the nature of product manufactured, suitable equipment either manually operated or operated semi-automatically (Electrical or steam based) or fully automatic machinery shall be made available. List of equipments and machinery recommended is indicated in Part II A. Proper Standard Operational Procedures (SOPs) for cleaning, maintaining and performance of every machine should be laid down.The licensee shall maintain batch-manufacturing record of each batch of Ayurvedic, Siddha and Unani drugs manufactured irrespective of the type of product manufactured (classical preparation or patent and proprietary medicines). Every licensee is required to provide facility for quality control section in his own premises or through Government approved testing laboratory. The test shall be as per the Ayurveda, Siddha and Unani pharmacopoeial standard.

No products derived from traditional systems may be manufactured without a licence from the State Drug Control Authorities. Patent and proprietary medicines derived from the traditional systems must contain ingredients, which are mentioned in the recognized books of the above systems, as specified in the Drugs and Cosmetics Act. A special committee and an advisory board advise the government for Ayurvedic, Siddha and Unani drugs. Pharmacopoeia committees have been constituted to prepare pharmacopoeias for all these systems (Rajagopalan, 1991) .In 1993, an expert committee appointed by the Indian government developed guidelines for the safety and efficacy of herbal medicines which were intended to be incorporated into the Drugs and Cosmetics Act and rules. It was proposed that no new herbal medicines other than those authorized by the licensing authorities be allowed to be manufactured or marketed, except for those mentioned in and manufactured in compliance with the formulae given in the "authoritative" books for Ayurveda, Siddha and Unani herbal medicines. A manufacturer of a new herbal medicine must include safety data and appropriate efficacy data in the marketing authorization application. Herbal preparations are defined as natural products in which the predominant active constituents are of plant origin. A classification for herbal medicines was proposed depending on their market availability, and the nature of the herbs:

- Category 1: already in use for more than 5 years
- Category 2: in use for less than 5 years
- Category 3: new medicines.

Depending on the nature of herbs and market availability, different requirements exist for submission of clinical trial data and toxicity data[2].

It is hoped that assessment of these traditional remedies could become the basis for a future classification of herbal medicines, as well as for evaluative studies on their efficacy and safety, and their potential use in national health care systems in different parts of the world.

2.2. *Approval for new herbal drugs*

Applications submitted to the DCG(I) for permission for clinical trials in respect of new drug applications (NDA) and abbreviated new drug applications (ANDA) are often referred to outside agencies like the Indian Council of Medical Research (ICMR) and the Department of Biotechnology, Government of India (DBT) for review. This arrangement often leaves very little control with regard to the time runs. In order that the applicant is enabled to complete the investigations in the shortest possible time, it is imperative that adequate infrastructure for fast track clearances is created in the Central Drug Standards Control Organisation (CDSCO) and also to Ayurvedic Yoga Unanai Sidhha and Homeopathy (AYUSH). The process of evaluation and review of applications of new drugs needs close collaboration with Indian Technical institutions. Industry, consumers , consumer organizations , foreign governments and international organizations. Thus, there should be chemists/ pharmaceutical technologists/chemical engineers to review areas connected with manufacture, in-process control, packaging, stability, purity and similar parameters of the product. Biotechnology-based and genetically engineered drugs are getting introduced with greater frequency. Pharmacologists/toxicologists should be there to evaluate the short term and long terms effects, including teratogenic and carcinogenic effects, in laboratory animals. Pharmacovigilence activities which includes Post marketing surveillance, Adverse Drug Reaction Monitoring etc. is also a critical functioning of Drug Regulatory Agency[10].

3.0. *General Guidelines for Methodologies on Research and Evaluation of Traditional Medicine (WHO)*

During the last decade, use of traditional medicine has expanded globally and has gained popularity. The challenge now is to ensure that traditional medicine is used properly and to determine how research and evaluation of traditional medicine should be carried out. Governments and researchers, among others, are increasingly requesting WHO to provide standards, technical guidance and information on these issues. Since 1991, WHO has developed and issued a series of technical guidelines such as *Guidelines for the assessment of herbal medicines; Research guidelines for evaluating the safety and efficacy of herbal medicines; and Guidelines for clinical research on acupuncture.*

3.1. Guidelines for the assessment of herbal medicines
(WHO Expert Committee on specifications for pharmaceutical preparations. Thirty-fourth report.

Geneva, World Health Organization, 1996:178–184 (WHO Technical Report Series, No. 863). These guidelines were finalized at a WHO Consultation in Munich, Germany, 19–21 June 1991.

For the purpose of these guidelines, herbal medicines are defined as follows: Finished, labeled medicinal products that contain as active ingredients aerial or underground parts of plants, or other plant material, or combinations thereof, whether in the crude state or as plant preparations. Plant material includes juices, gums, fatty oils, essential oils, and any other substances of this nature. Herbal medicines may contain excipients in addition to the active ingredients. Medicines containing plant material combined with chemically defined active substances, including chemically defined, isolated constituents of plants, are not considered to be herbal medicines. Exceptionally, in some countries herbal medicines may also contain, by tradition, natural organic or inorganic active ingredients, which are not of plant origin. The past decade has seen a significant increase in the use of herbal medicines. As a result of WHO's promotion of traditional medicine, countries have been seeking the assistance of the Organization in identifying safe and effective herbal medicines for use in national health care systems. In 1991, the Director-General of WHO, in a report to the Forty-fourth World Health Assembly, emphasized the great importance of medicinal plants to the health of individuals and communities. The objective of these guidelines is to define basic criteria for the evaluation of quality, safety and efficacy of herbal medicines and thereby to assist national regulatory authorities, scientific organizations and manufacturers to undertake an assessment of the documentation/ submissions/ dossiers in respect of such products.

Assessment of quality

Pharmaceutical assessment

This should cover all-important aspects of the quality assessment of herbal medicines. It should be sufficient to make reference to a pharmacopoeia monograph if one exists. All procedures should be in accordance with good manufacturing practices.

Crude plant material

The botanical definition, including genus, species and authority, should be given to ensure correct identification of a plant. The active and characteristic constituents should be specified and, if possible, content limits should be defined. Foreign matter, impurities and microbial content should be defined or limited.

Plant preparations

Plant preparations include comminuted or powdered plant materials, extracts, tinctures, fatty or essential oils, expressed juices and preparations whose production involves fractionation, purification or concentration. The manufacturing procedure should be described in detail.

Finished product

The manufacturing procedure and formula, including the amount of excipients, should be described in detail and identification of an active principle (e.g. "chromatographic fingerprint") The WHO Certification Scheme on the Quality of Pharmaceutical Products Moving in International Commerce should be applied.

Stability

The physical and chemical stability of the product in the container in which it is to be marketed should be tested under defined storage conditions and the shelf life should be established.

Assessment of safety

This should cover all relevant aspects of the safety assessment of a medicinal product. Reported side-effects should be documented according to normal pharmaco-vigilance practices.

Toxicological studies

Toxicological studies, if available, should be part of the assessment. Literature should be indicated as above.

Documentation of safety based on experience

As a basic rule, documentation of a long period of use should be taken into consideration when assessing safety. This means that, when there are no detailed toxicological studies, documented experience of long-term use without evidence of safety problems should form the basis of the risk assessment. However, even in cases of drugs used over a long period, chronic toxicological risks may have occurred but may not have been recognized.

Assessment of efficacy

This should cover all-important aspects of efficacy assessment. A review of the relevant literature should be carried out and copies provided of the original articles or proper references made to them. The pharmacological and clinical effects of the active ingredients and, if known, their constituents with therapeutic activity should be specified or described. As many herbal remedies consist of a combination of several active ingredients, and as experience of the use of traditional remedies is often based on combination products, assessment should differentiate between old and new combination products. Identical requirements for the assessment of old and new combinations would result in inappropriate assessment of certain traditional medicines. In the case of traditionally used combination products, the documentation of traditional use (such as classical texts of Ayurveda, traditional Chinese medicine, Unani, Siddha) and experience may serve as evidence of efficacy. An explanation of a new combination of well-known substances, including effective dose ranges and compatibility, should be required in addition to the documentation of traditional knowledge of each single ingredient. Each active ingredient must contribute to the efficacy of the medicine. Clinical studies may be required to justify the efficacy of a new ingredient and its positive effect on the total combination.

Product labels and package inserts should be understandable to the consumer or patient. The package information should include all necessary information on the proper use of the product.

3.2. Research guidelines for evaluating the safety and efficacy of herbal medicines

(Adapted from Research guidelines for evaluation the safety and efficacy of herbal medicines.Manila, World Health Organization Regional Office for the Western Pacific, 1993:35–40.)

It include details of Guidelines for toxicity investigation of herbal medicines, consisting of Acute toxicity test, Long-term toxicity test . Expected period of clinical use Administration period for the toxicity study (Single administration or repeated administration

Observations and examinations

Observations and examinations should be performed mainly on the following items (from 1 to 6):

1. General signs, body weight and food and water intake
2. Haematological examination
3. Renal and hepatic function tests.
4. Other function tests, like histopathological and blood chemical analysis.

4.0. Government Responsibility to strengthening herbal research

It is clear that the herbal industry can make great strides in India, with the co-operation between drug regulatory authorities, scientists and industries. However, standardization of methods and quality control data on safety and efficacy are required for proper understanding of the use of herbal medicines. It is the basic responsibility of the Government to ensure that drugs to be used by the public meet the established standards of quality, safety, biodiversity and efficacy. In India, the import, manufacture, sale and distribution of drugs and cosmetics in India is regulated under the Drugs and Cosmetics Act 1940, and Rules 1945. Standards of identity, purity, freedom from toxicity and strength in respect of every medicine and related products used for diagnosis, prophylaxis and treatment of diseases in human beings or animals have to be specified. Under the Act, distinct statutory functions and responsibilities have been assigned to Central and State Governments.

The quality of drugs has to be closely monitored so that drugs of doubtful quality are not manufactured. A drug is of acceptable quality under the Act not only if it meets the finished product specifications but also more importantly if it is manufactured in a plant complying with GMPs. Though the Drugs Controllers of the states are empowered to licence the manufacture and sale of drugs in their respective states under the Act, the DCG(I) in order to ensure uniform implementation of Rules, is enjoined with the responsibility of coordinating their activities and decisions[3]. For a manufacturer intending to export drugs, a GMP certificate under the WHO Certification Scheme on the quality of pharmaceutical products moving in international commerce is the generally accepted. The WHO Certification Scheme is a mechanism by which the importing country is in a position to ascertain whether it has been manufactured in accordance with internationally accepted GMPs. With the growth of the pharmaceutical industry, there has been considerable impetus to research and development activity on drugs. A number of medicines are now exported. This requires proper regulation so that safety, efficacy and quality issues are attended to in a globally accepted manner in the country.

References

1. Anonymous. 1999. IUCN *Criteria Review.* Provisional Report: Draft of the proposed changes and reconsiderations. Species. 31-32: 43-57.

2. Chakravarty , B K 1993. Herbal Medicines, Safety and efficacy guidelines. *The Regulatory affair journals* . 4:699-701.

3. Government of India website link file (cdsco.nic.in).

4. IUCN. 1994. *Red List Categories.* As Approved by the 40th Meeting of the IUCN Council. IUCN, Gland, Switzerland. 22pp.

5. Jayasuriya DC 1990 The regulation of medicinal plants - a preliminary review of selected aspects of national legislation.Unpublished Report.

6. Mehta , V 2002. *Indian herbal drug industry-Future prospects*–Website Information).

7. Rajagopalan TG 1991. Traditional Herbal Medicines around the Globe: Modern Perspectives. The Indian Perspective. Proceedings of the 10th General Assembly of WFPMM, Seoul, Korea, 16-18 October. *Swiss Pharma* 1991;13(11a):63-67.

8. Rao, C K , Geetha, BL and Suresh , G 2003. *Red list of threatened vascular plant species in India* , Ministry of environment and forest , Government of India.

9. Walter, K.S. and Gillett, H.J. (eds.) 1998. 1997 *IUCN Red* List *of Threatened Plants.*Compiled by the World Conservation Monitoring Centre. IUCN, Gland, Switzerland and Cambridge, UK. Lxiv+862pp.

10. WHO India website link (whoindia.org.in).

11. World Health Organization Regional Office for the Western Pacific. Research guidelines forevaluating the safety and efficacy of herbal medicines. Manila, World Health Organization Regional Office for the Western Pacific, 1993.

12. World Health Organization. *Quality control methods for medicinal plant materials.* Geneva,World Health Organization, 1998.

13. World Health Organization. Regulatory situation of herbal medicines: a worldwide review. Geneva,World Health Organization, 1998 (unpublished document WHO/TRM/98.1; available onrequest from Traditional Medicine (TRM/EDM/HTP), World Health Organization, 1211 Geneva27, Switzerland).

14. World Health Organization. *WHO monographs on selected medicinal plants.* Vol. I. Geneva,World Health Organization, 1999.

-ΔΔΔ-

8

Peste des Petits Ruminants (PPR) and Strategies for its Containment in WTO Regime

S C DUBEY AND D MONDAL

Division of Animal Health
Central Sheep and Wool Research Institute
Avikanagar Via-Jaipur, Rajasthan - 304501

By nature the Peste des petits ruminants (PPR) which is also referred as goat plague, has always been noticed as a highly contagious disease of small ruminants (SRs) caused by a Morbillivirus close to Rinderpest virus (RPV) and is becoming more widespread hence designated as an emerging plague of SRs in post RP era. PPR is characterized by classical signs of alimentary tract disorders (ATD's) and respiratory tract disorders (RTD's) exhibited mainly in form of fever, necrotic stomatitis, lacrimation, pneumonia, foetid enteritis and high mortality.

RP, the most feared livestock disease of previous century, is currently on the verge of eradication but another member of the same group of viruses causing PPR perhaps has emerged as a major worrying ailment for the farmers in many parts of Asia and Africa. PPR was once thought to be only an African problem but now it has been recognized in much of the Middle East and Indian subcontinent where it is causing alarming losses and there is a great risk that it could spread into the central Asian countries. **Peter Roeder, Animal Health Officer, responsible for viral diseases at FAO and the Secretary of the Global Rinderpest Eradication Program has stated that its becoming very clear that once a group of countries (like countries of Indian**

subcontinent) are undertaking livestock development and production program involving small ruminants, the PPR has be taken extremely seriously. PPR impact has been seriously realized in three international development projects based on sheep in Afghanistan and Jordan and goat development project in Bangladesh. In light of the statement of Peter Roeder, experiences in a few major projects and extent of PPR in indigenous SR population, it is prime time to protect the interest of poor shepherds of Indian subcontinent.

The Virus

Molecular studies of the morbilliviruses, in the (family *Paramyxoviridae*) revealed that type virus of the group is human Measles Virus (MV) but it also includes important animal pathogens such as RPV of cattle, buffalo and wild bovidae, PPR virus of SRs, CD Virus of domestic dogs and wild carnivore species, Phocine Distemper Virus (PDV) of seals and two viruses isolated from cetacean species- Dolphin morbillivirus (DMV) and Porpoise Morbillivirus (PMV)[1]. The PPR virus has the following characteristics[2]:

- May survive at 60°C for 60 minutes.
- Stable at pH from 4.0 to 10.0.
- Killed by alcohol, ether, and detergents as well as by most disinfectants (eg, phenol, sodium hydroxide) etc.
- Long survival time in chilled and frozen tissues.

Sequencing studies have shown that, despite being **only one serotype** of PPR virus, there are **different genetic lineages** of each virus. Studies on lineage of virus allows tracing of disease outbreaks outside their known geographic area. In recent decades, PPR emerged in South India and later spread to North India. Studies on the genetic relationships of the viruses from nineteen independent PPR outbreaks facilitated identification between viruses of distinct geographical origin by nucleotide sequencing followed by PCR amplification. Four separate lineages of the virus were identified and the virus isolates from Asia over the past 2 years were all of one lineage, which had not previously been identified in Africa or Asia[3].

Further, phylogenic analysis of virus isolates from 27 outbreaks in Asian and Middle Eastern countries, reported between1993 to 2000, and two recent outbreaks from the African continent were compared with the prototype African strain. Of the four known lineages of PPR virus the isolates detected to have their lineage as below. The geographical source of the new lineage 4 viruses is unknown although it is most closely related to African lineage 1. The possibility that the circulation of RPV, a related virus of cattle, would have masked its earlier presence PPR in northern India[4].

PPR virus	Lineage	Comments
West African isolate	1&2	Prototype PPR virus
Burkina Faso isolate	1	Isolated -1999
East African isolate	3	Isolated from Ethiopia -1996
Arabian & Southern India isolate	3	
Yemen isolate	3	Isolated -2001
Middle East, Asian subcontinent Nepal & Bangladesh Isolate	4	Most closely related to African lineage1
Kuwait	4	Isolated-1999

Epidemiology

(a) *Hosts:* Although goat and sheep are the natural hosts of PPR but goat appears to be more susceptible and suffer a more severe clinical disease than sheep[5]. PPR also occur in other species, particularly in captive wild ungulates from three families i.e. Gazellinae *(Dorcas gazelle)*, Caprinae *(Nubian ibex)* and Hippotraginae *(gemsbok)*. The American white-tailed deer *(Odocoileus virginianus)* has also been infected experimentally[6]. Cattle, buffaloes, camels, and pigs are also susceptible to infection but do not exhibit clinical signs and are unable to transmit the disease to other animals. PPR is not infectious to humans.

(b) *Distribution:* PPR is endemic in sub-Saharan Africa since several decades, it was first described in Ivory Coast in the 1940s[7]. The virus has been circulating in parts of sub-Saharan Africa for several decades and in the Middle East and southern Asia since 1993. Although, PPR in India, was first confirmed in March 1987 in sheep[8] but suspected of having RP is now believed that many outbreaks in India previously attributed to RP were actually PPR. The virus was isolated four more times by 1992, and major epidemics occurred in the state of Andhra Pradesh in 1994-1995 and 1997-1998[9]. In Rajasthan nearly 50% of susceptible population is all the time under risk[10].

PPR has also been reported in Sudan, Kenya, Uganda and Ethiopia. The Arabian Peninsula, the Middle East, and the rest of the Indian Subcontinent reported PPR incidents during 1993-1995. The disease has remained endemic in these areas. It has also occurred in Turkey in 1996, Iraq in 2000, Iran in 1994, Bangladesh in 1993 & 2000, and Nepal in 1995[4]. Studies clearly indicate that the disease is also endemic in our neighboring countries viz., Pakistan, Nepal and Bangladesh.

(c) *Transmission:* PPR virus survives poorly outside the host. Spread of infection to new areas is largely by close contact between infected and susceptible animals[11] particularly during the movement of infected animals. There are several means of transmission between animals[6]:

- Inhalation of aerosols produced by sneezing and coughing of infected animals.
- Outbreaks are more frequent during the rainy season or the dry, cold season[2].
- Direct contact with ocular, nasal or oral secretions.
- Direct contact with feces.
- Fomites such as bedding, water and feed troughs.
- No carrier state is known to exist.
- Besides, PPRV is also present in semen and embryos and probably transmited[12] viral infection through gonads.
- Feeding milk from infected animals to kids or lambs can spread infection.

(d) *Seroepidemiology:* Acute febrile viral infection or sometimes non-febrile non-clinical infection (in endemic areas) of PPR in goats and sheep elicits humoral antibody response that is easily detected by specific immune assays. A monoclonal antibody-based competitive enzyme-linked Immunosorbent assay has been in use to detect the presence of serum antibodies[13]. An observation on prevalence of antibodies to PPRV in 7,411 serum samples of small ruminants covering major states of the country is presented in Table-1 and graphical presentation of seroepidemiology exclusively of semiarid Rajasthan is depicted separately (Graph-1&2).

Table 1. Seroprevalence of PPRV in sheep and goat of India.

Indian States	Number of positive samples/total number of samples (Percentage of positive samples)						Reference
	Sheep serum		Goat serum		Total		
AP	114/278	41%	20/56	35.7%	134/334	40.1%	13
Assam	---	---	0/54	---	0/54	---	13
CG	---	---	45/102	44.1%	45/102	44.1%	13
Delhi	**3/10**	**30%**	**52/187**	**27.8%**	**55/197**	**27.91%**	14
Gujarat	8/10	0.80%	14/43	32.6%	22/53	41.5%	13
Haryana	---	---	8/57	14%	8/57	14%	13
	192/347	**55.33%**	**65/152**	**42.76%**	**257/499**	**51.50%**	14
HP	15/97	15.5%	34/84	40.5%	49/181	27%	13
J & K	11/27	40.7%	8/21	38.1%	19/48	39.6%	13
Karnataka	19/40	47.5%			19/40	47.5%	13
Kerala	**76/166**	**45.78%**	**5/536**	**0.93%**	**81/847**	**9.5%**	15
M P	0/3	---	9/30	30%	9/33	27.3%	13
MH	216/428	50.4%	388/536	72.4%	604/964	62.7%	13
Meghalaya	---	---	3/145	2.1%	3/145	2.1%	13
Orissa	¾	0.75%	11/70	15.7%	14/74	18.9%	13
Rajasthan	60/306	19.6%	4/25	0.16%	64/331	19.3%	13
Farm	**87/411**	**21.15%**	**17/68**	**25%**	**104/479**	**21.71%**	10
Field	**278/502**	**55.37%**	**266/480**	**55.41%**	**544/982**	**55.39%**	10
Tamil Nadu	0/19	---	19/37	51.4%	29/56	51.8%	13
UP	87/244	35.7%	252/1,017	24.8%	339/1,261	---	13
Uttaranchal	12/44	27.3%	128/630	20.3%	140/674	20.8%	13
Total					**2539/7411**	**34.2%**	

Serum samples tested by Singh et al , 2004 at IVRI, Mukteshwar for PPRV antibodies were derived from a number of geographical regions with in the country but, in light of the serosurveillance report of Rajasthan by Dubey et al, 2006 the results of the states with small sample size, may not be taken as a true representation of the target population of a particular agro-eco zones. The difference in result clearly indicates the need of survey including sizable population for better understanding of disease status in an area.

Screening of the 7,411 sera samples, revealed 34.2% prevalence of PPRV in small ruminants in India. The prevalence of antibodies to PPRV was noted to differ between species (i.e. sheep versus goats), age groups and geographical regions. The distribution and prevalence of antibodies to PPRV among various age groups of animals indicated that goats were exposed at an earlier age than the sheep, suggesting that goats may be more susceptible to infection with PPRV. The findings may be correlated with variations in the sheep and goat husbandry practices within different geographic regions, the topography of different states and the socio-economic status of individual Indian farmers[13].

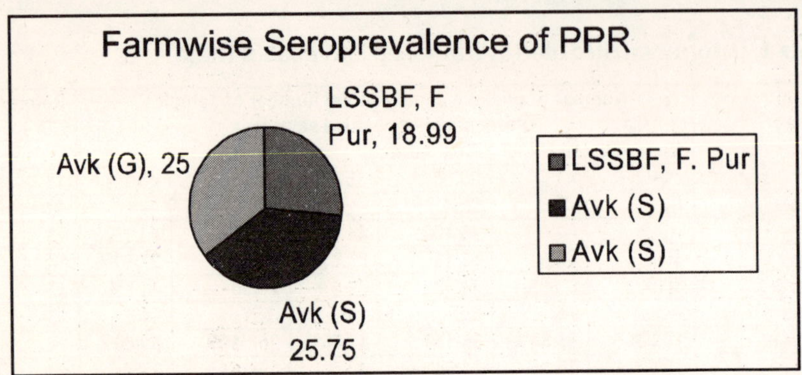

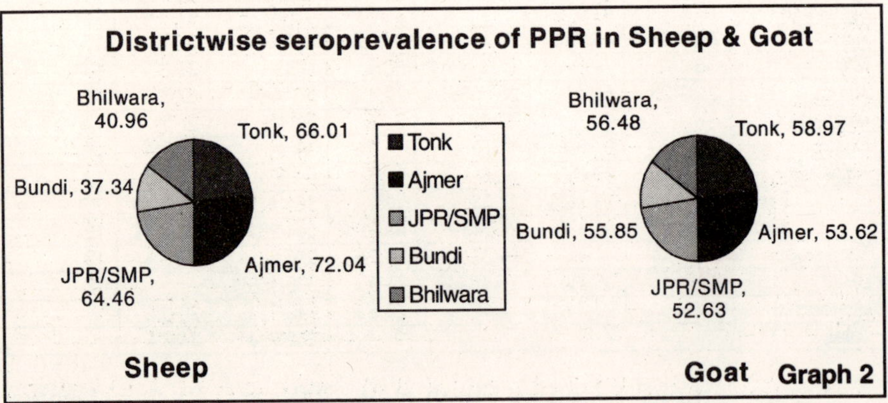

Graphs 1 & 2. Seroconversion in farm & field flocks of small ruminants of semi-arid Rajasthan.

(e) *Disease profile:* It is a fact that Morbilliviruses are highly pathogenic in their natural hosts and the economic consequences of epizootics of RP and PPR can be devastating. According to Singh et al, 2004 the disease outbreaks were more severe in goats than sheep and that the frequency of disease outbreaks was greater between the months of March and June (51.7%) as compared to other periods of the year. Analysis of another epidemiological study[16] of the 30 villages of semi-arid Rajasthan revealed, that most of the outbreaks were reported during the cooler months of the year (November to April- 63 cases) along with rare/nil reports during dry hot or humid hot months (May to October - 09 cases). It was also observed that flocks having mixed population of sheep and goat (32) or goat alone (31) were maximum affected then flocks exclusively of sheep (9) and hence reconfirm established observation of higher susceptibility for PPRV in goats than sheep[17,18,1]. Clinical signs and lesions in sheep were comparatively milder than goats. Observations indicated that goat played key role in PPR precipitation and spread in this area, where flocks with mixed population of sheep and goats are reared as normal animal husbandry practices.

Morbidity and mortality recorded by various authors in different states & some other countries are presented in Table 2. Further, morbidity and mortality can be as high as 100% and 90%, respectively when associated with other diseases such as capripox[4].

Table 2. Morbidity and mortality due to PPR in some Indian states & abroad.

Indian state	Species	Morbidity (%)	Mortality (%)	References
A. P.	Goat & sheep	49	High	19
W. Bengal	Goat	30.7	24	20
Rajasthan	Sheep	53	37.5	16
	Goat	54.7	36	
U.P.	Shep	52.9	29.3	21
		39.2	34.47	22
	Goat	84.3	60.3	21
		34.5 (abort-39)	41.3	22
Punjab	Sheep	57	32	23
	Goat	51	28	
H. Pradesh	Sheep& goat	26.8	18	24
Abroad				
Bangladesh	Goat	100	100	25
Ethiopia	Sheep	76	19	26
Nigeria	Sheep & goat	50	50	27
Saudi Arab	Sheep & goat	43	50	28

(f) *Disease economics:* PPR infection is one of the most important constraints to the increased production of small ruminants Earlier RP use to pose a serious threat to cattle, buffalo and wild ruminants but a global eradication campaign lead to eradication of the RP with possibility to declare zero RP status by 2010. PPR causes death in more than 50% of the affected animals due to high fever, pneumonia, diarrhoea and dehydration. The annual loss due to PPR in the small ruminants in India is approximately Rs.180 crores[29].

In another economic evaluation of PPR in Rajasthan[16] revealed a loss of Rs. 85.13 crore.

The loss figures were arrived considering latest population figures of SRs in Rajasthan state[30]. On the basis of the serosurveillance results the probability of outbreak in a year is assumed to be 50% and susceptible population of SRs is considered to be only 45% in view of presence of PPRV specific antibodies in 55% SR population of the state[16].

The total mortality loss due to PPR in SRs of the state amounts to Rs. 85.13 crores, of which nearly Rs. 66.86 crore in goat sector and Rs. 18.27 crores in sheep sector. The data indicated that the PPR disease is responsible for a loss of 2.31% of total value of SR comprising of 2.65% in case of goats & 1.58% in case of sheep. The calculations consider only the loss due to mortality and do not take into account other production losses that may arise due to the disease.

Parameters	**Goat**	**Sheep**
A Population in lakh (Raj.)	168	100.5
B % of adults	60	65
C % of young's	40	35
D Probability of outbreak (%)	50	50
E Susceptibility (%)	45	45
F Adult morbidity (%)	50	36
G Adult death rate (%)	18.5	12.7
H Value of each adult (Rs.)	2000	1500
I Lamb morbidity (%)	51	48
J Lamb death rate (%)	43	43
K Value of each young (Rs.)	750	500
L Loss due to mortality in cr. (adults) (AxBxDxExFxGxH)	41.98	10.08
M Loss due to mortality in cr. (youngs) (AxCxDxExIxJxK)	24.88	8.18
N Total loss in cr.	66.86	18.27
O Total mortality loss of SRs in cr.	**85.13**	
P Total value of adults (in crore) (AxBxH)	2017.08	980.26
Q Total value of youngs (in crore) (AxCxK)	504.27	175.94
R Total value in crore(P+Q)	2521.35	1156.21
S Total value of SRs in crore	**3677.56**	
T % loss due to PPR (N/R*100)	2.65	1.58
U % loss due to PPR in SR (O/S X 100)	**2.31**	

Clinical Features

Although, susceptibility to PPRV infection rises with age but the disease is fatal in the young animals[11]. The clinical signs imitate those of RP, but changes can occur faster. Specific clinical signs are outlined in various text books include sudden onset of the disease with high fever, severe depression and loss of appetite, watery nasal discharge progressing to profuse muco-purulent discharge with catarrhal exudate occluding the nostrils, respiratory distress and sign of bronchopneumonia, conjunctivitis with sometime catarrhal conjunctivitis, necrotic lesions in the mouth (dental pad, hard palate, cheeks), progressing to erosions of epithelium, severe diarrhoea, dehydration, weight loss and abortion.

According to[31] the prognosis of acute PPR is usually poor; especially when lesions do not resolve within 2 to 3 days or when extensive necrosis and bacterial infection give the animal's breath an unpleasant, fetid odor. Poor nutrition, stress of movement, and concurrent parasitic and bacterial infections worsen clinical signs. Dubey et al, 2004-06 observed two types of clinical signs infield flocks of semiarid Rajasthan. The disease was categorized as typical PPR (with ATDs and RTDs) or atypical PPR on the basis of the clinical signs exhibited by the affected sheep and goats (Table-3). Young animals (4 to 8 months) often have more severe disease.

Table 3. Symptoms observed in PPR affected animals

	Typical PPR		**Atypical PPR**
	ATD's	RTD's	
Clinical signs	**Common** – fever, anorexia, grinding of teeth, frothy salivation, dry muzzle, oral lesions inform of erosive stomatitis, glossitis (Fig-1), gingivitis, hemorrhages on hard palate, halitosis, abdominal pain, soiled perineal region (Fig-2) with foul smelling pasty feces or blackish diarrhoea and progressive dehydration. Convalescent animals showed poor body condition and debility.	Mucopurulent oculo-nasal discharges with opacity (Fig-3), high fever, severe congestion of nasal passage, nasal block leading to open mouth breathing (Fig-4) in some cases, occasional coughing, abnormal respiration and panting.	Nasal discharges, coughing some times, soiled perineal region with foul smelling pasty feces or blackish diarrhoea with possibility of reoccurrence. Mortality rare
	Rare - Abortion, stillbirth		No abortion

In typical cases of PPR, flocks were having history of one or the other stress born predisposing factors like cold spell, winter/spring rains (particularly after shearing), flocks kept under open sky roof, poor grazing availability during winter/cooler months and physiological stress (gestation & lambing) would also have played significant role in further aggravation of the situation alone or in combination, in favoring disease precipitation in partially undernourished animals. Studies at CSWRI reveled that over stocking rate is >70 times on public range land[32] and in turn might be facilitating pathogen transmission while in close contact. The gravity of under feeding in winter season in semi arid Rajasthan increases[33] as forage yield of range land use to be lowest (140kgDMha^{-1}) in winter than summer (193kg DMha^{-1}) and monsoon (490kg Dmha^{-1}). Higher number of clinical cases recorded in winter indicates favorable season for virus circulation through aerosols generated by coughing/sneezing and copious nasal discharge on or around thick fecal bed would have been possible more freely than in dry hot or humid hot months (May to October) of the year when thermo labile nature of the virus would have limited its transmission.

It is needless to emphasize that increased energy requirement of pregnant ewes and does in late pregnancy during winter and early lactation would have further deteriorated the health status inviting disease precipitation by picking up PPR virus, circulating in about 45-50 % SR population virgin to PPRV[16] and hence causing typical PPR along with sporadic abortion in a few animals. Further unhygienic habit to pile up fecal matter certainly would have played its role in transmission of primary agent (PPRV) and secondary bacterial pathogens responsible to attack alimentary and respiratory tract in sick SRs.

Post-mortem findings

Lesions are usually seen in the digestive and respiratory tract but can also be seen in other systems as well. In acute cases, the dominating PM findings includes dehydrated carcass with faecal soiling, necrotic lesions in the mouth/tongue and nose, congestion of the ileo-caecal valve, engorgement and blackening of the folds in the cecum, colon and rectum ('zebra striping'), enlarged spleen, oedematous lymph nodes and pneumonia.

Diagnosis

Differential diagnosis of PPR should be carried out with other infections producing similar clinical signs such as RP, Pasteurellosis, Orf, Pox, CCPP, BTD, Heart water, FMD, NSD, GI Parasite and plant or mineral poisoning. RP, in its classical form, is easy to identify clinically, however, clinical observations for both the diseases should always be confirmed by a laboratory.

Samples should be collected in acute phase of disease and be shipped fresh under cold chain (not frozen) within 12 hours of collection[31]. Samples include blood in EDTA anticoagulant (Blood & anticoagulant should be mixed gently), clotted blood or serum, mesenteric lymph nodes, spleen, lung, tonsils, pieces of the ileum and large intestine, swabs of serous nasal and lachrymal discharges. Ideally, samples should be collected from several animals in an outbreak. Epidemiological and clinical details should be provided with the samples, and each sample bottle should be marked carefully with permanent marker. Details of each sample's origin should be recorded for submission to the laboratory. As per OIE guidelines the following tests may be used to detect antigens and routine serology.

Immuno capture ELISA and counter immuno-electrophoresis (CIEP) are most rapid test for detecting viral antigen. AGID is very simple and inexpensive and gives results within 1 day, but not sensitive to mild forms of PPR. PCR and culture for isolation in lamb kidney or African green monkey cell cultures is confirmatory. Further, indirect fluorescent antibody test have been used to identify PPRV antigen in various tissues of organs of infected animal[34].

Routine serological tests include competitive ELISA (Sensitivity 99.4% Specificity 94.5%) and virus neutralization test, which is a prescribed test for international trade. However, Cross-neutralization with RP virus must be completed. Other serologic tests have been described, but not widely adopted includes CIEP and AGID test.

Treatment

There is no treatment for PPR however, mortality rates may be decreased by the use of drugs that control the bacterial and parasitic complications. According to OIE, use of specific antibiotics i.e. oxytetracycline and chlortetracycline are recommended to prevent secondary pulmonary infections. However, in absence of specific guidelines lot of animal use to die by the time effective treatment is practiced by the field veterinarians. Moreover high mortality in PPR affected animals is also observed in far of remote villages, where availability of veterinary services & medicines are not readily available. In view of the prevailing situations in semiarid Rajasthan, Dubey et al, 2004-06 developed and implemented a line of treatment including palliative ORS for PPR affected animals (using household ingredients), treatment to check secondary complication due to ATDs & RTDs and guidelines for day-to-day hygienic management of the affected flocks.

Palliative ORS/drench for sick animals (a most important step) - All sick animals were offered this preparation daily before treatment once or twice as per need. Green fodder was advised to all sick animals after wilting as source of vitamin A and nutrients availability. It was observed that palliative ORS drench helped in rehydration skin fold test, as source of energy and as electrolytes balancing factor together with mineral mixture offered orally in combination with antibiotic separately and hence played a life saving role.

Of these ingredients, barley flour is known for its nutritional quality, soothing and laxative effect on gut. In Ayurvedic system of medicine Haldi (Termeric, *Curcuma longa)* is known for its antiseptic, antiallergic, stomachic, blood purifier, and antihemorrhagic properties. Gur (Jaggery) added as source of energy and minerals whereas chaach (whey) as demulcent to mitigate effect of hemorrhagic enteritis in particular. Oral washing and application of Haldi + sweet oil paste facilitated prompt healing of oral lesions. Treatment given in chronological order helped in recovery even the animals with very severe oral lesions along with associated ATDs and RTDs showed improvement within 24 hours and survived.

Generally, liver extract + Vitamin B complex is used to overcome the anorexia, G.I. disorder and general weakness. Inj. Meloxicam is known as a safe logical partner of antibiotic therapy and it has long acting anti-inflammatory, analgesic and anti-exudative property. In general it is injected 4-5 times to facilitate complete recovery from RTDs to avoid reoccurrence of pneumonia. In acute sickness PPR born leukopaenia and lymphopaenia leads to immunosuppression that in turn facilitates further complications through secondary bacterial pathogens[35]. Ciprofloxacin + Tinidazole combination is known for its bactericidal action by providing broad spectrum coverage to mitigate adverse effect of bacterial pathogens responsible to exaggerate RTDs and ATDs due to secondary bacterial infection.

Sanitation and daily management instructions were given to harness the harmful traditional practices. Flock owners of Rajasthan believe that fecal bed (Fig-8) provide warmth to sheep and goat during winter, whereas, it is a known fact that unhygienic or contaminated surroundings due to loose feces and nasal droplets facilitates fast transmission of pathogens. Therefore, daily cleaning with lime dusting (once at 15 days interval) was advised. Cleaning of nasal opening of all sick animals with moist cloth or piece of gunny bag was advised to allow ease in respiration. Washing of mouth and nasal opening with potassium permanganate solution (1:10000) or boiled neem leaves water after cooling was advised along with application of Haldi (termeric) + sweet oil paste into mouth daily for 4 –5 days. Further, during night sheep and goat flocks of Rajasthan that are generally kept in open coral and hence invariably get cold exposure / stress particularly after sudden climatic change/rains/hail storm etc. and hence leading to precipitation of one or the other disease. Flock owners were therefore advised to provide temporary thatched shed using available dry grass/ panipula / bajra kadwi etc. to mitigate the impact of cold stress.

Although, the line of treatment and day to day care suggested, includes a numbers of activities for daily operations but careful implementation checked mortality with immediate effect. The line of treatment and suggestions worked out durin this study can be used to handle the acute PPR cases in sheep and goats. The cost of full course of treatment per animals was not more than Rs. 20-22 in this area.

Prevention

In the past, the RP vaccine was used to prevent PPR, however, this practice has now been phased out to avoid confusion during retrospective serologic studies. A homologous PPR vaccine is now available and gives strong immunity. There are a few genetically engineered recombinant vaccines undergoing limited field trials[2]. A combined recombinant vaccine to protect against RP and PPR has also been prepared. Currently a new marker vaccine for RP using reverse genetics technology is also being developed.

In India an effective Vero-cell line based live attenuated indigenous freeze-dried vaccine has been developed and validated using extensive field trials. The vaccine has a shelf life of more than one year at 4°C and provides immunity for three years. The vaccine is safe, potent and acceptable for use even during pregnancy. The vaccine has been developed at Indian Veterinary Research Institute (Mukteshwar campus) and Tamil Nadu Veterinary and Animal Science University,

Chennai. The technology for vaccine production has already been transferred to a few biological product units of the state AH Departments[36] and also to a private organization for commercial production and marketing. The vaccine would help in reducing losses due to morbidity and mortality and thus improve economic condition of poor sheep and goat farmers in the country. The vaccine costs less than Rs. 1/-.

PPR a potent biological weapon

Indian livestock population consists of 283.5 million bovine and 184.2 million small ruminants. The SR sector offers great potential for exports if it is completely free from trans-boundary animal diseases. Under the trans-boundary animal diseases, RP, FMD and PPR are the main diseases of global concern. The RP in old days was used as a biological weapon to devastate livestock based socio-economic infrastructure of the target country by introducing a group of Gray steppe oxen. These animals were known for their innate resistance to RPV with slow spread of virus without appearance of clinical signs. As sequel to widespread infection of RPV in local animal population, use to result in no transport, untilled fields, starving peasant and overthrown government. Similarly, use of PPRV as a potential biological weapon cannot be totally ruled out in current competitive world of global market trends where disease free livestock is prerequisite to enhance export potentials. Emerging concept of trans-boundary and zoonotic disease free livestock is the basic need of today's International trade to meet out hazard analysis & critical control point (HACCP), codex alimentarius and food and drug administration (FDA) requirement of WTO[37] and hence disease like PPR need to be harnessed by taking due measure in form of stamping out or vaccination or using a combination of both to check the possibility of use of PPRV as a devastating weapon to SR and there products.

Control through eradication

The livestock sector offers great potential for exports if it is completely free from trans-boundary and zoonotic animal diseases of global concern like RP, FMD, PPR, brucellosis, JD and Leptospirosis etc. Now an indigenous attenuated PPR virus vaccine is available that has proven effective in field trials, providing immunity for at least three years. Secondly, for effective implementation of control measures a proper and rapid diagnosis technique is also available.

Today RP, provably remains only in a few small areas and with international commitment, could soon be eliminated completely. Now, with the experiences that we have accumulated through the eradication efforts of RP, it is most possible that PPR may also be brought to its knees. PPR has proved to be difficult in many countries, especially when the disease is first introduced. If we decided to move from RP to PPR it would be very important to discriminate between PPR and RP, because RP has sometimes also been recognized to affect sheep in the Indian subcontinent with signs similar to PPR. Moreover, PPR is an important disease in its own right hence differentiation is essential particularly with RP, at this stage where the RP eradication program represents an international emergency.

Unfortunately, unlike the situation with RP, coordinated eradication of PPR is not a likely prospect in the immediate future but progressive control of the disease is possible. Vaccination, combined with common sensible actions such as safeguarding flocks from PDFs and unhygienic management practices may provide good control. However, a few points should be kept in mind while preparing action plan for PPR eradication program (PPREP) in view of WTO regime:

1) The turnover rate of small ruminant populations is, much faster than that of larger livestock, which means that vaccination has to be used more intensively and more frequently.
2) All sero-surveillance should be carried out using indigenously developed monoclonal antibody-based diagnostic kits of IVRI, Mukteshwar for the detection of PPRV antigen or antibody for uniformity of results.
3) Assessment of disease status through serosurveillance and disease surveillance for the trans-boundary animal diseases would be mandatory for efficient disease reporting, monitoring and forecasting system.
4) Initially a program of progressive control, could be taken up in certain areas to declare them free from PPR. In view of disease profile PPREP can be initiated through a participatory approach among SAARC countries.
5) Financial support should be invited from world agencies to handle PPREP.
6) Differentiation of PPR from a variety of diseases that cause similar ATD's and RTD's for example RP, FMD, BTD, pneumonic

mannheimiosis / pasteurellosis and CCPP should be regularly carried out by authentic laboratory following OIE/IVRI guidelines.

References

1. Quinn P J, Markey B K, Carter M E, Donnelly W J and Leonard F C. 2002. *Veterinary Microbiology and Microbial Diseases.* Blackwell Science. Ltd.

2. OIE. 2002. (Office International des Epizooties/World Organization for Animal Health). Peste des petits ruminants. Technical disease card database. Apr 22.

3. Shaila M S, David S, Forsyth M A, Diallo A, Goately L, Kitching R P and Barrett T. 1996. Geographic distribution and epidemiology of peste des petits ruminants viruses. Virus Research, Volume 43, Issue 2, August, Pages 149-153.

4. Dhar P, Sreenivasa BP, Barrett T, et al. 2002. Recent epidemiology of peste des petits ruminants virus (PPRV). Vet Microbiol Aug 25;88(2):153-9.

5. Radostits, O.M., Gay, C.C., Blood, D.C and Hinchcliff, K.W. 2000. Veterinary Medicine- A text book of Diseases of Cattle, sheep, pigs, goats and horses, 9th Edition, pp. 1077-1079. Saunders, London.

6. Saliki J T. 2002. Quoted in website : PPR update April 7, 2003.

7. Gargadennec L and Lalanne A. 1942. La peste des petits ruminants. Bulletin of servs Zootech. Epizoot. Afr. Occid. Fr. 5:16-21.

8. Shaila, M.S., Purushothaman, V., Bhaskar, D.A., Venugopal., K. and Venkatesh, R.A.1989. Veterinary Record, 125: 602.

9. Taylor WP, Diallo A, Gopalakrishna S, et al. 2002. Peste des petits ruminants has been widely present in southern India since, if not before, the late 1980s. Prev Vet Med. Jan. 22;52 (3-4):305-12.

10. Dubey S C, Sonawane G G, Tripathi S, Saravanan P and Singh R K. 2006. Seroprevalence of PPR in small ruminants of semiarid Rajasthan. Ind. J Ani. Sc. (In press).

11. Ozkul, A., Akca, Y., Alkan, F., Barrett, T., Taner, K., Dagalp, S.B., Anderson, J., Yesilbag, K., Cokcaliskan, C., Gencay, A. and Burgu, I. 2002. Prevalance, distribution and host range of peste des petits ruminants virus in Turkey, Emerging Infectious Diseases, 8:708-712.

12. Dubey S C, Mondal D and Tripathi S. 2006. Reproductive tract diseases (RTD's) and gonadal transmitted diseases (GTD's) of small Ruminants Proc. "National Seminar on Innovations and Recent Advances in Reproduction for Augmenting Small Ruminant Production" 28-30 Dec. 2006. Avikanagar.

13. Singh R P, Sarvanan P, Singh R K and Bandopadhaya S K. 2004. Rev.sci. tech. Off. Int. Epiz. 23 (3):807-19.

14. Anonymous. 2006. Seroprevalence of PPR in sheep & goat population in Haryana & Delhi states (personal communication).

15. Sunil kumar N S, Ravishankar C, Jayaprakasan V, Mini M and Krishnan Nair G. 2005. Seroprevalence of peste des petits ruminants in Kerala. Indian Vet. J., May;82:570-571.

16. Dubey S C, Mondal D, Tripathi S, Suresh A, Singh R K and Singh V K. 2004-06. Epidemiology, therapeutic measures and economics of PPR in small ruminants of semi-arid India. (Unpublished observations).

17. Mathur P B and Dubey S C. 1994. *Sheep and Goat Diseases.* ICAR Publication.

18. Dubey S C and Somvanshi R. 2002. *Bhed-Bakarion ke roag avum unka niyantran.* CSTT Pub. (Min. of HRD, GOI). Pp329-37.

19. Rajeswari R. 2000. PPR in small Ruminants in Andhra Pradesh. Ind Vet. J., 77(5): 373-75.

20. Saha A, Lodh C and Chakraborty A. 2005. Prevalence of PPR in goats. Indian Vet. J., June;82:668-69.

21. Kumar G S, Rathore B S and Malhotra M L. 1999. Ind. J A Scs. 69 (6):365-68.

22. Shankar H, Gupta V K and Singh N. 1998. Occurrence of pest des petits ruminants like disease in small ruminants in Uttar Pradesh. Ind. J A Scs. 68(1):38-40.

23. Dhand N K, Sharma C S, Sandhu K S, Sharma D and Singh J. 2002. Ind. J. Anim. Scis. 72(10):853-54.

24. Wadha D R, Prasad B, Verma S and Sharma A K. 2002. Occurrence of PPR like outbreak among migratory goat and sheep in Himachal Pradesh. Ind. J. Vet. Med. 22(2):105-106.

25. Islam M R, Samsuddin M, Rehman M A, Das P and Devan M L. 2001. Outbreaks of PPR in Black Bengal goat in Maymansing in Bangladesh. Bangladesh Veterinarians. 18(1):14-19.

26. Tibbo M, Woldemeskel M and Gopila A. 2001. An outbreak of respiratory disease complex in sheep in central Ethiopia. Trop. Anim. Hlth. Prod. 33(5):355-65.

27. Ezeibe M C O. 2005. Relative resistance of goat to PPR following sheep and goat transmission of infection. J. Sust. Agric. & Environ. 7(1):15-18.

28. Hocisawi F M T, Elzen EMEA, Md G E, Gameel A A, Al-Afaleg A I, Hegazi A, Basir A L. Vet. Bull. NO- 3153. (2005).

29. Bandopadhy, S.K. 2002. The Economics appraisal of a PPR control programme in India. In 14th Ann. Conf. Nat. Sem. on Mang. Of Viral diseases with emphasis on global trade & WTO regime. Ind. Vir. Soc. 18-20, Jan, Bangalore.

30. Indian Livestock Census -17th. 2003. All India summary Report- Livestock, Poultry, Agri. Machinary, Impliments and Fishery Statistics, GOI, MOA, DAH & D, New Delhi.

31. Saliki JT. 1998. Peste des petits ruminants. In: US Animal Health Association, Committee on Foreign Animal Disease. Foreign animal diseases: the gray book. Ed 6. Part IV. Richmond, VA: US Animal Health Assoc.

32. Sankhyan, S.K., Shinde, A.K. and Karim, S.A. 1999. Indian J. Anim. Sci., 69: 617-620.

33. Shinde, A.K., Sankhyan, S.K., Singh, N.P., and Verma, D.L. 2006. Small Rum. Res. (In press).

34. Pawaiya R V S, Misra N, Bhagwan P S K, and Dubey S C. 2004. Pathology and distribution of antigen in goat naturally infected with PPR virus. Indian Journal Animal Science 74: 35-40.

35. Rajak K K, Sreenivas B P, Hosamani M, Singh R P and Bandopadhayay. 2005. Experimental studies on immuno suppressing effect of PPR virus in goats. Comp. Immun. Mico, 28(4):287-96.

36. ICAR website, Success story, Vaccine for Plague (*Peste des petits ruminants* - PPR): A Disease of Small Ruminants.

37. Dubey S C, Shinde A K and Singh B N. 2006. *Vyavaharic Pashupalan* (Applied Animal Husbandry of Cattle, Buffalo, Sheep, Goat, Horse, Swine & Camel) IBDCo, Lucknow, 369 pages.

-ΔΔΔ-

9

Control of Capripox Virus Infections in India

D D KULKARNI

High Security Animal Disease Laboratory
Indian Veterinary Research Institute, Bhopal - 462 021

In India, sheep pox and goat pox occur throughout the regions where sheep and goats are reared. In coming years, due to openness in international trade, the small ruminants would be put to 'high production-stress' that would make them more vulnerable to attacks of pox and may also increase its severity, thereby necessitating the control and eradication with more urgency. Attempts to protect either goats with sheep pox vaccines or sheep with goat pox vaccines are largely unsuccessful. In India, sheep pox has been controlled mainly through the use of a live attenuated vaccine prepared from the Romanian strain of SPV, but an efficient vaccine is yet to be developed for goat pox. It is normally recommended that homologous vaccines should be used to protect goats effectively against goat pox.

Viruses belonging to the Capripoxvirus genus of the Poxviridae family are etiologic agents of important diseases of domestic animals in the developing world. These include lumpy skin disease virus, which causes disease only in cattle in central and southern Africa (will not be discussed in this presentation), sheep pox virus (SPV) and goat pox virus (GPV), which cause disease in sheep and goats in northern and central Africa, southwest and central Asia, and the Indian subcontinent, as far East as Bangladesh, China and Myanmar. Occasionally capripox spreads from Turkey into Greece[20].

India has an estimated 60 million sheep and over125 million goats. At least 17 Indian states have significant number of sheep and goats while eight of them have populations exceeding 10 million[3]. Sheep pox and goat pox occur throughout the regions in India where sheep and goats are reared. A joint survey on the epidemiology and economic impact of sheep and goat pox in Maharashtra state during 1997-98 was carried out by Bharatiya Agro-Industries Foundation, Pune with Australian collaboration, overall morbidity rates for kids, does and bucks were 67%, 58% and 75%, while the mortality rates were 53%, 45% and 58% respectively[3].

Historical Background

Goat pox and sheep pox are OIE notifiable diseases. Goat pox, a highly contagious viral disease of goats, was described in detail in ancient veterinary medical texts, and has been widespread since early times. Hansen reported goat pox in 1879 from Norway[44]. Later on it was observed during the First World War in Macedonia and became enzootic during 1926 with a mortality rate of 15%[8]. The sheep pox originated in Central Asia and then spread to many western countries. Many sheep pox outbreaks occurred during 1847–1866 in Great Britain and eradication was achieved in 1866, but only recently from continental Europe[7]. Sporadic outbreaks still occur in Eastern European Mediterranean islands, probably originating from imported animals[20]. In India, sheep pox was reported for the first time in 1931-32 and goat pox in 1936[43].

Host-range

Capripox viruses infect only ungulates, and most strains cause clinical disease in only one species. Under natural conditions, GPV is highly host specific, infecting only goats but there are strains that may cause disease in both sheep and goats. Kenyan and Yemen isolates as well as an Oman sheep isolate infect sheep and goats equally[14,26]. Usually, Middle East and Indian isolates of goat pox are host specific and do not infect sheep[13, 22, 49]. It is possible that the host preference shown by different strains is due to their adaptation to either goats or sheep in a restricted geographical area. However, it is envisaged that when sufficient isolates have been examined biochemically, no clear distinction will be possible between goat and sheep isolates, but rather a spectrum would emerge in which some strains have clear host-preferences while others would be less defined and would naturally infect whichever host they come into contact with[24]. The GPV may also produce lesions in rabbits and reindeer[32].

Exceptionally, it is observed that a few strains of GPV are pathogenic for human beings; however, many workers reject this fact. There is no conclusive evidence that capripox infects humans. A report from India[48] implying goat pox to cause human infection was merely based on clinical signs. There was no attempt to isolate the causative virus or perform serology on the convalescent serums of the three patients to differentiate the infection from contagious ecthyma, which is a known zoonotic agent that occurs worldwide. A report from Sweden[4] indicated that human infection occurred during an outbreak of goat pox. Although serological studies seemed to indicate that the apparent causative agent of the outbreak was not vaccinia or contagious ecthyma, no virus was isolated. Therefore, it cannot be said that goat pox virus caused human infection.

Probability factors

A survey conducted under Joint Indo-Australian project had shown that farm flocks, long distance migratory flocks and village flocks (in that order) were more likely to experience pox than the stall-fed flocks. Some other factors that would affect the occurrence of the disease were- variable and long incubation period (5–14 days), resistance of the virus (it may remain infective on the wool or hair of recovered animals or in surroundings for long periods), high level of contact between flocks ? through shared grazing and watering, camps and holding under a closed enclosures in a very humid and crowded atmosphere at night (this is particularly true for goats) and ?tendency of owners to dispose off sick animals in market and effecting new purchases without bothering for quarantine etc.

However, it has also been seen that the recovered animals have good immunity, there is relatively low turnover of animals as well as ?low level of introduction of new animals into the flock and there is no carrier state. These would be considered as favourable factors for control. Considering the risk and favourable factors, the average annual probability that a sheep flock would experience pox was calculated as 0.34% compared to 1.45% for goat flocks and 1.83% for mixed flocks[3].

Why it is necessary to control the capripox?

Infectious diseases jeopardize the optimum productivity. Besides mortality and other economic losses, they restrict the export of livestock and their products. Freedom from major animal diseases provides a significant advantage in export markets for animals and animal products. As per OIE, sheep and goat pox are among 17 'List-A' diseases that are transmissible with potential for widespread and rapid transmission and are of serious socio-economic consequence.

There are some 40 recognised breeds of sheep and 20 recognised breeds of goat in India, as well as larger populations of non-descript local animals. Although these sheep and goats have evolved in different geographical areas they share similar low input subsistence systems of rearing with very limited preventive health care. As a result of natural selection, many of these animals are highly resistant to diseases endemic in each area. Based on data from the Central Sheep and Wool Research Institute, Avikanagar, the overall morbidity and mortality due to sheep pox were much lower in fine wool sheep breeds (such as Nali, Chokla, and their crosses), than in meat sheep breeds (such as Malpura, Sonadi and their crosses)[51]. In coming years, due to openness in international trade, the small ruminants would be put to task for increase in their production potential. This 'high production-stress' would make sheep and goat more vulnerable to attacks of pox and may also increase the severity of the diseases, thereby necessitating the control and eradication with more urgency.

Although capripox is considered as one entity affecting sheep and goats universally, in India sheep pox is a serious disease whereas goat pox is limited to clinical signs in goats and mortality in kids. The research on sheep and goat pox in India has been fragmented. At present the disease is largely recognised by clinical symptoms but these are unable to reliably differentiate pox from other exanthematous conditions like contagious ecthyma. Owing to their intimate association of with their flocks, shepherds in India have a detailed knowledge of their animals, their health status and are able to provide detailed information about performance of their flocks and the occurrence of disease. However, due to negligence of veterinarians towards the migrating small ruminants, the lack of coordinated programmes and specific diagnostic tools, it is difficult to implement the control measures.

Transmission

Affected sheep and goats shed the virus at every stage of the disease. Virus is present in secretions and excretions of infected animals, including milk, and in scabs from skin lesions, but these are not considered to be important sources for transmission. Most transmission is by direct contact, but indirect contact and mechanical transmission by insects also occur. In endemic areas, spread occurs mainly in summer till late monsoon. According to ICAR Report (1998), the outbreaks have been recorded throughout the year, approximately 33% occurring in summer, 25% in autumn, 27% in winter and 16% in spring.

Short distance aerosol transmission from nasal secretions and saliva is an important method of spread[26] and transmission is most likely to occur by direct contact when animals are herded together at watering places, yards and markets. Environmental contamination of such sites is also important because of the resistance of the virus to inactivation. Mechanical transmission of virus by biting insects and by stable and houseflies feeding on exudates from skin lesions is suspected[53]. The practice of nomadic shepherds to migrate the sheep across various states (to utilize post-harvest agricultural residues as nutrition) tends to perpetuate the diseases over much of their range.

Control

Sheep pox is a major health hazard of sheep, which can be controlled mainly by vaccination. Various kinds of vaccines are available but the live attenuated vaccine is the best choice. Capri pox-free countries maintain their disease-free status by the restriction of imports of livestock and animal products from affected areas. In the case of countries remote from enzootic areas the radical slaughter of affected species and their movement restrictions, coupled with a ring vaccination of radius 25-50 km, results in elimination of disease[9].

In enzootic countries, annual vaccination of susceptible animals with a live vaccine controls the disease. In India, sheep pox has been controlled mainly through the use of a live attenuated vaccine, but an efficient vaccine is yet to be developed for goat pox[43].

Maternal immunity provides protection from capripox for up to 3 months[23]. Although maternal antibody will inactivate the vaccine, it is advisable to vaccinate all the stock over 10 days old. Ring vaccination is frequently practiced during outbreaks in enzootic areas, but usually only the species that are clinically affected are vaccinated.

Traditionally, some methods were used as prophylactic measures viz. *ovination*[31,46,47] (the inoculation of infected scab material as such or suspended in glycerin saline with the false notion that it prevents the infection), ***Sero-clavelization***[37,45] (administration of serum from convalescent sheep and/or anti-SPV serum either at different sites or as a mixture of both prepared just before use, at only one site), and *Sensitized virus*[53] (the incubation of a mixture of potent anti-SPV serum and SPV at 15–18° C for 3 days) inoculation, etc. All these methods have several disadvantages and are not recommended now.

Inactivated vaccines

Various inactivating agents have been used with variable success in inactivating different strains of SPV, which include-formaldehyde, ethanol, butanol, ethyl-ether, aniline dyes, crystal violet, brilliant green, boric acid, heat, sodium merthiolate, phenol, b-propiolactone (BPL), hydroxylamine, N-acetyl ethyleneimine (AEI) and ultra-violet irradiation. However, formaldehyde, b-propiolactone, hydroxylamine, ethyleneimine etc. have been found to be more effective[5]. Among the adjuvants, Aluminum hydroxide has been employed more commonly and saponin, oils are also used[5]. Recently aluminum chloride was reported to be the most effective adjuvant for SPV[54].

Various inactivated vaccines were used by employing various strains of SPV.[11,12,14,15,18,27,34,49,56] With such preparations, disadvantages like short-term immunity, poor shelf-life, clogging of needles and syringes in cold seasons, large dosage and transportation, occasional necrosis, mortality[41] etc. have been observed. There was a poor correlation between neutralizing antibodies and immune status of vaccinated sheep.[52] Therefore, the use of killed vaccines are discouraged owing to their temporary protection.[29]

Live virus vaccines for sheep pox

A number of SPV strains viz., Perego, Bakirkoy, Algerian, SGPV, Kenyan, Pendik, SPV/RH, Mauritanian, Roumanian-Fanar, Kazakhstan, Chitinsk, Mongolian (USSR), NISKI, Karnal, RM/65, Mathura SP8, Jaipur, Ranipet, Hyderabad etc. have been used with variable success[5]. The passage level at which each of the above strains got attenuated varies with the strains, virulence and cell system used. Among the cell cultures; lamb testes, kidney and thyroid, calf kidney and fetal muscle, fetal sheep kidney, lung and skin tissues and continuous cell lines like BHK21 and Vero have been the most commonly employed systems for attenuation[6]. Live vaccine containing SPV adsorbed onto aluminum hydroxide gel induced strong immunity as early as 15 days post-vaccination and maintained for about two years[2,35] whereas an Indo-China strain induced immunity only for 8 months[58]. These vaccines with shelf life of 10 months under refrigeration had around 83% protection and no post-vaccinal adverse reactions. Similarly, oil-adsorbed virus induced complete protection, whereas glycerin saline vaccine produced severe reactions[41]. In addition, immunity to sheep pox is predominantly cell mediated[9]. Live vaccines have been found safe in pregnant ewes and result in normal lambing. There was no shedding of virus from the vaccinated ewes. Annual vaccination of adult sheep and the vaccination of lambs (born to protected ewes) at 2 months of age provide the best immunity[38]. In India, a live attenuated vaccine prepared from the Romanian strain of SPV is used to confer immunity in all types of sheep against sheep pox and is quite safe and effective[21].

Live virus vaccines for goat pox

In enzootic areas, both live attenuated and inactivated vaccines are useful in the prevention and control of goat pox, but inactivated vaccines give only short-term immunity.[33,36,59]

Various live attenuated vaccines have been used for goat pox with varying degrees of protection.[16,17,28,42] A subunit vaccine also appears to be of some use in the control of disease as revealed by higher neutralization indices in immunized goats.[10] Moreover, a single vaccine prepared from a strain of capripoxvirus that infects sheep and goats equally was found to be effective in controlling both goat pox and sheep pox for at least 12 months.[9,25]

Polypeptide vaccine

The VP3 protein of sheep pox virus was identified as immunogenic protein and induced significantly high level of humoral and cell mediated immune responses in rabbits[39]. Immunization of sheep with VP3 protein emulsified with Freund's incomplete adjuvant elicited 67% protection to sheep against virulent challenge after 21 days post-vaccination[40]. This needs further exploration for prophylactic purposes.

Combined vaccines

Combined vaccines are economical and can reduce the stress to animals. Several combined vaccines with inclusion of capripox were tried in the past and no impairment of immunity was reported by majority of workers. A mixed vaccination comprising of PPR (attenuated 75/1 strain) and capripox (RM/65) in Northern Cameroon has resulted in poor protection of vaccinated goats when challenged with virulent capripox, despite inducing neutralizing antibodies[30]. In India, live attenuated vaccines are currently in use for controlling PPR and sheep pox, while goat pox vaccine is under experimental field trial. A recently developed bivalent vaccine at IVRI Mukteshwar (comprising goat pox and *Peste des Petits ruminants* viruses) was found to be safe and induced protective immune response in goats against both the antigens. Goats immunized subcutaneously with 1ml of vaccine consisting of 10^3 $TCID_{50}$ of each of GPV and PPRV were monitored for clinical and serological responses for a period of 4 weeks post-immunization and post-challenge. All the immunized animals resisted challenge with virulent strains of either GPV or PPRV on 28th day post-vaccination. Single immunization with goat pox vaccine was found to confer protective immunity at least for 26 months as revealed by challenge studies[19].

Cross protection

Reports of cross-protection of sheep and goats against goat pox and sheep pox and other related diseases such as CPD are often contradictory and inconclusive[57]. In an experimental study in Cameroon, goats vaccinated with RM65 strain of SPV could not withstand challenge with wild strain of GPV and the failure of the vaccine to protect against virulent virus challenge was attributed to lack of protection between SPV and the wild strain of GPV used in the experiment. This can be explained by partial cross protection among sheep pox and goat pox vaccine viruses[55]. In India, attempts to protect either goats with SPV vaccines or sheep with GPV vaccines are largely unsuccessful[1,37]. It is normally recommended that homologous vaccines should be used to protect goats effectively against goat pox.

The following points have been recommended by FAO for sanitary prophylaxis -

- Isolation of infected herds and sick animals for at least 45 days after recovery
- Slaughtering of infected herd (as far as possible)
- Proper disposal of cadavers and products
- Stringent disinfection
- Quarantine before introduction into herds
- Animal and vehicle movement controls within infected areas

In India, the control of sheep pox has been made possible only due to effective vaccine. The goat pox has remained as a neglected area. It may be a long way to go for eradication of capripox from India but it is worth to set the target to do so in coming 10 years in the following manner-

- The state level disease investigation labs and vaccine institutes should be asked to take up identification of local goat pox strains and production of live attenuated vaccine to cater the local need. The sheep pox vaccine production be scaled up.
- At national level, attempts to take up research on goatpox vaccines including combined and recombinant vaccines should be encouraged.
- Sheep pox vaccinations should be done regularly every year on a larger scale and this should continue for at least for 5 years before assessing the impact.
- Sheep pox and goatpox vaccinations should be done simultaneously and regularly to all sheep and goats including the migratory flocks.
- The sheep and goat rearing persons should be encouraged to co-operate with disease control process. This is possible by educating them, avoiding animal movements during outbreaks, giving incentives, and propagating concept of stall-feeding goats.

References

1. Agrawal S. K. & Soman J. P. 1997. Assessment of immune efficacy of attenuated live goatpox vaccine against sheep pox and contagious ecthyma. Indian Vet. J. 74(12): 1016-1018.
2. Anandan R., Sundara Rajan S., Kannamani G. & Jayaraman M.S. Studies on live and inactivated sheep pox vaccines. Cheiron 1972; 1:42–55.
3. Anonym. (1998) Development and standardization of diagnostic tests and vaccines for the control of sheep and goat pox diseases in India and Australia. Final Report on ACIAR PROJECT No. PN 9451.
4. Bakos K. & Brag S. 1957. Studies of goat pox in Sweden. Nord. Vet. Med., 9:431-439.
5. Bhanuprakash V. Epidemiology of sheep pox infection in Karnataka, the analysis of viral protein profiles of field and vaccine strains and evaluation of live attenuated vaccines. PhD Thesis. Uni Agri Sci, Bangalore, Karnataka, India; 2001.
6. Bhanuprakash V., Indrani B.K., Hegde R., Kumar M.M.& Moorthy A.R.S. A classical live attenuated vaccine for sheep pox. Trop. Anim. Hlth. Prod. 2004; 36(4): 307–20.

7. Bhanuprakash V., Indrani B. K., Hosamani M. & Singh R. K. 2006. The current status of sheep pox disease. Comp. Immunol. Microbiol. Infect. Dis. 29(1): 27-60.

8. Blanc G. Melanidi C. & Stylianopoulo M. 1929. Variola of goats in Greece. Arch. Inst. Pasteur Hellenique, 2(1): 129-133.

9. Carn V.M. 1993. Control of capripoxvirus infections. Vaccine, 11(13): 1275-1279.

10. Carn V. M., Timms C. P., Chand P., Black D. N.& Kitching R. P. 1994. Protection of goats against capripox using a subunit vaccine. Veterinary Record, 135(18): 434-436.

11. Celliker A. & Arik F. Effect of Merthiolate on sheep poxvirus. Pendik Vet Kont Arastirma Ensti Derg 1968; 1:52–6.

12. Celliker A. & Arik F. Aluminium hydroxide-adsorbed sheep pox virus vaccine. Br Vet J 1962; 118:159–61.

13. Datta S. & Soman J. P. 1991. Host range and physico-chemical characterization of 'Ranchi' strain of goat-pox virus. Indian J. Anim. Sci. 61(9): 955-957.

14. Davies F. G. 1976. Characteristics of a virus causing a pox disease in sheep and goats in Kenya with observations on the epidemiology and control. J. Hyg; 76(2): 163–71.

15. Das S. K, Mallick B. B. 1984 Susceptibility of various cell culture systems to sheep poxvirus strains. Indian J. Exp. Biol; 22:205–8.

16. Davies F.G. & Mbugwa G. 1985. The alterations in pathogenicity and immunogenicity of a Kenya sheep and goat pox virus on serial passage in bovine foetal muscle cell cultures. J. Comp. Path. 95(4): 565-572.

17. Dubey S. C. & Sawhney A. M. 1978. Live and inactivated tissue culture vaccines against goat pox. Indian Vet. J., 55:926-927.

18. Goyal S.M, & Singh I.P. 1975 Inactivation of sheep pox virus by formaldehyde, hydroxylamine and N-acetyl ethyleneimine. Indian J Anim Sci; 45(10): 748–53.

19. Hosamani, M. S., Singh, S. K. Mondal, B., Sen, A. Bhanuprakash, V. Bandyopadhyay, S. K., Yadav, M. P. & Singh, R. K. 2006 A bivalent vaccine against goat pox and Peste des Petits ruminants induces protective immune response in goats. Vaccine 24(35-36): 6058-64.

20. House J. A. 1992. Sheep and goat pox. In Veterinary diagnostic virology: a practitioners guide. 217-219; [244].

21. Jadhav K.M, Pandey A.K and Radadia N.S 1989. Field observations on sheep pox tissue culture vaccine. Indian Vet. J. 66: 908–912.

22. Kitching R. P. 1983. Progress towards sheep and goat pox vaccines. Vaccine, 1: 4-9.

23. Kitching R. P. 1986. Passive protection of sheep against capripoxvirus. Research in Veterinary Science, 41(2): 247-250.

24. Kitching R. P. 1994. Sheep and goat poxviruses. In: Webster RG, Granoff A, eds. Encyclopedia of Virology. London: Academic Press, 1160-1165.

25. Kitching R. P., Hammond J. M. & Taylor W. P. 1987. A single vaccine for the control of capripox infection in sheep and goats. Res. Vet. Sci. 42(1): 53-60.

26. Kitching R. P. & Taylor W. P. 1985. Transmission of capripoxvirus. Res. Vet. Sci., 39(2): 196-199.

27. Kulshreshtha A. K, Pathak P. N, Moghe M. N, & Tanwani S. K. 1991 Studies on comparative efficacy of different sheep pox viruses as vaccines. Indian Vet J; 68(6): 506–9.

28. Mahmood M.A, Ahmad M, Afzaal S, & Khan M. S, 1993. Experimental immune response of Iranian goat pox vaccine virus strain against prevalent virus in Pakistan. Pakistan Vet. J., 13(3): 125-127 [222].

29. Manual OIE. Manual of standards for diagnostic tests and vaccines. 4th ed.; 2000.

30. Martrencher A, Zoyem N. & Diallo A. 1997. Experimental study of a mixed vaccine against PPR and capripox infections in goats in Northern Cameroon. Small Ruminant Res; 26(1–2): 39–44.

31. Munro D.A. 1947. Protective inoculation with dry pegion pox and sheep and goat pox vaccines. Indian J Vet Sci; 17:193–200.

32. Nandi S, & Rao T.V.S, 1997. Goatpox - a major threat to the goat industry in India. Indian Farming, 47(3): 26-27.

33. Pal J. K. & Soman J. P. 1992. Further trials on the inactivated goat pox vaccine. Indian J. Virol. 8(2): 86-91.

34. Pandey A. K, & Kumar S. 1974. A note on the use of Russian strain of sheep pox virus for preparing sheep pox vaccine. Indian J Anim Sci. 44 (12): 1015–6.

35. Pandey A. K, Malik B. S. & Bansal M. P. 1969. Studies on sheep pox virus. II. Immune and antibody response with cell culture adapted virus. Indian Vet J. 46: 1017–23.

36. Prasad I. J, & Datt N. S. 1973. Observation on the use of live and inactivated vaccines against goat pox. Indian Vet J. 50:1-10.

37. Prasad G, Garg S. K, & Singh I. P. 1986. Passive transfer of immunity against sheep pox. Indian Vet Med J. 10(2): 78–81.

38. Precausta P., Kota F. & Vellut G. 1979. A new freeze-dried living virus vaccine against sheep pox. Comp Immunol Microbiol Infect Dis. 1(4): 305–19.

39. Rai A, Singh U. S, Goel A. C, Pandey K. D, Mishra S. C & Gupta B. K. 1986. Characterization of an immunogenic viral protein of sheep poxvirus. Indian J. Virol. 1(2): 120–6.

40. Rai A, Goel A. C., Pandey K. D., Mishra S. C. & Gupta B. K.1986. Immunogenicity of a virion polypeptide of sheep poxvirus in sheep. Indian J. Virol. 2(1): 11–15.

41. Ramani K. & Sundara Rajan S. 1976. Comparative efficacy of different sheep pox vaccines. Indian Vet J. 53: 313–8.

42. Ramyar H, Hessami M, & Ghaboussi B, 1974. Goat pox: immunogenicity of vaccine virus modified in cell culture. Recueil de Medecine Veterinaire. 150(2): 131-133.

43. Rao T. V. S. & Bandyopadhyay S. K. 2000. Comprehensive review of goat pox and sheep pox and their diagnosis. Anim Health Res Rev; 1:127–36.

44. Rafyi A. & Ramyar H. 1959. Goat pox in Iran, serial passage in goats and the developing eggs and relationship with sheep pox. J. Comp. Path. Therap. 69: 141-147.

45. Red'Ko A.S. 1945. Convalescent serum for control of sheep pox in lambs. Vet Moscow; 1:21.

46. Sabban M.S. 1955 Sheep pox and its control in Egypt using a desiccated live virus vaccine. Am J Vet Res; 16:209–13.

47. Sapre M.V, Kalia O.P. 1980. Incidence of sheep pox outbreak in Rajasthan. Auto-vaccination on large scale. Wool Woolens India; 17(3): 38–40.

48. Sawhney A.N, Singh A.K & Malik B.S, 1972. Goat-pox: An anthropozoonosis. Indian J. Med. Res. 60: 683-684.

49. Sharma S.N & Dhanda M.R. 1970 Studies on sheep and goatpox viruses-immunization trials. Indian J Anim Sci; 40:626–39.

50. Sharma S. N. & Dhanda M. R. 1972. Studies on sheep- and goat-pox viruses. Pathogenicity. Indian J. of Anim. Hlth. 11:39-46.

51. Sharma M. M., Uppal P. K., Lonkar P. S. & Mathur P.B. 1986. Epidemiology of a sheep-pox outbreak in mutton and fine wool type sheep at an organized farm. Indian J. Anim. Sci. 56: 1183–1186.

52. Singh B. K, Pandey A. K. & Das S. K. 1983. Serological response in sheep vaccinated with attenuated and inactivated sheep pox vaccines. Indian J Comp Microbiol Immunol Infect Dis. 4(2): 90–6.

53. Singh I. P., Pandey R. & Srivastava R. N. Sheep pox: a review. 1979. Vet. Bull. 49(3): 145–54.

54. Solyom F, Perenlei L, Roith J. 1982. Sheep pox vaccine prepared from formaldelhyde inactivated virus adsorbed to aluminium hydroxide gel. Acta Microbiol Acad Sci Hung. 29(2): 69–75.

55. Toth A. 1937. An outbreak of sheep pox. Allatorv Lapok. 60:151–2.

56. Uppal P. K. & Nilakantan P. R. 1967. Observations on the use of live and inactivated virus vaccines against sheep pox. Indian Vet. J. 44:815–27.

57. Wang S.H. & Jiang H. X. 1988. Vaccination of sheep against sheep pox with an attenuated goat pox vaccine. Chinese Journal of Veterinary Medicine, 14(12): 36-38.

58. Wynohradnyk V. & Cirstet J. 1956. Vaccination against sheep pox with live attenuated virus adsorbed on aluminium hydroxide. Annu. Inst. Pat. Igiene. Anim; 6:195–204.

59. Yadav M. P., Pandey A. B., Negi B. S., Sharma B. & Shankar H. 1986. Studies on inactivated goat pox vaccine. Indian J. Virol. 2(2): 207-221.

-ΔΔΔ-

10

Situation of Bluetongue in India and Strategies for its Prevention and Control

A B PANDEY

All India Network Programme on Bluetongue Disease
Centre for Animal Disease Research and Diagnosis
Indian Veterinary Research Institute, Izatnagar – 243 122 (U.P)

INTRODUCTION

Bluetongue (BT), an arboviral disease is caused by the bluetongue virus (BTV), which naturally infects domestic (sheep, goats, cattle) and wild ruminants (deer, pronghorn antelope, bighorn sheep), camelids and some other herbivores such as elephants. Disease is characterized by the catarrhal stomatitis, rhinitis, enteritis and lameness. Bluetongue is transmitted by several species of Culicoides (biting midges). It is almost exclusively a disease of sheep, although cattle and goats show sub clinical infection. Cattle are the main mammalian reservoir of the virus and play a very important role in the epidemiology of the disease. The frequency of BTV infection in cattle is generally higher than in sheep. Exotic sheep are more susceptible than indigenous and cross-bred sheep. Shifting from host factors, however, outbreaks do occur in indigenous breeds due to the prevalence of highly virulent strains in Indian sub-continent. In non-domestic ruminants, disease can vary from an acute haemorrhagic disease with high mortality, as observed in white-tailed deer, to an inapparent disease as seen in the North American elk (Cervus canadensis). The reports on BT in wild life are scanty in India. But the possibility of these animals to contract the disease is high as wild ruminants particularly deer are highly susceptible to the BT.

BT can cause spectacular disease outbreaks and thus placed in one of the 15 diseases classified in 'List A' by the Office International des Epizooties (OIE, 2006).

The name "bluetongue" was given by Spreull[35] to describe the distinctive cyanotic tongue in some severely affected sheep. It was first reported in South Africa after Merino sheep were introduced from Europe in the late eighteenth century.

CAUSATIVE VIRUS

BTV is a member of the Orbivirus genus, currently one of nine genera classified in the family Reoviridae. Within the Orbivirus genus, 14 groups are differentiated on serological basis. There is significant immunological cross-reactivity between members of the BT and Epizootic Hemorrhagic Disease (EHD) serogroups[11]. The BTV is quite stable at 25°C. The slow freezing of virus at -10° or -20°C results in the loss of infectivity and hence the blood containing virus should be stored at 4°C without any substantial loss of infectivity for a year. The tissue culture virus can be stored in 50% buffered lactose- peptone solution at - 20°C without any significant loss of titer after 8 years. The virus is resistant to ether, chloroform and sodium deoxycholate and is inactivated beyond the pH range of 6-8.

The virion is about 86 nm in diameter with icosahedral symmetry (Hewat *et al.* 1992) and is composed of 80% proteins and 20% nucleic acid of ten segments of double stranded RNA. The base composition of genome has 42.4% G+C value. Each of these genome segments code for one of the ten viral proteins. Seven of these proteins are 'structural', combining to form the viral particles and three non-structural proteins, NS1, NS2, NS3, are detected within the infected cells. They are non-enveloped and composed of three protein layers. The outer layer contains two proteins, VP2 and VP5. VP2 is primarily responsible for the induction of type-specific neutralizing antibodies and its variable sequence resulted in the generation of 24 recognized serotypes of BTV. These are also involved in cell attachment and penetration of virus during initiation of infection. This genetic diversity of bluetongue virus is a consequence of both drift and reassortment of individual gene segments sequel to selective pressure of antibodies. Middle and inner layers, which form a bi-layered icosahedral core particle, are composed of two major proteins, i.e. VP7 (middle layer) and VP3 (inner layer) and three minor proteins- VP1, VP4, VP6 and within this are the ten species of double-stranded RNA.VP7 is a major determinant of serogroup specificity.

Coding of viral protein by Genome-segment

Protein	Genome segment
Structural	
VP1 (Polymerase, inner core)	L1
VP2 (Nonglycosylated, neutralizing, Haemagglutinating outer capsid)	L2
VP3 (Conserved, group specific, inner core)	L3
VP4 (Conserved, enzymatic, inner core)	M4
VP5 (Indirectly neutralizing, outer core)	M6
VP6 (Conserved, enzymatic, inner core)	S9
VP7 (Conserved, group specific, outer core)	S7
Nonstructural	
NS1 (Conserved, group specific, inner core)	M5
NS2 (Conserved, phosphorylated, inner core)	S8
NS3 (Cell membrane, release of virus)	S10

VECTOR

Midges of the genus Culicoides are biological vector for BTV[22]. Culicoides transmit BTV to and from susceptible animals, which have become infected by feeding on viraemic vertebrates. After a replication period of 6–8 days, and following its appearance in the salivary gland ('extrinsic incubation period'), the virus can be transmitted to a vertebrate host during a blood meal.

Of approximately 1400 species of Culicoides worldwide, less than 20 are considered actual or possible vectors[22]. There is evidence for the involvement of novel Culicoides vector(s) (*C.obsoletus* and or *C.pulicaris*) and for survival of BTV for significant period in the absence of adult vectors i.e. over wintering.

In India, although disease is endemic but detailed study on vector species responsible for disease and virus-vector relationship are scanty. BTV has been isolated from Culicoides in India and China, but the species were not identified. Recent studies on vectors indicated that *C.oxystoma* is found to be most prevalent. Other vectors responsible for transmission include *C. clavipalpis, C. anophelis* and *C. imicola* in West Bengal, *C. anophelis* and *C. actoni* in Uttar Pradesh, *C. simiklis* in Uttaranchal and *C. orientalis* in Gujarat[27].

EPIZOOTIOLOGY

In India, the disease was first reported by Sapre[32] from Maharashtra state in flock of sheep and goat. Since then, several outbreaks of BT have been reported in sheep from almost all the states of the country. Now, BT has endemic status in India. A serological survey has indicated the presence of bluetongue virus (BTV) antibodies in cattle and buffalo in several states in India. However, clinical BT has not been frequently observing in cattle and or buffaloes. The distribution of BTV serotypes and their competent vector spp has been presented in the Table-1.

BTV has evolved a life cycle where alternate cycles of virus replication in vertebrate and invertebrate hosts are essential for virus persistence; consequently Culicoides spp acts as principal vector in transmission of BTV. There is no evidence of vertical transmission of the virus in the invertebrate host. Observations on the placental transmission of virus in the vertebrate host are contradictory[31]. In addition, there is no convincing scientific evidence that natural BTV infection crosses the placenta to cause fetal infection and production of persistently infected adult animals[11]. There is little evidence of direct or indirect contact transmission in either host, other than rare instances of seminal transmission (here male should be viraemic and semen should contain red and white blood cells) in vertebrates[25]. The virus cannot be spread by meat, milk or dairy products, but can be through mechanical vectoring by means of instruments (blood transfer) during periods of viraemia.

The virus seems to be maintained in midges-cattle cycle as a primary cycle, and once a certain level of infection is attained, it spills over to initiate a secondary cycle in sheep. This generation occurs in late summer and early autumn when the vector populations are at a maximum. Thus, cattle act as both primary and reservoir vertebrate hosts[10]. Susceptibility of dogs for BTV infection has also been reported on inoculation of inadvertently contaminated canine vaccine with BTV, indeed pregnant bitches that received this contaminated vaccine typically aborted and died[2,3].

The presence of BTV year round or seasonal depends on the climatic conditions and therefore, the geographic distribution of BTV is categorized into three ecologic zone types: endemic, epidemic and incursion. Endemic zones are defined as typically tropical regions where BTV transmission occurs throughout the year and sub-clinical infection is common. In endemic zones, clinical disease generally occurs in newly introduced, immunologically naïve susceptible species. Epidemic zones include temperate areas where outbreaks occur seasonally, generally in the late summer when vector population is at peak. Incursion zones are areas that experience outbreaks infrequently when climatic conditions favour disease transmission by vectors as observed in greatest outbreak of Mediteranian basin.

India is a geographically vast country having diverse geo-climatic zones. Broadly, the climate of the country classified into tropical, sub-tropical and temperate with major part of it constitutes tropical region. Disease is often said to be 'endemic' in India although much of the major outbreaks encountered during late summer and early winter season i.e., September through November periods.

Table. 1. Distribution of BTV serotypes in Indian subcontinent and competent vector Culicoides spp.

Region	BTV Serotypes	Vector Species
Indian subcontinent	1-18, 20, 21, 23	C.imicola, C.oxystoma, C.clavipalpis, C.actoni, C.anophelis, C.orientalis, C.simiklis

The geographical boundaries of BTV are limited to the prevalence of competent vector spp. The northern- and southernmost boundaries of BTV have historically been 40° North and 35° South latitudes, respectively. However, it is now documented that BTV occurs up to 50° North (Figure 1). Global warming would be predicted to expand the existing northern and southern extremes of global BTV distribution. For instance, C. imicola, the established BTV vector, has recently extended its range into many of the more northern areas of south-central & south-east Europe. In exceptionally favourable years the virus will spread even further to 50° North. There is evidence that infected midges are carried on the wind for very long distances[33]. It has been postulated that the major epidemics of bluetongue, in regions where disease occurs only sporadically, can often be traced to windborne carriage of infected Culicoides from distant areas[12].

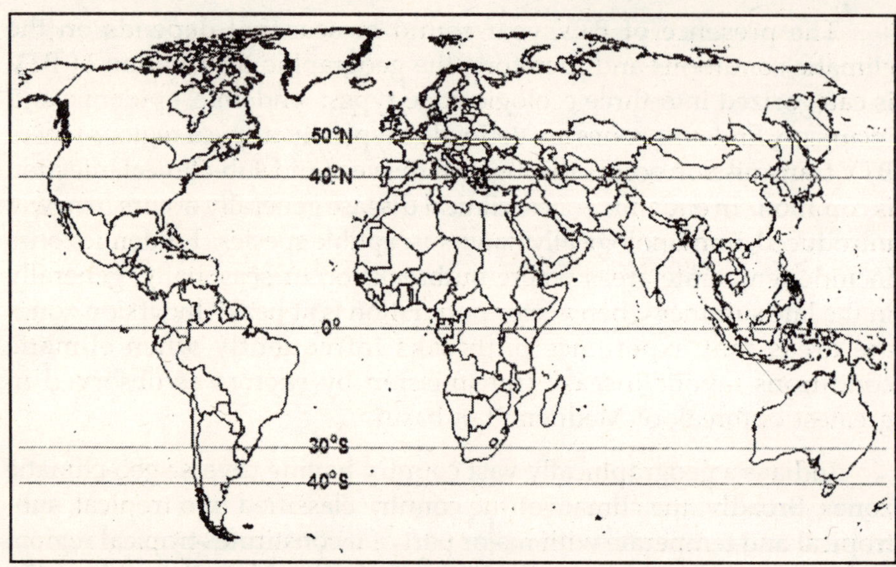

Fig. 1. Geographical boundaries of Blue Tongue virus: The early distribution of BTV and recent extension of its boundaries around the world shown by straight lines. This area includes tropical and subtropical areas between 40°North and 30°South of earlier period and between 50° North & 40° South of recent extension.

In India, so for 21 serotypes (except 19, 22 and 24) have been reported from different states, of which serotype 1, 2 and 23 are found to be most prevalent[20,21] (Table 1 & 2). The first outbreak of this disease, reported in 1964 in sheep and goat farm from Maharashtra state was based on clinical symptoms. Subsequently[5] further reported infection of sheep in Uttar Pradesh. It was reported the occurrence of disease in this country[37]. Henceforth several outbreaks of BT have been recorded from different states of India in indigenous, crossbred and exotic sheep. The virus was isolated and identified as serotypes 3, 4, 9, 16 and 17 from Himachal Pradesh and serotypes 1 and 4 from Haryana[14]. Subsequently, reported isolation of virus serotypes 9 and 18 from Maharashtra. In southern India, serotypes 1, 4, 5, 6, 7, 11, 12, 14, 15, 16, 17, 20 & 23 in Tamilnadu[15] and serotype 1, 2, 12, 16, 17 & 20 in Karnataka were most commonly seen.

ECONOMIC IMPORTANCE AND IMPACT

Bluetongue is an economically important disease of ruminants for several reasons. It is inflicting great economic losses to livestock industry in India and different parts of the world. These losses are attributed

Table 2. Distribution of various BTV serotypes in India (Sreenivasulu *et al.*, 2003)

States	Species	Virus Isolation	Antibody detected by Sero-surveillance
Andhra Pradesh	Sheep	2, 9, 15	4, 12-14, 17-19
	Cattle	-	6, 12
Tamil Nadu	Sheep	1,2, 3, 16, 23	1, 4-7, 11, 12, 14-17, 19, 20
Karnataka	Sheep	18, 23	1, 2, 12, 16, 17, 20
	Cattle	-	1, 14, 16
Maharashtra	Sheep	1, 2-4, 8, 9, 16, 18	-
Gujarat	Buffalo	-	1, 15, 17
	Cattle	-	2, 12, 20
Madhya Pradesh	Sheep	18	-
Uttar Pradesh	Sheep	9, 18, 23	-
Haryana	Sheep	1, 4	14
	Cattle	-	1, 2, 8, 12, 16
Himachal Pradesh	Sheep	3, 9, 16, 17	4
Jammu & Kashmir	Sheep	18	-
West Bengal	Sheep	15	-

primarily to mortality and high morbidity, including abortion, milk, meat, and fleece losses in sheep. In endemic areas vaccination is a recurring cost. Now in India lot of trials going on to produce effective vaccine by pouring huge amount of money. In the recent years, several countries have shown interest in the germ plasm of Indian breeds of cattle and other animals for genetic and molecular studies and for some animal products such as fetal calf serum. But, inclusion of BTV in the OIE 'List A' has direct consequence that continuingly affecting our global trade, particularly to Canada, New Zealand, and the European Union, which are classified by the OIE as bluetongue free and parts of Australia, which have been regionalized as bluetongue-free[11]. It has been estimated that in the late 1970s, the ban on US cattle semen exports resulted in an annual loss of $24 million[12]. Worldwide losses due to BTV have been estimated to US$ 3 billion/ yr[36].

India may be the only country which experiencing severe outbreaks due to the prevalence of highly virulent strains. Table.3 depicts the severity of outbreaks from 2000 to 2004 in different months in India.

Table 3. Month wise distribution of bluetongue outbreak in different parts of India, 2000-2004 (http://www.oie.int)

2000

Year (Month)	Species	Total outbreaks	Total cases	Deaths	States (Outbreaks)
April	Ovine	2	90	2	Karnataka
May	Ovine	2	32	5	Karnataka (1), Rajasthan (1)
September	Ovine	2	12	4	Karnataka
October	Ovine	259	770	49	Karnataka
November	Ovine	131	7724	470	Karnataka (91) Andhra Pradesh (40)
December	Ovine	30	719	129	Karnataka (28), Andhra Pradesh (2)

2001

Year (Month)	Species	Total outbreaks	Total cases	Deaths	States (Outbreaks)
January	Ovine/Caprine	6	No data	No data	No data
March	Ovine/Caprine	10	-	-	-
July	Ovine/Caprine	3	-	-	-
August	Ovine/Caprine	7	-	-	-
September	Ovine/Caprine	26	-	-	-
October	Ovine/Caprine	28	-	-	-
November	Ovine/Caprine	406	-	-	-
December	Ovine/Caprine	224	-	-	-

2002

Year (Month)	Species	Total outbreaks	Total cases	Deaths	States (Outbreaks)
January	-	3	33	-	Karnataka
February	-	3	32	-	Karnataka
March	-	1	10	-	Karnataka
May	-	2	9	-	Assam
August	-	1	6	-	Tamil Nadu
November	Ovine	2	44	-	Karnataka
December	Ovine	1	20	-	Karnataka

2003

Year (Month)	Species	Total outbreaks	Total cases	Deaths	States (Outbreaks)
January	Ovine/Caprine	3	250	25	Karnataka
August	Ovine	43	6664	326	Karnataka (1), Andhra Pradesh (42)
September	Ovine	304	12181	1964	Andhra Pradesh (303), Pondicherry (1)
October	Ovine	21	719	139	Andhra Pradesh
November	Ovine/Caprine	13	189	64	Karnataka (1), Andhra Pradesh (10), Tamil Nadu (2)
December	Ovine/Caprine	7	44	5	Karnataka (3), Andhra Pradesh (4)

2004

Year (Month)	Species	Total outbreaks	Total cases	Deaths	States (Outbreaks)
January	Ovine	1	30	4	Karnataka
April	Ovine	1	6	3	Karnataka
June	Ovine	5	26	2	Andhra Pradesh
July	Ovine	22	214	18	Karnataka (13), Andhra Pradesh (5), Tamil Nadu (4)
August	Ovine	19	351	38	Karnataka (13), Andhra Pradesh (6)
September	Ovine	75	865	103	Karnataka (3), Andhra Pradesh (72)
November	Ovine	8	35	2	Karnataka (2), Andhra Pradesh (6)
December	Ovine	152	7723	843	Karnataka (2), Andhra Pradesh (5), Tamil Nadu (145)

Annual outbreaks of bluetongue in India (2000-2004)

Year	Jan.	Feb.	Mar.	Apr.	May.	Jun.	Jul.	Aug.	Sep.	Oct.	Nov.	Dec.	Total
2000	0	0	0	2	2	0	0	0	2	259	131	30	426
2001	6	0	10	0	0	0	3	7	26	28	406	224	710
2002	3	3	1	0	2	0	0	1	0	0	2	1	13
2003	3	0	0	0	0	0	0	43	304	21	13	7	391
2004	1	0	0	1	0	5	22	19	75	0	8	152	283

Factors involved in expression of disease

Among sheep, fine wool and mutton breeds are very susceptible. Furthermore, sheep that are native to tropical and subtropical regions of the world where BTV is endemic are comparatively resistant to BT, whereas European breeds such as the Merino are highly susceptible. Nutritional status, immune status, and age also influence the severity of BT in individual sheep, as can environmental stresses such as high temperature and ultraviolet radiation.

i. *Virus factors :* Field serotypes of BTV in endemic areas exhibit remarkable genetic heterogeneity and virulence, even amongst strains of particular serotype that co-circulate in animal perpetuation[29]. This considerable genetic variability of BTV is reflected by differences in phenotypic properties of each virus strain, including their virulence to susceptible ruminants. For example, India possesses extremely infectious strains and suffers severe BT whereas, in eastern Australia, only non-virulent strains are present.

ii. *Host factors :* BTV naturally infects domestic and wild ruminants, camelids, and some other herbivores such as elephants. It is well documented that sheep and white-tailed deer are the species that are most susceptible to BT than any other ruminant species. Cattle are less susceptible to disease.

iii. *Vector factors :* Culicoides vectors are critical to the survival and transmission of BTV, as infection is not contagious and there is no credible evidence of long-term maintenance of BTV in ruminants. Thus, BTV infection occurs only where competent vectors are present[22]. Furthermore, BT disease usually occurs during late summer and early autumn when number of insect vectors is highest in BTV-endemic areas. The central role of the insect in BT epidemiology ensures that prevalence of the disease is governed by ecological factors, such as high rainfall, temperature, humidity and soil characteristics, which favor insect survival. The insect vectors of bluetongue virus breed in moist conditions in a variety of habitats, particularly damp, muddy areas, fecal and plant matter. In many parts of the world therefore, the disease has a seasonal occurrence[9]. Vectors have nocturnal feeding habits, preferring still, warm conditions, pastures and open pens[33]. A single midge bite is sufficient to infect a susceptible ruminant host[23,24].

The duration of viraemia in the infected vertebrate is an important factor in the transmission of BTV to competent midges. BT is no longer considered a persistent infection of ruminants, especially cattle[17,18]. Viraemia in most of the infected cattle is reported for less than 4 weeks with fewer than 1% exceeding 8 weeks. The exceptional and maximum viraemia reported in sheep is 54 days[16]. But normally, viraemia lasts for 6-10 days and rarely persists for longer than 14 days in affected sheep and in cattle for about 7-28 days. This duration of BTV viraemia in domestic ruminants has been a critical issue in international trade and placement of trade barriers. The OIE currently recognizes a 100-day infective period based on PCR observation for the movement of animals across the country.

PATHOGENESIS

Pathogenesis of BT disease in sheep depends on the virulence of the virus strains as well as the nutritional status, immune status, age of the animal and environmental stresses such as high temperature and ultraviolet radiation. On the bite of infected vector, the virus multiplies primarily in the regional lymph node which leads to primary viraemia where all the cells of the blood promiscuously become infected[18]. But, the cell associated viraemia occurs primarily in the RBCs and platelets, although dissemination of the virus occurs mainly through mononuclear cells[12]. Further multiplication of the virus occurs mainly in the endothelial cells including periendothelial and pericytes of capillaries, which is main contributory factor in the development of pathological lesions in sheep. This multiplication of virus in the leukocytes leads to panleukopenia and in smaller blood vessels lead to micro-vascular endothelial injury[13].

Endothelial injury leads to increased capillary permeability resulting into oedema of various parts of the body especially in face region, larynx and lungs[30]. It is responsible for generalized vascular thrombosis leading to multi-visceral necrosis, skeletal and cardiac infarction development, and bleeding tendency due to consumption of various clotting factors and platelets with severe hemorrhage, which is especially more so in sheep and white tailed deer [19]. Affected animals frequently suffer pharyngeal or esophageal myo-degeneration leading to dysphagia, or regurgitation, and animal may succumb to aspiration with the development of bronchopneumonia. In severe cases, death of the animal occurs either from pulmonary edema, bronchopneumonia, cardiac insufficiency or any of these combinations.

CLINICAL SIGNS AND POSTMORTEM FINDINGS

After an incubation period of 4-7 days, fever up to 103° - 105°F develops and persists as long as the viraemia is present in the body (approx. up to 6-8 days). Within 24-48 hrs onset of fever, animal starts showing nasal discharge, salivation, hyperemia of nasal buccal mucosa and discharge may become mucopurulent and blood stained. In next few days, there will be edema of lips, tongue, face, ears, inter-mandibular space giving "monkey face appearance", hyperemia may become more intense with petechiation, and cyanosis of tongue in few cases which later may become swollen and protruded (Stott, 1998). During 5-8 days following onset of fever, there will be necrotic ulcers on cheeks, tongue, gums and they usually heal slowly with diphtheritic membrane formation. Respiration may become laboured and dysentery may occur during this period. When mouth lesions begin to heal, coronitis may occur (at the end of febrile period) which is characterized by the coronary swelling and streaky hemorrhage in the periople owing to affection of underlying sensitive laminae. Loss of body condition, muscle wastage and torticollis may develop as a late sign[25]. Animal may exhibits stiffness, reluctance to move, arching of back and recumbency. Affected areas of skin show break in the wool or much of the fleece being tender or cast. Death due to pulmonary edema and/or cardiac insufficiency occurs at any stage up to a month post-infection or prolonged convalescence. Mortality may vary from 5-70%.

In utero transmission of the virus in pregnant animals leads to abortion, foetal death or lamb/calf born small, underdeveloped with or without malformations. Various reported malformations include hydranencephaly, cerebral anomalies like cerebral dysplasia, excessive gingival, enlarged tongue, anomalous maxillae, dwarflike build and rotations and contractures of the distal extremities[10].

Postmortem examination of animals revealed all the lesions attributed mainly to "micro-vascular endothelial damage". Grossly, there is congestion, oedema, hemorrhage and ulcerations of digestive and respiratory mucosae (mouth, pharynx, larynx, oesophagus, fore stomach, intestine, tracheal mucosa)[10]. Principally, necrosis of cardiac muscle involving the papillary muscles of left ventricle and petechial to ecchymotic hemorrhages in the tunica media near the base of the pulmonary artery, are almost considered as pathognomonic findings[28]. Sub-epicardial and sub-endothelial haemorrhages, particularly involving the left ventricle, are common. Hyperaemia and edema of lungs are also consistent findings. Lymph nodes and spleen show hypertrophy. Kidneys commonly congested and there may be petechial hemorrhage in mucosa of the urinary bladder, urethra and vulva or prepuce. Hoof laminae and coronary band show congestion, swelling and streaky hemorrhages.

DIAGNOSIS

A. *On the basis of clinical signs:* A presumptive diagnosis of BT in affected sheep can frequently be made, based on clinical signs and lesions.

B. *Laboratory diagnosis:* BTV infection usually be subclinical and laboratory confirmation, either by isolation and identification, serology or by nucleic acid detection methods, is required[7].

i. Samples to be collectd

1. Blood for virus isolation is collected from febrile animal in anticoagulants, such as heparin or EDTA and kept cool (4°C), but not frozen. Since the virus is highly cell associated, it coexists with the antibody in the circulatory system, and so upon de-freezing, RBCs will undergo lysis with release of virus particles which are then immediately neutralized by the antibodies.
2. Spleen and lymph nodes are preferred tissues for virus isolation or antigen/genome detection by ELISA/PCR. They should be chopped finely and preserved in glycerol phosphate buffer with pH 7.6 and stored at 4°C, but should not be frozen.
3. Serum for serological tests. Preferably paired samples should be collected at acute and convalescent phases and should be stored at -20°C.

ii. Virus isolation in culture

a. *Sheep inoculation:* not economical method, so not routinely practiced.

b. *Intravenous inoculation of embryonated chicken eggs:* It is the second most sensitive technique for isolation of BTV. It is a prescribed test for international trade. Bluetongue virus may be detected in the inoculated eggs by antigen/nucleic acid detection methods.

c. *Intracerebral inoculation of newborn mice:* not routinely practiced.

d. *Tissue culture method:* Cell culture BTV isolates are identified and confirmed by tests based on group-specific antibodies using fluorescent or enzyme conjugates. Isolates placed in the BT sero-group are typed by virus neutralization tests using serotype-specific antisera. Commonly, baby hamster kidney (BHK-21) and African green monkey kidney (Vero) cell cultures are used.

iii. Serotyping of BTV by virus neutralization

Virus neutralization is the 'gold standard' test for serotype determination of virus isolate using serotype-specific neutralizing antibodies directed against VP2. Neutralization tests are type-specific for the currently recognized 24 BTV serotypes and can be used to serotype a virus isolate, or can be modified to determine the specificity of antibody in sera. The following techniques may be used:

a. Microtitre neutralisation: most widely used test.

b. Plaque reduction

c. Plaque inhibition

d. Fluorescence inhibition

iv. Serogrouping of viruses

This is based on the reactivity of group-specific BTV protein VP7.

The following techniques are used:

a. *Immunocapture ELISA:* With BT serogroup-specific monoclonal antibody (MAb) to avoid cross-reactivity between members of the BT and EHD serogroups[1]. It is rapid, sensitive, most widely used test

b. *Immunofluorescence:* not widely used.

c. *Immunospot test:* not widely used

d. *Indirect peroxidase-antiperoxidase test:* not widely used.

v. Serological tests

a. *Competitive or blocking ELISA:* It is rapid, sensitive, can be easily standardized and most widely used. It is prescribed tests for international trade.

b. *Agar gel Immunodiffusion Test:* This test to detect anti-BTV antibodies is simple to perform and. Since 1982, the test has been the OIE standard testing procedure for the international movement of ruminants. However, one of the disadvantages of the AGID used for BT detection is its lack of specificity.

c. *Complement Fixation Test:* not widely used.

d. *Fluorescent antibody technique (FAT)* has been found useful for the detection of group specific antigen, but, not widely used.

vi. Nucleic acid detection techniques

a. *Polymerase chain reaction (PCR):* Primer-directed amplification of viral nucleic acid has revolutionised BT diagnosis[4,8,38]. Results to date indicate that PCR techniques may be used, not only to detect the viral nucleic acid, but also to 'serogroup' orbiviruses and provide information on the serotype and possible geographic origin (topotype or genotype) of BTV isolates. Nested PCR (nPCR) assay increasingly is used, because it is highly sensitive and specific. It is the only prescribed test for detection of diseases of livestock that OIE recommended for international trade. Ruminants whose blood is negative by "nPCR" assay pose no threat for inadvertent movement of BTV by trade. The "real time" PCR technology is faster than traditional PCR methods, and is less prone to problem of contamination.

b. *Differential diagnosis:* BT should be differentiated from foot and mouth disease, vesicular stomatitis, sheep pox, peste des petits ruminants, pneumonia, contagious ecthyma, photosensitization, copper deficiency, oestris Ovis, Ulcerative dermatosis, and Rift Valley Fever.

DISEASE PREVENTION AND CONTROL

Once established in an area, it is impossible to actively eradicate bluetongue virus. The virus will circulate, generally subclinically, in cattle and other ruminants and in midges. In countries marginally suitable for virus persistence, the virus may be maintained for several years before dying out. The interruption of BTV cycle is the most effective method of prevention and control. This can be achieved by following methods.

i. *Movement restrictions, testing and hygiene:* In disease free countries, controlled importation of susceptible animals, serological testing of animals during quarantine period in vector-free area and importation of semen from bulls shown to be serologically BTV negative should be practiced. In disease prevalent countries, restriction of animal movement, slaughtering and burning of infected stock and **serological surveillance** of area round infected zone constitute main strategies to contain the disease.

ii. *Isolation and identification of virus types prevalent:* Typing of the virus isolates prevalent in a particular state or area would facilitate in selection of virus types to be incorporated in the vaccine and thus the disease control.

iii. *Vaccination:* Of several vaccine options available, namely live attenuated, killed or recombinant, only attenuated virus vaccines are in current use in several countries. In India, its use is still awaited.

a) *Live attenuated vaccines:* It is the only current licensed vaccines for BT. In South Africa, for example, they have been used for over 40 years and are known to induce an effective and lasting immunity. However, they must be type-specific, and therefore multiple vaccinations may be needed. Most disadvantages of live attenuated vaccines are possible reversion to virulence and transmission through insect bites can occur apart from inducing teratogenesity and congenital malformations in the fetuses[17,18].

b) *Recombinant vaccines:* There are several options for the development of recombinant BTV vaccines, including live virus delivery of BTV antigens and the virus-like particles generated in infected insect cells by recombinant baculoviruses expressing the four major BTV proteins, VP2, 3, 5 and 7. These vaccines need further studies on duration of immunity and ability to suppress viraemia following challenge. None of the recombinant vaccines are currently licensed and are not under use.

c) *Inactivated vaccines:* Although the efficacy of inactivated virus vaccines has been investigated in some laboratory studies, they do not appear to have been used in the field. However, in view of their being safe, attempts are being made to develop and commercialize such vaccine under All India Network Programme on this disease Ensuring complete inactivation is often problematic but that can be ensured by testing the safety in proper manner. Further studies are required on duration of immunity and protective dose.

iv. *Control of vectors:* Spraying nearby animals with **long-acting insect repellent**, housing animals at night, **reducing insect breeding sites** by spraying or improved drainage and spraying buildings with insecticide should be undertaken[24]. **Disinfection of premises is not appropriate since virus is insect transmitted.**

References

1. Afshar A., Thomas F.C., Wright P.F., Shapiro J.L. & Anderson J. 1989. Comparison of competitive ELISA, indirect ELISA and standard AGID tests for detecting bluetongue virus antibodies in cattle and sheep. Vet. Rec., 124, 136–141.
2. Akita G.Y., Ianconescu M. and MacLachlan N.J., 1994. Bluetongue disease in dogs associated with contaminated vaccine. Vet Rec; 135:283-284.
3. Akita GY, Ianconescu M, MacLachlan NJ, Osburn BI, Greene RT, 1994. Bluetongue disease in dogs associated with contaminated vaccine. Veterinary Record, 134(11):283; 5 ref.
4. Bandyopadhyay, S.K., Kataria, R.S. and Tiwari, A.K. 1998. Detection of bluetongue virus genome segment 6 sequences by RT-PCR. Indian Journal of Experimental biology, 36: 1034-1037.
5. Bhambani, B.D. and Singh, P.P. 1968. Bluetongue disease of sheep in India. The Indian Veterinary Journal, 45: 370-371.
6. Caporale V., Calistri P., Goffredo M., & Meiswinkel R. 2003. The distribution of Culicoides imicola in Italy. Application and evaluation of current Mediterranean models based on climate. J. Vet. Med. B., 40, 132–138. (6).
7. Clavijo A., Heckert R.A., Dulac G.C., Afshar A., 2000. Isolation and identification of bluetongue virus. Journal of Virological Methods, 87:13-23.
8. Dangler C.A., De Mattos C.A., DeMattos C.C. & Osburn B.I. 1990. Identifying bluetongue virus ribonucleic acid sequences by the polymerase chain reaction. J. Virol. Methods, 28, 281–292.
9. DeMaula CD, Leutenegger CM, Jutila MA 2002. Bluetongue virus-induced activation of primary bovine lung microvascular endothelial cells. Vet Immunol Immunopathol; 86:147-157.
10. Erasmus BJ, 1990. Bluetongue virus. Virus infections of ruminants, 227-237; 31 ref.
11. Gibbs EP and Greiner EC 1994. The epidemiology of bluetongue. Comp Immunol Microbiol Infect Dis; 17:207-220.
12. Gibbs EPJ, Greiner EC, 1988. Bluetongue and epizootic hemorrhagic disease. The arboviruses: epidemiology and ecology. Volume II, 39-70; 169 ref.
13. Howerth EW, Greene CE and Prestwood AK 1988. Experimentally induced bluetongue virus infection in white-tailed deer: coagulation, clinical pathologic, and gross pathologic changes. Am J Vet Res; 49:1906-1913.

14. Jain, N.C., Sharma, R and Prasad, G. 1986. Isolation of bluetongue virus from sheep in India. Veterinary Record, 119: 17-18.

15. Janakiraman, D., Venugopalan, A.T., Ramaswamy, V and Venkatesan, R.A. 1991. Serodiagnosis evidence of prevalence of bluetongue virus serotype among sheep and goat in Tamil Nadu. Indian Journal of Animal Sciences, 61: 497-498.

16. Koumbati M, Mangana O, Nomikou K, Mellor PS, Papadopoulos O, 1999. Duration of bluetongue viraemia and serological responses in experimentally infected European breeds of sheep and goats. Veterinary Microbiology, 64(4): 277-285.

17. MacLachlan NJ 1994. The pathogenesis and immunology of bluetongue virus infection of ruminants. Comparative Immunology, Microbiology and Infectious Diseases, 17(3/4):197-206; 59 ref.

18. MacLachlan NJ. 1994. The pathogenesis and immunology of bluetongue virus infection of ruminants. Comp Immunol Microbiol Infect Dis; 17:197-206.

19. Mahrt CR and Osburn BI 1986. Experimental bluetongue virus infection of sheep; effect of vaccination: pathologic, immunofluorescent, and ultrastructural studies. Am J Vet Res; 47:1198-1203.

20. Mehrotra, M.L. 2000. Status of bluetongue disease in India. Indian Veterinary congress. 18-19th February, 2000. pp. 187.

21. Mehrotra, M.L., Shukla, D.C., Singh, K.P., Khanna, P.N. and Saikumar, G. 1995. Isolation of bluetongue virus from an outbreak of Morbilli virus infection in sheep and goat. Indian Journal of Comparative Microbiology Immunology and Infectious Diseases, 16: 135-136.

22. Mellor PS, 1990. The replication of bluetongue virus in Culicoides vectors. Bluetongue viruses, 143-161; [In Current Topics in Microbiology and Immunology, 162]; 101 ref.

23. Mellor PS, Boorman J, Baylis M, 2000. Culicoides biting midges: their role as arbovirus vectors. Annual Review of Entomology, 45:307-340.

24. Mellor PS 2000. Replication of arboviruses in insect vectors. J Comp Pathol; 123:231-247.

25. Office International des Epizooties, 1998. Supporting document for the OIE International Animal Health Code chapter 2.1.9 on bluetongue. OIE Ad hoc working group on bluetongue, September 1998.

26. Pandey, A.B., Yadav, G., Nandi, S. and Singh, R. 2004. Serosurveillance of bluetongue disease in different parts of India. XI Annual Conference of IAAVR and 4th Indian Veterinary congress, Held at IVRI, February, 27-28, 2004.

27. Pandey, A.B., Nandi, S., Singh, R., Samanta, S., Audarya, S.D. and Suresh, I. 2005.Third Annual Scientists' Meet, 1-2 Sept. 2005 held at Water and Land Management Institute, Aurangabad.

28. Parsonson IP 1990. Pathology and pathogenesis of bluetongue infections. Curr Top Microbiol Immunol; 162:119-141.

29. Pierce CM, Balasuriya UB and MacLachlan NJ 1998. Phylogenetic comparison of the S10 genes of field and laboratory strains of bluetongue virus from the United States. Virus Res; 55:15-27.

30. Pini A 1976. Study on the pathogenesis of bluetongue: replication of the virus in the organs of infected sheep. Onderstepoort J Vet Res; 43:159-164.

31. Roberts DH, 1990. Bluetongue: a review. State Veterinary Journal, 44(124):66-80; 40 ref.

32. Sapre, S.N. 1964. An outbreak of bluetongue in goats and sheep in Maharashtra state, India. Veterinary Reviews, 15: 69.

33. Sellers R.F. 1981. Bluetongue and related diseases. In: Gibbs EPJ ed. Virus Diseases of Food Animals. London, UK: Academic Press, 567-584.

34. Spreull J. 1905. Malarial catarrhal fever (bluetongue) of sheep in South Africa. J Comp Pathol Therap; 18:321-337.

35. Spreull, J. 1905. Malarial catarrhal fever of sheep. Veterinary Record, 14: 629-633.

36. Tabachnick W.J., MacLachlan N.J. and Thompson L.H. 1996. Susceptibility of Culicoides variipennis sonorensis to infection by polymerase chain reaction-detectable bluetongue virus in cattle blood. Am J Trop Med Hyg; 54:481-485. Verwoerd DW, Erasmus BJ, 1994. Bluetongue. In: Coetzer JAW, Thomson GR, Tustin RC, eds. Infectious Diseases of Livestock with special reference to Southern Africa. Cape Town, South Africa: Oxford University Press, 443-459.

37. Uppal, P.K. and Vashudevan, B. 1980. Occurrence of bluetongue in India. Indian Journal of Comparative Microbiology Immunology and Infectious Diseases, 1: 18.

38. Wade-Evans A.M., Mertens P.P.C. and Bostock C.J. 1990. Development of the polymerase chain reaction for the detection of bluetongue virus in tissue samples. J. Virol. Methods, 30, 15–24.

-ΔΔΔ-

11

Foot and Mouth Disease Control Strategies for Livestock Development in India

S N SINGH

Biovet Private Limited
560, C-Block, 1st Main, AECS Layout
Kundanahalli, Bangalore - 560 037

The Indian dairy farming and agriculture business has grown to Himalayan heights to put the country as number one milk producer in the world today with above 70 million tons record production. The country has a huge livestock population of cattle, buffaloes, sheep and goats to the tune of 450 million heads. The production of livestock products like milk, meat, wool, hides and other byproducts have also increased tremendously in recent years.

Improvement in the genetic potential of the livestock by means of cross breeding have made the cross bred animals, apart from increasing stressful productivity management practices have made animals more susceptible to diseases like Foot and Mouth disease, infectious bovine rhinotracheitis, haemorrhagic septicaemia, black quarter, brucellosis. Most of these diseases an be controlled by systematic vaccination and monitoring.

The export market of livestock products is ever increasing. The increase is particularly significant to other Asian countries, countries of the Middle East and Europe. The world trade organization (WTO) plays a significant role in determining trade policies, it would be of utmost importance that export products be free of important infectious diseases. especially those listed by the OIE. A stringent monitoring and control policy should be implemented to prevent spread of these diseases (FAO / OIE / WTO).

India has followed the OIE proposal for the eradication of Rinderpest disease. Sero surveillance and disease diagnosis was vigorously followed along with compulsory vaccination of all animals. It should be noted that the OIE would recognize India to be Rinderpest free soon. An immune belt has been created along the borders of the endemic areas.

Many of the developed nations are free from most of the diseases listed in OIE particularly FMD and hence their livestock products are better accepted worldwide. The prevalence of Foot and Mouth Disese in India is a major trade barrier.

Foot and Mouth disease is an acute infection caused by a virus. The virus belonging to the picornaviridae family, is the smallest virus known so far. The disease is characterized by formation of blisters, followed by ulcers on the mucosa of the. mouth cavity and on the skin of the feet, hence the name: "Foot and Mouth Disease" . Recent event in Foot and Mouth Disease control 2006 FAO Meet, Cyprus clearly defines the following important considerations.

1. Lack of early warning of emergence of new antigenic types of Foot and Mouth Disease type A has contributed to the scale of the subsequent regional epidemic in the I.R. of Iran, and Turkay
2. Incursion of an African type A virus into the Mediterranean region has occurred for first time in Egypt in 2006, leading to a widespread and severe epidemic in the naïve animal production.
3. Regional or national vaccine banks do not currently exist in the countries of the middle east and that there is often a prolonged lead time before delivery of vaccine from commercial supplier.
4. That delay in diagnosis of the new virus incursions has resulted from the use of diagnostic methods and reagents that did not sensitively detect emergent viruses of different type or antigencity.
5. There is a need to identify the extent of biosecurity measures to prevent new farm infections given the cost and impact of culling and vaccination programmes, given the widespread dissemination of the type A epidemics in Turkey and Egypt, and the type O epidemic in the northern Europe in 2001 which occurred during periods of cool and humid conditions which favored virus transmission.

6. Significant quantities of animal products are brought by air traffic into European countries every day by passengers from Foot and Mouth Disease endemic countries in Africa and elsewhere.

7. There is an increasing trade-driven movement of livestock commodities from Foot and Mouth Disease-endemic areas in Asia

Animals Susceptible for FMD

All cloven-footed animals including cattle, buffaloes, sheep, goat, pigs, elephants and other ruminants are susceptible to the virus. For obvious reasons, the disease is more important in cattle, buffaloes, bulls and bullocks in India.

Disease Spread

The infected animal is the main source of infection. Infection may spread either through direct contact or by indirect means, the infected feed and fodder, infected utensils, and infected means of transportation or, through carrier cattle attenders. Infections has also been reported to travel through air. However, at most times, the wide spread of infection results from congregation of animals in cattle fairs, cattle markets or large-scale transportation of agricultural produce in bullock carts.

The incubation period for Foot and Mouth disease is as short as 48 to 72 hours and as long as 10 to 14 days. On an average, it varies from 3 to 7 days.

Seven immunologically distinct serotypes of Foot and Mouth disease viruses have been reported worldwide. There is no cross-immunity between serotypes, immunity to one does not confer immunity against any of the other types. Four serotypes 0, A, Asia-1 and C are the reported serotypes in India.

Disease Economics

Besides the acute stage of the disease, characterized by the formation of ulcers in the mouth, feet and udder, the virus of foot and mouth disease exhibits its pathology in some of the vital hormonal glands, which control the metabolic processes of the body. Disordered functioning of heat regulating centres leading to panting is one example. The disturbance in physiological process of lactation leads to a significant reduction in milk yield. In milch animals lesions on teat and udder can lead to mastitis, which may damage the teats and thereby affect the milk yield on permanently.

The economic losses to the livestock industry attributed to Foot and Mouth disease are large. The OIE / FAO / APHCA place a massive significance in their attempts to eradicate. Foot and Mouth disease world wide.

Direct losses to livestock sector are due to

1. Abortion in 25 % pregnant animals.
2. Reduction in meat production by 25 % in endemic area.
3. A drop in milk production by 50 %.
4. A reduction in wool production by 25 % in affected sheep.
5. Mortality rate of up to 5.5 % of the affected cases.

Indirect loss to livestock sector are due to

a. Loss of production function during the acute phase of disease.
b. Loss of milk yield on a permanent basis.
c. Loss of breeding capacity including abortions.
d. Loss due to reduced draught capacity in working bullocks.
e. Interference with food production programme.
f. Loss in cattle trade both national and international.
g. Loss resulting from temporary cessation of A.I. programmes.
h. Loss in flesh in meat-animals.
i. Mortality in young calves due to heart failure.
j. Flare up of inter-current infections like Theileriosis and Anaplasmosis.

This adage is very relevant in the case of Foot and Mouth Disease than other diseases. In countries where Foot and Mouth Disease is wide spread regular programmes of large-scale vaccination using Foot and Mouth Disease vaccines are being followed.

Prevalence

Foot and Mouth disease is enzootic in Africa, Asia. South America and parts of Europe.

The disease has been reported from various parts of the world except Japan, New Zealand, Australia, Canada and the United State of America. Foot and Mouth disease is a reportable disease in most countries and attempts are made by the FAO/OIE to collect data on the prevalence of the disease in various countries. The identification of the various virus serotypes is based upon complement fixation test, liquid phase blocking ELISA and recently by nucleic acid recognition method acid recognition method (Polymerase chain reaction). Overall it has been found that the serotype '0' and the serotype 'A' occur more frequently than the other serotypes. The disease is endemic in India.

Strategy to control

Sero-type spredominantly occuring in the country are mainly type '0' (70%)followed by Asia-l and type , A '. There has been no report or minor occurrence of type 'C' outbreak(s). In Punjab, Uttar pradesh and Maharastra in 1998 were predominantly by Type '0'. The various serotypes of foot and mouth disease virus are antigenically distinct and do not cross react. Depending upon the prevalence of the type of the virus causing disease, the vaccine used in the area are determined. If a single type of virus is seen prevalent, a monovalent vaccine (with only one type of the virus antigen) is used. If two types seen, bi-valent vaccine with two types of antigens are used. In India, a tetravalent vaccine with antigen from type A,O,C and C and Asia-1 are used. To prevent antigenic drift, vaccines usually with more than one strain are used in manufacture.

Hoechst Roussel Vet maintains a reference collection of the vaccine strains of FMDV. By comparing outbreak strains with the vaccine strains could be identified. A repertoire, an appropriate vaccine strains could be identified. A repertoire of antibodies is also developed to determine shared neutralizing epitopes, thereby giving an indication of the potential value of vaccine strains in helping control the outbreak. Linkages are being established within India, with IVRI, the United Kingdom Institute for Animal Health, at Pirbright and other institutes globally via the Internet.

In 1951-52 over 900,000 outbreaks were reported in Europe, the European countries finally eradicated the disease and from 1992 the member countries of European Economic Community (EEC) are adopting, a non-vaccination and stamping-out policy. This had largely come about by the maintenance of solid vaccination coverage because the European FMD commission in 1957 accepted the systematic vaccination would be necessary for number of years to reduce the incidence of the disease so as to make other measures like slaughter policy an economically feasible.

FMD Vaccines

International standards for FMD vaccines can be founc! in the British Veterinary Pharmacopoeia, British Veterinary pharmacopoeia Codex, European Pharmacopoeia (Veterinary) (1993) and the OIE Manual of Standards for Diagnostic Tests and Vaccines. National Veterinary Authorities usually exercise control of the use of FMD vaccines. Indian Veterinary

Pharmacopoeia is being planned to release shortly. The FAO/OIE is formulating International standards for safety, potency and antigenic mass requirements for the various vaccine strains. The dosage of the vaccine mass depending upon the epidemiological situation of the area is also being worked out.

History of FMD vaccine development

The three critical elements of FMD vaccine production are antigen production, virus inactivation and the addition of suitable adjuvants. Historically, the original source of FMDV for vaccine production was clinically derived material, such as infected cattle tongues in 1926.

In 1951, Franked described a new technique for the production of FMDV on an industrial scale in tongue explants. It was the FMD vaccine made in this system that was used in the Netherlands in the first of the highly successful mass annual prophylactic vaccination campaigns to be carried out in Europe.

The advantages of this production system were its simplicity, low/ lack of cellular protein contamination of the virus harvest and the fact that adaptations of the virus to the culture system was not required.

A significant development in FMDV antigen production was the transition to tissu e culture methods of virus growth. Initially, small-scale production in roller bottles using primary calf kidney cells was instigated in Italy in 1963.

However, following the introduction in 1964 of a continuous cell line derived from baby hamster kidney fibroblasts (BHK 21) that supported the growth of FMDV, this system gained wide acceptance in FMDV vaccine production.

Later a variety of monolayer systems were devised to increase culture vessel surface area. and thus productivity.

However, the greatest scale-up capacity for FMDV production was the advent of technology, which exploited the ability of BHK-21 cells to grow in deep suspensions culture in fermentors (bioreactors) that are used widely now days. Vaccine Development, production and selection application needs further development alternative potency methods, early detection evaluating different DIVA tests for susceptible spaces mainly ruminants. Both 'carrier' to be seen as a major challenge for production and control. Therefore, Foot and Mouth Disease vaccine development quality control quality assurance including NSP antigen and 3D & A analysis is important to understand

Antigen production

Commercially available FMD vaccines are still based upon inactivated whole virus particles, mostly grown in BHK-21 in a battery of fermentors 100 1200 liters capacity located in strict containment area under controlled air conditions.

Virus growth in cell culture system is followed by a series of treatments to clarify, inactivate, purify and concentrate the viral harvest.

During FMD antigen production, temperature and pH have to be closely controlled because of thermal instability and the low tolerance of the virus to pH conditions outside the ranges 7.0 - 8.0. Whole virus particle (146S) content is critical to the potency of the final product and measurement of 146S is used for vaccine formulation calculations.

Inactivation

Inactivation is one of the most critical steps in the production of FMD vaccine. Initially formaldehyde was used to inactivate alum adsorbed virus. Inprocess evaluation of this system proved to be difficult and it has mostly been superseded by the use of first order kinetics inactivants of the aziridine group of chemicals, most recently binary-ethylene-imine (BEl). Ideally this procedure is performed twice in separate inactivation vessels.

Good Manufacturing Practices (GMP)

In recent years, lhere has been a move away from end product testing towards the philosophy of in-process control. This policy has been encouraged at HR Vet. Through the promotion of Good Manufacturing Practice (GMP). In process inactivation controls are performed upon the virus harvest by tissue culture titration in sensitive cells, spectrophotometeric analysis or serological assays.

Further most inactivation concentration and purification by ultrafiltration or precipitation with polyethylene glycol (PEG) could yield a final virus product with a concentration factor of upto 1000 fold. Post inactivation purification reduces the non viral protein component of the antigen harvest, which is important in the reduction of potential hypersensitivity reaction in vaccinated animals.

Adjuvants

Inactivated whole virus vaccines against FMD are formulated as mono or polyvalent products with suitable stabilizers, buffers and adjuvants to enhance their potency. In aqueous formulations, the inactivated viral antigen is adsorbed to aluminium hydroxide and further adjuvanted with saponin.

Such vaccines are used successfully world wide for the immunization of ruminants.

However, commercial aqueous vaccines have not been successful in immunizing pigs (reactions at the site of injection were observed), and concentrated, inactivated antigens formulated as oil adjuvanted vaccines have been used widely in this species.

Oil adjuvanted FMD vaccines are also used in cattle, particularly in South America. Improved formulations have reduced the local reactions initially seen in this species.

Advantages are claimed for the use of oil-adjuvanted FMD vaccines in cattle in the areas of duration of immunity and the ability to immunize calves. Simple water-in-oil preparations can be made by the emulsification of the antigen in aqueous solution with light mineral oil and an emulsifying agent. Silverson and Ystral on-line pumps are used for the emulsification process and to ensure stability of the emulsion under field conditions.

Alternatively, a more easily injectable formulation can be made by further emulsification in a second aqueous phase to produce a stable water emulsion [double oil emulsified (DOE)]. There are several reports of the successful experiment use of these DOE FMD vaccines in cattle and pigs.

Following the completion of the blending process and addition of suitable preservatives, the vaccine bottled should be subjected to prescribed in vitro sterility test,safety I innocuity and potency tests in cattle, as described in the European Pharmacopoeia (Veterinary).

Safety tests are performed *in vivo* using the whole vaccine inoculated into susceptible animals and *in vitro* using eluted antigen inoculated onto sensitive cell culture.

Minimum potency assurance required is assessed by a variety of serological and I or animal challenge procedures.

FMD vaccines have a shelf life of one year if stored at 4^0C.

Production capacity

The production capacities have been increased tremendously to meet the demands of country as well as India current production capatcity is 80 million doses in totality and need to increase the capacity to 200 million doses by 2008.

Use of FMD Vaccines

In order to establish satisfactory immunity, it is usual to give a primary course of two inoculations with an interval of 2-4 weeks.

Re-vaccination may be given at 4-12 month intervals depending upon local epidemiological conditions and the quality of the vaccine. Therefore, the primary vaccination course may be delayed until four months of age in the offspring of regularly vaccinated mothers, although there is some evidence that calves can respond at one month old or younger.

The Role of Vaccination in FMD Control Strategy

Prophylactic

The successful control of FMD in countries with endemic or epizootic disease has often been based upon the regular use of inactivated whole virus vaccines as part of a regional FMD control policy.

The short lived nature of protective immunity in cattle following vaccination compared to FMD infection has led to the need to vaccinate annually or bi-annually, and even thrice a year in areas with a high risk of exposure to the virus.

Antigenic variation within a serotypc has lIIadc it commol1 praclicc to include more than one strain of a particular serotype in FMD vaccines.

Mass prophylatic vaccination against FMD, usually practiced only in the cattle population, is the first step towards controlling FMD in endemic areas.

The aim of this policy is that, over a period of years the load of FMDv in the environment will be reduced as the number of outbreaks, and therefore animals, with clinical disease will fall.

Obviously, good veterinary services are essential to maintain the vaccination campaign and monitor disease status in the country.

If the level of immunity to FMD in the target population in excess of 75% is achieved, the disease, should be adequately under control so that extra measures, such as importation control, quarantine and stamping out foci of infection, can be effective.

An example of the successful implementation of these policies was the reduction in outbreaks of FMD in Europe from 30000/year in 1965 to less than 1000/year by 1975.

It is extremely important that an antigenically appropriate vaccine should be used. It is essential that the antigenic relationship between field isolates and the vaccine strains in use should be ascertained regularly. The next stage in FMD control is to stop mass prophylactic vaccination and, by means of stringent surveillance, rapid diagnosis and importation control, a state of freedom from infection could be achieved. This is the current situation for Uruguay and the European Union countries.

Emergency

General vaccination is recommended for countries where the disease is enzootic, or where the threat of an outbreak is very great. If an outbreak occurs, a booster vaccination with the relevant serotype will increase the resistance of the population.

The process of stamping out of infection is difficult under Indian conditions because of social reasons.

Mass vaccination coverage to 80% of the animal population will reduce the incidence of foot and mouth disease in endemic areas. A generation of a vaccine should be advocated to contain the disease.

Committed people, a proven vaccine, a good delivery system and effective vaccination coverage along with active support from the farmers, Governmental decision makers, government research institutes, nongovernmental agencies, and manufacturers of vaccine would effectively control foot and mouth disease in India.

References

1. Anon 1996. Report from the OIE World Reference Laboratory for Foot and Mouth Disease Bulletin office International des Epizooties, 108:804.
2. Anon 1996. Report from the OIE World Reference Laboratory for Foot and Mouth Disease Bulletin office International des Epizooties, 108:867.
3. Carpenter. WC, Rai D.V., Samuel A.R. & Hofner M.C, 1996. Comparison of a radioactive and nonradioactive method for sequencing foot and mouth disease virus isolates Revue Scientifique et Technique, office International des Epizooties, 15: 875-882.
4. Donaldson, A.I. & Kihm, U, 1996. Research and technological developments required for more rapid control and eradication of foot and mouth disease Revue Scientifique et Technique, Office International des Epizooties, 15:863-873.
5. Donaldson A.I., Kitchinc R.P. .& Barnett, P.V. 1996. Foot and mouth disease. In. Manual of Standards for Diagnostic Tests and Vaccines, Lists A and B Diseases of Mammals, Birds and Bees, PP 47-56, paris, Office international des Epizooties.
6. Donaldson, A.I. 1997. Foot and mouth disease in Taiwan Veterinary Recorded, 140:407.
7. Donaldson, A.I. 1997. Risks of spreading foot and mouth disease through milk and dairy products Revue Scientifiique et Technique, Office International des Epizooties, 16: 117 -124.
8. Donaldson, A.I. 1997. Global eradication of foot-and mouth disease considerations and proposals. In: Empress Livestock Programme Rome, FAO.
9. Donaldson, A.I. & Have,P. 1997. Criteria to consider and operational requirements for implementing emergency (ring) vaccination Report of the Session of the Research Group of the Standing Technical Committee of the European Commission for the Control of Foot and mouth diseases, Kibbuttz Ma'ale Hachanisha, Appendix 29: 201-207.
10. Ferris, N.P. 1997. Development and use of the ELISA in the control of foot and mouth disease. Proceedings, International Symposium on Diagnosis and Control of Livestock Diseases Using Nuclear and Related Techniques Towards Disease Control in the 21 sl Century, Vienna. Salf JS & Iloot MC. (1997) Virus titres in the development of foot and mouth virus persistence in cattle. Report of the Session of Research Group of the Standing Technical Committee of the European Commission for the Control of Foot and Mouth Disease, Kibbutz Ma'ale Hachamisha, Appendix 4:45-51.

-ΔΔΔ-

12

Emerging and Re-emerging Caprine and Ovine Diseases in India

S NANDI AND R S CHAUHAN

Centre for Animal Disease Research and Diagnosis
Indian Veterinary Research Insitute, Izatnagar, U.P.

Small ruminants namely sheep and goats constitute a substantial portion of livestock wealth in India and play an important role in the national economy. Animals are vulnerable to a variety of infectious emerging and non-emerging diseases. The emerging diseases are those the incidence of which have recently been increased and appeared likely to continue increasing. The reemerging diseases are those which had formerly been controlled but is now increasing again. In the recent years, a number of emerging and reemerging diseases affecting the small ruminants have been reported causing enormous economic losses. Out of many emerging diseases, PPR, sheep pox, goat pox, BT, enterotoxaemia and brucellosis are notable one. So, an in depth knowledge of the emerging and reemerging diseases including the pathogenesis, clinical symptoms, diagnosis, treatment and control is the need of the hour so that the management of the diseases can be made in a very effective and efficient manner.

PESTE DES PETITS RUMINANTS (PPR)

It is a severe rinderpest (RP) like disease of small ruminants viz., sheep, goat and wild ruminants characterized by pyrexia, catarrhal inflammation of ocular and nasal mucous membrane, erosive stomatitis, enteritis and pneumonia. The disease is also called goat plague, kata, pseudorinderpest and stomatitis pneumo-enteritis complex. The current and standard name of the disease was chosen by the International Workshop of the disease in Ibadan, Nigeria, 1980.

PPR first reported in 1942 in Ivory Coast by Gargadennec and Lalanne. For long time, PPR was confined to West Africa, but now it exists in East Africa, from sub-saharan Africa to Arabian Peninsula, the middle East, Bangladesh, Nepal, Pakistan and India. For the first time in India PPR virus isolated from TN[65] and confirmed it by cDNA probes. Subsequently, outbreaks in sheep and goat were reported from other parts of TN, AP, Karnataka, Maharashtra, Rajasthan, UP, HP and WB during 1992-1995 [20]. It was earlier confined to Southern India and is now widely distributed in whole Indian subcontinent [45].

The etiological agent of PPR is a virus under the genus *Morbillivirus* and family *Paramyxoviridae.*

The PPR virus and RP virus are the two candidates within the same family and genus causing disease in small and large ruminants respectively but are not rigidly species specific. Both can infect large and small ruminants. In nature, these viruses maintain a separate and independent existence with a distinct epidemiology. Generally, PPR virus infects small ruminants and RP virus causes disease in large ruminants (Radostits *et al.,* 2000).

Susceptible Hosts

PPR is a rinderpest like disease of sheep and goats. Clinical signs and pathological lesions are confusingly similar to those of RP. Both sheep and goats are susceptible to PPR but goats are more susceptible than sheep in field outbreaks. The young, adult and both the sexes of sheep and goats are susceptible[70,60]. Experimentally, cattle have been infected without clinical signs and resisted experimental challenge inoculation with virulent RP virus suggesting PPR virus conferring cross protection against RP virus in animals. PPR in wildlife has also been reported. The disease has also been reported in a variety of wild animals viz., dorcas gazelle (*Gazella dorcas sp.*) , dama gazelle (*Gazella dama*) nilgai (*Bosephalus tragocamelus*), Nubian ibex (*Capra ibex nubiana*), red deer (*Cervus elaphus*), Laristan sheep (*Ovis orientalis laristanica*), gemsbok (Oryx gazelle) etc[27].

Mode of Transmission

Natural infection occurs mainly through direct contact with infected sheep or goat. Secretions, excretions and faeces contain high concentration of virus and spread of the disease take place through inanimate objects contaminated with it like RP. The disease is also transmitted through ingestion of infected materials or inhalation. There is no carrier state in animal but wild ruminants may play a role in the spreading of the disease[26]. Most of the disease outbreaks occur in wet months when rainfall causes goats to hurdle together enhancing the air borne transmission. The virus survives for long time in nature if high relative humidity persists.

Pathogenesis

After entry into the body, the virus invades the retropharyngeal lymph node and both alimentary and respiratory tract mucosa are affected. Inflammation of alimentary tract mucosa causes enteritis and diarrhoea, dehydration and death in young animals. Pneumonia and respiratory distress are due to inflammation of respiratory tract mucosa. Young goat may die due to anoxic anoxia. The disease may be further complicated by secondary infections. Bacteria like *E. coli* and *Pasteurella sp.* and protozoa like coccidia have been found to be associated with PPR to aggravate the condition. Animals may die due to severe pneumonia and death. Recovered animals may develop life long immunity[54].

Clinical Signs

The clinical signs of acute PPR in goats and sheep is fever (39-41°C) after an incubation period of 6 days [29]. There is profuse serous and nasal and ocular discharges which may turn from mucopurulent to purulent. Some may show signs of conjunctivitis. There is increased respiration rate (120/minutes) with extended head and mouth breathing. Signs of pneumonia become prominent and animal may die due to respiratory distress. The lesions are found in alimentary tract mucosa. Necrotic lesions are evident in lip, buccal mucosa (stomatitis), gum (gingivitis), dental pad, palate and tongue[47]. Diarrhoea is most commonly observed after 3-4 days and faeces contain mucous and blood[13]. Due to diarrhoea, emaciation and dehydration set in, temperature goes down, animals become exhausted and finally death occurs. Experimental PPR infection led to gradual decline in plasma volume, serum sodium and potassium values in goats. There were no significant changes in chloride, bicarbonate, phosphate, urea, total protein and albumin in the disease. The virus has destructive effect on lymphocytes.

Lesions

These are mostly alike in the two diseases. The most prominent pathological lesions such as erosions in the lips, dental pad, tongue, soft and hard palate and cheeks, esophagus, pillars of the rumen, abomasums and intestine, conjunctivitis and gastroenteritis are noticed. There is pronounced necrosis of lymphocytes in lymph nodes, tonsil, Payer's patches and spleen. Erosive lesions in large intestine along Payer's patches give rise to **"Zebra marking"**. The distinguishing pathological features between RP and PPR is the frequent occurrence of primary broncho-interstitial pneumonia in the latter disease. This is due to the affinity of the virus to bronchial epithelial cells and alveolar macrophages which are infected and destroyed[66].

Histopathologically, there is degeneration of epithelial cells at the edge of erosions of the oral mucosa in experimentally infected sheep and goat. Lymph nodes showed depletion of lymphoid cells, infiltration of macrophage in the germinal centres and sinuses and pyknosis and karyorrhexis of lymphoid cells. Lymph nodes in the head regions are haemorrhagic, necrotic and there is depletion of lymphoid cells. Spleen exhibited extensive haemorrhages, necrosis and hemosiderin deposits. Interstitial pneumonia with occasional eosinophilic nuclear inclusion body in the pneumocytes are noticed. In the lungs , there are many intranuclear and I/cytoplasmic inclusions and alveolar giant cells. There are degenerated villi in the small intestine, depletion of lymphoid cells in the Payer's patches and eosinophilic nuclear inclusions in the epithelial cells. Sections of the small and large intestine had ulcerated mucosa and submucosal haemorrhages[7].

Diagnosis

It is based on clinical signs and haematological observations and laboratory tests.Laboratory diagnosis of the PPR is done either by serological test or recently developed non serological tests. Various techniques have been applied in the serological diagnosis of PPR in sheep and goats. These include AGPT, CIE, VNT, ELISA (indirect, competition and immunocapture) [48,16,22]. Differential electrophoretic profile in the 'N' protein, nucleic acid hybridization using either radio-labelled or biotinylated c-DNA probes and RT-PCR using P and F gene specific primers have been used for precise, correct and reliable diagnosis of PPR from RP[22,25]. However, ELISA using MAbs against either 'N', 'H' or recombinant 'N' protein could differentiate PPR from RP[37,21]. Immune electron microscopy using ferritin conjugated antibodies can differentiate PPR from RP but expensive and sophisticated instrument is necessary. Serum neutralization test using homologous and heterologous viruses and sera could differentiate PPR from RP but the test is time consuming, cumbersome and requires viable viruses in the field sample.

Control

PPR is now endemic in India. There are lot of outbreaks reported in the country throughout the year. Legislative control of animal movement and slaughter of affected animals is not feasible. The only effective way to control the disease is preventive vaccination. Besides, strict sanitation and hygienic measures are to be adopted in flock. The new animal from disease prone area should not be introduced in a flock. Sick animals purchased from market should not be allowed to mingle with healthy animals without keeping in quarantine for a finite period.

Vaccines

Tissue culture rinderpest (TCRP) vaccine has been proved to be effective against PPR as it confers satisfactory level of cross immunity. Vaccine should be administered at 3-4 months of age when there is little or no maternal antibodies. The vaccine can be used in an actual outbreak to protect the animals after keeping them segregated from the infected ones for a week or two because there is a risk that some animals already may be incubating the disease may succumb. Furthermore, some researchers do not suggest to vaccinate the animals against PPR with TCRP vaccine as there is a chance that PPR virus may mutate and result in evolution of a new variant of PPR virus. So, both the inactivated and live attenuated vaccines against PPR have been developed. Inactivated PPR vaccine is not suitable as immunity following its use is short lived. Vero cell adapted PPR vaccine developed by IVRI and TNVASU have been proved to be effective against PPR in sheep and goat[64]. Vaccinia virus recombinants expressing the fusion (vRF) or the haemagglutinin (vRH) and a double recombinant expressing both the genes (vRFH) of RP virus provided complete protection to goats against PPR. The recombinant vaccine using capripox viruses for expression of the fusion and HA proteins of RP virus also exerted satisfactory level of immunity in goats to be protected against PPR.

GOAT POX

It is a malignant disease of goats characterized by high rise of body temperature and generalized pock lesions [35].

Distribution

Goat pox has been reported from all parts of the world. Various workers have reported the outbreaks of the disease from different states of India. It is widely prevalent in different states of India[67,58]and may attain epidemic proportion with mortality rate of 80% in kid and ~50% in adults[62].

Etiology

The pox virus is one of the largest and most complex of all viruses. There is no nucleocapsid conforming to either of the two types of symmetry found in most of the viruses hence it is called a complex virion. The virus is brick shaped and there is irregular arrangement of tubules on outer membranes. The causative agent of the goat pox is double stranded DNA containing virus belonging to the genus *Capripox virus*, subfamily *Chordopoxvirinae* and family *Poxviridae*. The other members of the genus *Capripox virus* are lumpy skin disease virus and sheep pox virus[44]. Capripox viruses are antigenically and genetically very close to each other. It has been reported that it is possible to infect goats with sheep pox virus and sheep with goat pox virus[68].

Susceptible Hosts

The capripox virus are peculiarly host adapted and are pathogenic only for sheep or goats. Although the Kenyan sheep and goat pox virus strain is the mildest producing only a local lesions in both sheep and goats, many strains of Mediterranean regions and from India appear highly adapted to either sheep or goats. Virus strains will produce local reactions when inoculated into the other species and there is no natural cross infection between the two, appears to occur under field condition. A unique property of most of the sheep pox and goat pox virus strains is the high level of host adaptation[8].

Mode of Transmission

The method of transmission of goat pox virus has been the subject of much speculation. The usual mode of transmission is either through contact with infected animal or inhalation[35]. The virus may gain entry through wound and abrasion. The greatest concentrations occur in the skin from 10-14 days and the titre then falls but virus can be detected up to 28 days in necrotic tissue or scabs[18].The virus may also be transported to other places by dog, cat etc. The disease may be transmitted through food and water mechanically.

Pathogenesis

The incubation period is 7-8 days. After getting entry into the body, the virus reaches the blood circulation via lymphatic ducts and is then carried to the skin and target organs. There is development of lung lesions leading to bronchopneumonia. The virus is present in conjunctival and nasal secretions. There is appearance of papules on the mucous membranes. The presence of virus in the skin papules produce irritation and the affected animals rub their body on other animals or walls. The virus may pass through placenta to the foetus and pregnant animals may abort[56].

Clinical signs

The clinical signs of goat pox virus infection vary with different hosts and in different geographical areas. Goats of all ages can be affected. The disease is more severe in young animals. The first signs of disease are rhinitis, conjunctivitis and pyrexia (40-42°C). Animals stand with an arched back, have poor body coat and have inappetance. One to two days later, cutaneous lesions develop (0.5-1.5 cm in diameter) often accompanied with lesions on the external nares, lips and within the mouth. The lesions on the skins are largely confined to the areas where the wool or hair is shortest such as on the head, neck, ears, axilla, groin, peritoneum and under the tail. Lesions present on the tongue and gums tend to ulcerate and high mortality rates occur when lesions develop in the respiratory and alimentary tract. Post-mortem examination shows multiple consolidated areas (0.5-2.0 cm in diameter) in the lungs. Secondary bacterial pneumonia is the common cause of death. Similar lesions are also visible in the liver, kidneys, abomasums and other organs. The appearance of the skin lesions varies and in some outbreaks the lesions are obviously vesicular, they coalesce and involve large areas of the body, especially the lower abdomen. While lesions may be restricted to predilection sites, they often involve the whole of the body in both young and old animals. Some lesions heal *in situ*, but the majority form necrotic plugs, which are shed leaving a shallow 'punched out' ulcer. This heals leaving a depressed permanent scar. When an epidemic occurs in a completely susceptible flock with no previous contact with the virus, the morbidity may be as high as 75% and mortality 50%. Mortality in young animals may approach to 100%[35].

Lesions

Pock lesions appear in all the parts of the body e.g. nostrils, face, ear, lip, cheek, snout, feet, thigh, abdomen, neck, teat and udder. Papules are formed within 1-3 days. The papules have raised dark brown centres and these changes into vesicles with creamy coloured fluid in 2 days. Skin nodules of varying sizes with thick grey materials are formed within 9-12 days. On 12 days onwards, the nodules regress forming reddish brown crusts which fall off leaving a granulation tissue underneath on 20th days. Lesions are found in the trachea with accompanying catarrhal pneumonia. Lesions occasionally reach abomasums causing haemorrhagic enteritis.

On histological examination the changes comprise of haemorrhages in the epidermis and edema of subcutis. There is ballooning degeneration of epithelial cells, coagulative necrosis with pyknotic nuclei and infiltration of mononuclear cells are evident. The intracytoplasmic inclusion bodies are observed in the epidermal cells. Lung parenchyma reveals pneumonic changes characterized by congestion, exudation, thickening of interlobular septa and infiltration of inflammatory cells[12,62].

Diagnosis

The disease is diagnosed on the basis of

a) *Clinical signs :* Pock lesions on the skin throughout the body.

b) *Histological examination :* I/C inclusion bodies in epidermal cells.

c) Isolation of virus in lamb or kid kidney, testicle cell culture, BHK21 cell line, Vero cell line or chorio-allantoic membrane.

d) *AGPT :* It gives precipitation band in 24-48 hours[57].

e) *CIE :* It gives precipitation band in 2-3 hours[59]

f) *FAT :* Direct FAT is used to detect the presence of pox virus in the oedema fluid. Indirect FAT is employed to determine the antibody status of a flock.

g) *ELISA :* Immunocapture ELISA can be used to detect the goat pox virus in scab suspension. Indirect ELISA is employed to screen the serum samples for the presence of antibodies against goat pox virus in a flock[58].

h) *Polymerase chain reaction (PCR) :* Recently PCR using primers based on the attachment gene has been successfully adopted to detect the capripox virus genomic DNA in skin scab samples. The amplified product could be visualized in phenol-chloroform extracted DNA samples as well as samples which were boiled in water bath for 10 minutes only[56].

Control

a) Import of animals from disease prone areas are to be prohibited.

b) The destruction of affected flocks or the quarantine of the sick animals from healthy ones.

c) Strict sanitary measures are to be adopted.

d) Live attenuated goat pox vaccine adapted and attenuated in cell culture (lamb/kid kidney, testicle, BHK21, Vero etc.) are also available in the market giving solid lifelong protection when administered S/C or I/D @ 1 ml/animal.

SHEEP POX

Sheep pox causes a severe and highly contagious disease in sheep which is listed in group A disease of the OIE[14]. It is defined as a malignant pox of sheep characterized by fever and generalized development of pock lesions. Lesions first appear as vesicles which later turn into pustules on the exposed parts of the body. There is high mortality rate in lambs.

Distribution

The disease is in existence since very long in Asia and Europe. It is prevalent in many areas of middle East countries, Africa, Asia and Mediterranean countries. There are lot of disease outbreaks in sheep flocks in different states of India around the year [67,55].

Etiology

The sheep pox virus is one of the largest and most complex of all viruses. There is no nucleocapsid conforming to either of the two types of symmetry found in most of the viruses hence it is called a complex virion. The virus is brick shaped with irregular arrangement of tubules on outer membranes. The causative agent of the sheep pox is double stranded DNA containing virus belongs to the genus *Capripox virus*, subfamily *Chordopoxvirinae* and family *Poxviridae*.

Susceptible Hosts

The sheep are naturally susceptible. The susceptibility also depends on the breed and age of animals. Younger sheep are more susceptible over older ones. In young lambs, the disease may flare up in epidemic proportion. Merino breed is more susceptible over indigenous one.

Transmission

The disease spreads by direct or indirect contact with contaminated objects. In infected pens, virus in the scab can remain viable for as long as 6 months. Infection can occur through either the respiratory or the cutaneous route. Mechanical transmission by arthropods may occur. Affected sheep shed the virus at every stage of the disease up to 8 weeks after the lesions have healed. Merino sheep are most susceptible whereas Algerian sheep are comparatively resistant. On inhalation, the virus enters the lungs where from it enters the circulation by lymphatic channels and is then carried to skin and mucous membranes. The virus may pass from the infected ewe to the foetus through placental blood and the lambs may be born with pox lesions[34,69].

Pathogenesis

The incubation period varies from 4 to 14 days. On entrance, there is multiplication of the virus and resultant viraemia. The virus get distributed throughout the body system and generalized skin infection takes place. There is infiltration of cells and necrosis of the epidermal layers. Prickle cell layers of the skin proliferate and this extends to the corium. Vesicles thus appear are multilobular in nature. Vesicles turn into scabs. Intracytoplasmic acidophilic inclusion bodies are present in the infected epithelial cells. The virus may affect mucous membranes of the pharynx, trachea and abomasums. There is formation of vesicles and ulcers but inflammatory changes are occasionally noticed in intestinal and respiratory tract mucosa[56].

Clinical signs

The clinical signs of sheep pox depends on the breed of sheep, the virulence of the virus strain, the nature of secondary infection and organ involved. The disease may appear usually in three different clinical forms. (a) Malignant form (b) Mild (benign) form and (c) Abortive form.

(a) *Malignant form :* This is the most common form. It is characterized first by appearance of marked general distress with elevated high temperature (106-107°F), rough staring coat, loss of appetite and numerous areas of erythema all over the body surface. The eruptions are more prominent on the cheeks, nostrils, lips, ears, eyelids, vulva, armpits, thighs, chest, mammary glands, testicles and under surface of the tail. The lesions pass through the typical pock stages. The vesicular stage which may be haemorrhagic with a tendency to generalize, is followed by the development of pustules which later turn into scabs. Healing is slow and may take more than a month. Approximately 35-40% of the affected cases develop secondary pneumonia characterized by respiratory distress, cough and thick mucopurulent nasal discharge. Such cases are invariably terminated fatally. Profuse watery diarrhea of foetid nature with sub-normal body temperature before death is almost a constant feature. Affected animals may also die of disease due to pyaemia or septicaemia caused by secondary bacterial complications[40].

(b) *Mild (benign) form :* Adult Algerian and indigenous sheep are usually affected with this form and eruptions are confined around eyes, lips and nose.

(c) *Abortive form :* Generalization is rare and mortality is low (5%). Ewe may abort foetus with pock lesions. Lactating ewe may develop the signs of mastitis due to lesions in the udder.

Pathology

The characteristic papules, vesicles, pustules and scabs are noted on cutaneous surface. Numerous circular nodules grayish white in colour and comparatively of smaller diameter (4-6 mm) than those of the skin are always present throughout the tissues of the lungs. Consolidation of apical, cardiac and ventro-lateral borders of the diaphragmatic lobes with marked oedema is present in about 35% cases. The reddish fluid oozes out on cutting the oedematous lungs. The pleura show no gross lesions. The trachea shows petechial haemorrhages with few ulcers and purulent exudates. The petechial haemorrhages are also found on epicardium and endocardium of the heart. The liver is found enlarged and acutely congested. The kidneys are usually affected with white infarcts and subscapular haemorrhages and showed intense congestion of both medulla and cortex. Adrenals are also congested. The mediastinal, mesenteric and prescapular lymph glands are usually enlarged and oedematous[62].

Diagnosis

Similar to goat pox.

Control

Vaccines being the most economical and efficient way to control the sheep pox. A variety of vaccines both inactivated and live attenuated have been developed since the beginning of the century by a number of researchers employing aluminium hydroxide, alum and saponin as adjuvant and formalin, merthiolate and BPL as inactivating agent. The route of administration of most of the vaccines is either I/D, S/C or I/M and elicited an immune response lasted for 6 months to 1 year. However, in India now-a-days live attenuated tissue culture sheep pox vaccines are mainly produced by two manufacturers, viz., IVRI, Izatnagar and Biomed, U.P.

Sheep pox is enzootic in India. About 250 outbreaks affecting about one third of total sheep population are reported in India every year causing enormous losses. As sheep rearing is mostly the livelihood of small and marginal farmers in the rural areas, the importance of the disease is clearly revealed. So, to control the disease, the prudent approach is the regular immunization of all the susceptible flocks with a potent and efficacious vaccine. Besides, proper monitoring and stringent surveillance of the disease should be carried out in regular interval to get a clear picture of the disease in India. Lastly, proper management, hygienic condition, sanitation and balanced ration should be provided along with the other measures to keep the disease under control[56].

BLUETONGUE

It is an infectious, noncontagious arthropod borne viral disease of domestic and wild ruminants namely sheep, goat, cattle, camels, llamas, deer and antelopes[39]. This is predominantly a disease of sheep but occasionally cattle and goats are subclinically affected. The disease is characterized by high fever, catarrhal stomatitis, rhinitis, enteritis and lameness due to inflammation of the coronary bands and sensitive laminae of the feet. Bluetongue (BT) is now established as endemic disease of animals in India[53,73]. Antibodies against a number of serotypes of bluetongue virus have been detected in sheep, goat, cattle, and buffalo in different parts of the country. The BT disease has been included in OIE list A which has resulted in strict quarantine measures and expensive and costly testing. Keeping in view the importance of the disease, All India Network Programme on Bluetongue has been started in the year 2001 with 10 centres including IVRI, Izatnagar is the co-ordinating centre. Different centres are working on different lines encompassing the repository, typing, sero-surveillance and development of vaccines[6].

Distribution

The disease is prevalent in many countries of the world. In the Indian subcontinent, the disease was first reported from West Pakistan (now Pakistan) in 1959 in a flock of Ramboullet sheep imported from USA. In India, the disease was first reported among sheep and goats of Maharashtra state based on clinical signs and detection of bluetongue virus (BTV) specific antibody in sera of recovered animal[63]. Later BT was reported almost all the states of India on the basis of virus isolation in chicken embryo or cell culture system. BTV has also been isolated from *Culicoides sp*[32].

Etiology

The disease is caused by a double stranded RNA virus belonging to the genus *Orbivirus* and family *Reoviridae*. The virus genome of 10 ds RNA segments of varying lengths which code for the 7 structural and 3 non-structural proteins. The BTV is non-enveloped and spherical. The virions contain two icosahedral protein shells termed as inner core and indistinct outer capsid. The characteristic ring like configuration leads to the name of this Orbivirus (Orbis meaning ring or circle). The structural proteins are numbered as VP1 to VP7 and the nonstructural proteins are designated as NS1, NS2, NS3 and NS3A. The outer capsid of BTV consists of two major polypeptides namely VP2 and VP5. VP2 is unique for each of the BTV serotypes, where as VP5 revealed an intermediate level of conservation. The variability of VP2 and VP5 protein reflects the role of these proteins in the induction of serotype specific neutralizing antibody. The outer capsid proteins are associated with the virulence and cell adsorption and VP2 is a haemagglutinating protein[44]. The BTV core is composed of 2 major polypeptides (VP3 and VP7) and 3 minor polypeptides (VP1, VP4 and VP6). VP7 has been identified as the soluble group specific antigen and conserved in all the serotypes of the bluetongue virus and serum samples suspected for bluetongue are screened for antibodies against this protein in competition ELISA[46].

Susceptible Hosts

Although all the domesticated and wild ruminants are susceptible to bluetongue disease, it is basically a disease of sheep. Young sheep under 1 year are more prone to infection. British and Merino sheep are more susceptible than African sheep. European mutton breeds are more susceptible than the Merino. Frank clinical cases of bluetongue were not detected in indigenous (Nali and Sondi) and cross bred sheep indicating that they are comparatively more resistant to bluetongue virus. The disease remains in subclinical form in cattle. The degree of susceptibility of goats, although variable, is markedly less than that of sheep. White tailed deer and other wild ruminants may be naturally infected[30, 38, 3].

Mode of Transmission

Bluetongue virus is transmitted by *Culicoides sp.* and is only enzootic in areas where adults of competent vector species are present for most of the time of the year, thus maintaining a continuous series of virus. The species of culicoides which are known to act as vectors for bluetongue virus are most active in the temperature range of 18-29°C and almost inactive below 10°C or above 30°C. There are in excess of 1000 species of culicoides worldwide, only six are important and known to be capable of transmitting the virus. These include *C. variipennis, C. imicola, C. filvus, C. actoni, C. wadai* and *C. nubeculosus*[31].

The geographic and seasonal distribution of vectors is critical for both the spread and maintenance of the BTV in the field. At ambient temperature of 25°C, productive infection of the salivary glands and the ability to transmit the virus, takes approximately seven days after ingesion of the original viraemic blood meal. Only females of the culicoides vectors take blood meal from vertebrate hosts prior to egg laying. White tailed deer and antelope are considered to be reservoir of infection and thus act as carrier of virus from one season to other[41,43].

Pathogenesis

On entrance through the bite of the vector, the virus multiplies in the blood stream and leads to a primary viraemic stage. The virus has an affinity for the endothelium, precapillary arterioles, venules and red blood cells. Small vessels underlying stratified squamous epithelium, particularly the oral mucous membrane, skin and coronet of the hoof are mostly affected. The virus is also detected in reculoendothelial cells of lymph nodes especially those draining tissue of the head. The characteristic lesions appear due to damage of blood vessels. It produces hyperemia, haemorrhages and oedema of various tissue. The virus has got both abortigenic and teratogenic characteristics[39,51].

Clinical Signs

The disease is divided into 3 forms (a) acute (b) subacute and (c) chronic. Naturally occurring acute bluetongue disease in sheep has an incubation period of less than a week (2-4 days in experimentally infected sheep). There is high rise of temperature (105-106°C), nasal discharge, salivation and lacrymation. The nasal discharge is mucopurulent and often blood stained and the saliva is frothy. Swelling and edema of the lips, gums, dental pad and tongue may be evident. There may be ulceration of the lips, dental pad, gum and tongue. Necrotic ulcers develop on the lateral aspects of the tongue which is swollen and purple in colour hence the disease is called bluetongue. Foot lesions including laminitis and coronitis and manifested by lameness and recumbency appear when the mouth lesions begins to heal. The appearance of a dark red to purple band in the skin just above the coronet is considered to be an important diagnostic sign. The twisting of head and neck to one side (wryneck) may be seen in few cases about 12th day. Oedema of the face, jaws and the ears and hyperemia of the wool less skin may be present. Diarrhoea and dysentery may supervene. Rapid shallow respiration is not uncommon and pneumonia may be evident[52].

Subacute or subclinical form is noticed in cattle and goats. There is a mild to moderate fever, hyperemia of the mucosa and conjunctiva. Some cattle may develop syndrome like that seen in severely affected sheep. This form of the disease in cattle appears to be a hypersensitivity reaction in a previously affected animal. The same hypersensitivity does not occur in sheep. Cattle and sheep infected during early pregnancy may cause abortion of congenital deformities which include hydranencephaly, microcephaly, curvature of the limbs, blindness and deformity of the jaw[5].

Course

In mild cases, the prognosis is generally favourable and recovery is usually rapid and uneventful with some loss of condition. In severe cases, the prognosis should be guarded. Animals which appear to be recovering satisfactorily may suddenly collapse and die. On the other hand, animals which have had severe symptoms may recover completely. The recovery is usually prolonged when animals are weak and debilitated because of varying degree of buccal lesions and muscular involvement. Lambs especially take a long time to recover and usually remain runts.

The mortality rate in bluetongue varies remarkably and the outcome is influenced by various factors. The pathogenicity of virus strains differs markedly and the sunlight is known to aggravate the clinical course of the disease. Under field conditions, The mortality rate usually varies between 2-30%. Death occurs mostly from 1-8 days after the appearance of visible signs. Bronchopneumonia often resulting from aspiration of rumen ingesta, is probably the most common cause of death[53].

Pathology

The severity and distribution of the lesions found in bluetongue infected sheep depend on the severity of the clinical symptoms and particular state of the disease at which death occurs. The typical pathological picture may be largely obscured by secondary complications such as bronchopneumonia or metabolic disturbances in animals dying late in the course of the disease.

The prominent lesions of digestive system are usually found in and around the mouth. The mucosa is oedematous, hyperemic, cyanotic and petechiae or ecchymoses may be present. Excoriations covered with grey necrotic tissue are evident on the lips, dental pads, anterior dorsum of the tongue and internal surface of the cheek. Hyperemia of the ruminal papillae, ruminal pillars and reticular folds is commonly observed. The remainder of the digestive tract may show congestion and catarrhal inflammation. In the respiratory system the nose is frequently partly occluded by a grayish brown scab composed of desquamated epithelium, inspissated serum and foreign debris. The trachea may contain white froth and the lungs may be oedematous and pneumonia is common as a result of inhalation of ingesta. In cardiovascular system, there is widespread hyperemia, oedema and haemorrrhages. Distinctive haemorrhages, ranging in size from 2 to 15 mm are consistently found at the base of the pulmonary artery and are generally considered as pathognomic of bluetongue. Subepicardial haemorrhages near the apex of the heart and subendocardial haemorrhages on the papillary muscles may be present. Petechiae are observed in the pericardium along with mild hydropericardium and hydrothorax. The lymph nodes, particularly those draining tissues of the head are commonly enlarged, oedematous and haemorrhagic. The spleen may be slightly enlarged and subscapular petechial haemorrhages may be present in the thymus. The lesions in the skeletal muscles mostly observed in thigh, shoulder, back and neck consist of petechial or ecchymotic haemorrhages and hyaline degeneration. The muscular degeneration continued even the subsidence of fever and is largely responsible for the severe emaciation characteristic of bluetongue. The kidneys are commonly congested with petechial haemorrhages on the mucosa of the urinary bladder, urethra, vulva or penile sheath[51,52].

Diagnosis : Presumptive Diagnosis

It is based on the history, clinical findings (oedematous swelling and cyanosis), seasonal occurrence and negation of spread by direct contact.

Confirmatory Diagnosis

The serological tests which are generally used for confirmatory diagnosis of bluetongue consist of agar gel immunodiffusion test, plaque inhibition test, neutralization test, fluorescent antibody test, complement fixation test, competition ELISA and dot-ELISA[1,46]. Besides, isolation of bluetongue virus in suitable system viz., sheep, chicken embryo or cell culture inoculation offer the alternative method of confirmatory diagnosis. Recently RT-PCR based on the structural protein and nonstructural protein gene specific primers, nucleic acid hybridization and nucleic acid sequencing offer the most sensitive ways to detect the viral genome in the clinical samples[17].

Different Diagnosis

The disease should be differentiated from the following diseases.

(a) *Foot and mouth disease* : This is a highly infectious and contagious disease and spread by direct contact while bluetongue is not transmitted by direct contact but require insect vector for its transmission. FMD produces vesicles on the tongue and feet but there is no bluish coloration of the tongue. In bluetongue a pink line appears on coronet which is not present in foot and mouth disease. In FMD there is no development of pulmonary oedema and pneumonia which is present in bluetongue disease.

(b) *Malignant catarrhal fever (MCF)* : There is extensive erosions on the buccal cavity, tongue, hard palate and fore stomachs but in bluetongue the lesions are not extensive in nature. MCF is sporadic in nature where as bluetongue is seasonal in nature. Corneal opacity due to severe panopthalmitis is found in MCF but not in bluetongue.

(c) *Photosensitization* : There is dermatitis of light skin on exposure to light and is not caused by infection. In photosensitization there is reddening and oedema whereas in bluetongue there is erosion, oedema, necrosis and crust formation. Cyanosis of tongue mucosa and pink line in coronary band which are characteristic features of bluetongue are absent in photosensitization.

Control

The following measures may be undertaken to prevent and control the bluetongue.

1. Grazing of the animals should be avoided in areas where there is lot of culicoides vectors.
2. Attempts should be made to curb down the vector population by spraying insecticides and good water management.
3. Good hygienic and sanitary conditions should be adopted to control the spread of the disease.
4. Movement of all the animals susceptible to bluetongue should be restricted.
5. Contact with wild ruminants should be avoided.
6. Strict quarantine of the affected and sick animals should be made.
7. Animals importation from bluetongue disease prone countries should be strictly avoided.
8. *Immunoprophylactic measures* : In India there is no vaccine available which is effective against BT. However, in foreign countries both live attenuated and inactivated vaccines are available and

routinely used to control the BT. In India, preliminary studies on the development and evaluation of inactivated (BEI, BPL and hydroxylamine) and adjuvanted (Aluminium hydroxide and saponin) BTV vaccines have shown the encouraging results in eliciting both humoral as well as cell mediated immunity in vaccinated animals which withstood challenge virus infection also.

ENTEROTOXAEMIA

Enterotoxaemia is an acute highly fatal disease of sheep and goats primarily of lambs and kids caused by the absorption of toxin produced by *Clostridium perfringens* type D in intestinal tract and characterized by endocardial haemorrhage, gastroenteritis, pulpy kidneys, nervous symptoms and death[9]. *Clostridium perfringens* is a Gram positive rod shaped, anaerobic, non-motile, capsulated, non-sporulating haemolytic bacterium. It is caused by the type D toxin produced by *Clostridium perfringens*. Besides, the toxin, production of hyaluronidase by the organism also contributes to its pathogenicity.

Pathogenesis

The organism is soil borne and is a common flora of the alimentary tract of all species of warm blooded animals. The organism proliferates rapidly under the slightest anaerobic condition in the intestine. The force-fed or well fed lambs are found to be more susceptible. Feeder lambs on rich clover pasture or feedlot on heavy grain ration are more susceptible. Adult sheep on corn fields or excessive feeding during hot weather and ewes on good pastures of grass or winter wheat are also susceptible to the disease. Very high protein diet may serve as a contributory factor while overeating on concentrates and lush pastures are important inciting factors. A sudden change from a poor to a luscious pasture, overeating on rich feed, grazing on fodder crops e.g. green wheat, green Lucerne etc. may lead to the disease. Sluggishness of the bowel movement (atony of the ileum) causing anaerobic condition is an important factor.

The toxins produced consist most of å (epsilon) factor with smaller quantities of á and â fractions. Due to the action of the toxins on the intestinal mucosa, the permeability and absorption of the toxins increase facilitating the entry into the circulatory system leading to rapid fatal toxaemia. The toxin also acts on intestine to cause diarrhea.

Due to the effect of the toxins on the kidneys, they become swollen and pulpy. If the toxins reaches the brain nervous symptoms develop[71].

Clinical Signs

In majority of the cases, the course of the disease is peracute and death is sudden. Sheep normal at night are found dead in the morning. Sometimes, sheep may walk with staggering gait and knuckling over the fetlocks. Other symptoms are excessive salivation, champing of jaws, hyperesthesia, rapid breathing, coma and finally death[50].

In some animals, progressive weakness, depression, excitement, convulsions, excessive salivation, champing of jaws, opisthotonus, circling and/or pushing against objects are observed. Diarrhoea is observed at any stage of the disease. Temperature may rise 2 to 3°F in animals with convulsive symptoms. Death usually occur within few hours but is generally preceded by violent galloping movement, twitching of the muscle and rolling of the eyes. In chronic cases, diarrhoea may linger and loss of condition precedes recovery [74].

Diagnosis

The presumptive diagnosis can be made on the history of overfeeding of grains. Postmortem lesions consists of serious effusions (large blotches) in the peritoneum, epicardium, endocardium, intestinal serosa, abdominal muscles, diaphgram and thymus gland. The kidney may appear as swollen reddish, soft and pulpy (pulpy kidney) due to hyperemia. There is congestion of lungs, trachea and abdominal lymph nodes. Haemorrhagic abomasitis and duodenitis are also noticed. The liver is friable, congested and spotted. The intestine is devoid of contents but the stomach is packed with food.

The laboratory diagnosis of the disease is based on the isolation of organism and detection of *Cl. Perfringens* toxin D by mouse inoculation test[33] and ELISA[72]. The differential diagnosis should be made with regards to HS, coccidiosis, listeriosis, plant or chemical poisoning and botulism[36].

Prevention and Control

The disease can be prevented by proper husbandry and minimizing the predisposing factors to a maximum extent. Lambs or sheep above 3 months of age should be vaccinated with 2.5 ml of *Cl. Perfringens* type D of formalized vaccine subcutaneously. A second vaccine with equal amount should be given 14 days after the first vaccination to ensure effective immunity. Usually the vaccine is given before the occurrence of the disease. Sheep which are vaccinated in the previous year need only one shot each in the subsequent years. Pregnant ewes can be vaccinated up to 3 weeks before lambing[33].

BRUCELLOSIS

Brucellosis is an important infectious disease of the reproductive system of sheep and goats characterized by inflammation of the genital organs, foetal membranes, abortion, sterility and formation of granulomatous lesions in various tissues. The causative agent is a small non-motile, non-spore forming, Gram negative coccoid. *Brucella melitensis* and *Brucella ovis* are the primary pathogen of goats and sheep respectively[28].

Pathogenesis

The disease is prevalent worldwide. The disease is transmitted usually by oral route through contaminated feed and water with aborted foetal content or foetal membrane, vaginal exudates and milk. It is also transmitted through inhalation, skin wound, conjunctiva, mammary gland and coitus. Congenital infection may also occur. Bacteria invade the mucosa of the digestive tract via the oral route and reaches the mesenteric lymph node to multiply there. They are carried to the general circulation through the thoracic duct resulting in bacteraemia. The organism localizes in the pregnant uterus and foetal membrane, lymph node, bone marrow, spleen, genital organs, joints and tendons. The organism usually appear in pregnant uterus at 2 to 4 months following infection, multiplies in the chorionic epithelium to spread between the chorion and uterine mucous membrane resulting in inflammation, degeration, autolysis and necrosis. Just prior to immediately after abortion the organism appear in mammary gland or its lymph nodes[4].

Geographical Distribution

Brucellosis of sheep caused by *Br. ovis* has been reported in most of the major sheep producing regions of the world. The flock prevalence of infection can be as high as 75% and as many as 60% of rams may infected[11].

Susceptible Hosts

Only sheep are affected although the ram is more susceptible than ewe. While tailed deer and goats can be infected only experimentally leading to epididymitis. The Merino and Merino derived cross breeds are less susceptible to disease than British breeds.

Transmission

The organism is present in the placenta, vaginal discharges and milk of infected animals. The organism can survive on pasture for several moths but transmission by fomites has no practical significance. The infected ram is the source of infection and perpetuates the disease in a flock. The rams excrete the organism in semen. The transmission between rams occurs via passive venereal infection and by direct ram to ram transfer. Passive venereal infection occurs from ewes that have been bred by an infected ram. Infection may result in early embryonic death and occasionally abortion or the birth of weak and poorly viable lambs[111].

Clinical Signs

The incubation period varies from 3 weeks to 6 months. Acute oedema and inflammation of the scrotum is noticed. The regression of the acute syndrome is followed after a long latent period by the development of palpable lesions in the epididymis and tunicae of one or both testicles. There is gradual loss of condition, birth of weak calves, decreased lactation, retained placenta and orchitis in male animals. Hygroma of knee joints results in arthritis and lameness. Infection of mammary gland results in interstitial mastitis and enlargement of supramammary lymph node. Abortion during late pregnancy (last 2 months) is the most obvious sign in goats and sheep[615,49].

Pathological Lesions

The placenta appears oedematous either focally or diffusely and covered with fibrin flakes. The cotyledons are dull in appearance and chorion may be thickened with a thick tenacious brown exudates. Mammary gland and testes may reveal mastitis and orchitis respectively with granulomatous lesions described as above. The foetus shows oedema in all parts and cavities of the body. Spleen and lymph nodes are swollen, hyperplastic and show necrotic foci[64].

Microscopically, there is infiltration of phagocytic cells and lymphocytes surrounded by fibrous tissue proliferation. The organisms are present in chorionic epithelial cells. In males, proliferation of fibrous tissue compresses or replaces the epididymis.

Diagnosis

The presumptive diagnosis can be made on the basis of symptoms and lesions. The definitive diagnosis is made by plate agglutinin test, CFT, STAT (1:40 and above positive), ELISA, milk ring test, FAT, isolation and identification of bacteria and demonstration of organisms in tissue sections using special stains[24,2,10].

Prophylaxis

The universally recommended vaccine is Elberg's Rev 1 which is effective in both sheep and goats Rev 1 is a living attenuated vaccine that produces a bacteremia which is cleared by 14 weeks in goats and a shorter time in sheep. Vaccination of animals at 3-8 months of age confers a high degree of immunity which lasts for > 4 years in goats and 2.5 years in sheep. Vaccination of pregnant goats and sheep will result in abortion and generally not recommended. A number of other vaccines are also available but none is fully effective.

Control

In order to control, systemic measures should be undertaken.

1) Serological testing of herds and immediate disposal of positive cases.
2) Proper zoo-sanitary practices and good management are of paramount importance.
3) Screening test should be carried out at regular intervals to eliminate the infected animals and disinfect the premises.
4) A measure of control can be achieved using scrotal palpation to detect infected rams with scrotal abnormalities followed by culling.
5) Serological tests are repeated at monthly intervals with culling of reactors until all rams are sero-negative.
6) Vaccination may be the most economical and practical means of controlling diseases in areas with a high incidence of infection[19,42].

References

1. Afshar, A., Thomas, F.C., Wright, P.F., Shapiro, J.L., Shettigara, P.T. and Anderson, J. 1987. J. Clin. Microbiol. 25 : 1705-1709.
2. Agarwal, G.S., Singh, S.V. and Batra, H.V. 1999. Indian J. Anim. Sci. 69 : 463-465.
3. Akita, G.Y., Glenn, J., Castro, A.E. and Osburn, B.I. 1993 J. Vet. Diag. Invest. 5:154-158.
4. Alton, G.G. 1990 Brucella melitensis. In : "Animal brucellosis" (Nielsen, K., Duncan, J.R., eds).
5. Anderson, G.A., Stott, J.L., Gershwin, L.J. and Osburn, B.I. 1987. J. Gen. Virol. 68 : 2509-2514.
6. Anon 2005. Annual Report of AINP on Bluetongue.
7. Aruni, A.W., Jayapal, G., Vairavasamy, K. and Ebenezer, D. 1999. Indian J. Anim. Sci. 69(9) : 690-692.
8. Bandyopadhyay, S.K., Gajendragad, M.R., Dhal, N.K., Gupta, A.R. and Yadav, M.P. 1984. Indian J. Anim. Sci. 54: 961-964.
9. Basavalingappa, B.S., Krishnamurthy, G.V., Isloor, S. and Mayanna, A. 2005. Indian J. Comp. Microbiol. Immunol. Infect. Dis. 26(1): 29-31.
10. Batra, H.V., Chand, P., Ganju, L., Mukherjee, R. and Sadana, J.R. 1989. Res. Vet. Sci. 46:143-146.
11. Blasco, J.M. 1997. Preventive Vet. Med. 31: 275-283.
12. Buller, R.M.L. and Palumbo, G.J. 1991. Microbiol. Rev. 80-122.
13. Bundza, A., Afshar, A., Dukes, T.W., Myers, D.J., Dulac, G.G. and Becker, S.A. 1988. Canad. J. Vet. Res. 52: 46- 49.
14. Carn, 1993. Vaccine.11: 1275-1279.
15. Chandra, M., Singh, B.R., Shankar, H., Agarwal, R.K. and Babu, N. 2005. Indian J. Anim. Sci. 75 :220-221.
16. Chandran, N.D., Kumanan, K. and Venkatesan, R.A. 1995. Trop. Ani. Hlth. Pro. 27(2) : 89-92.
17. Dash, P.K. 2000. M.V.Sc thesis submitted to Deemed University, I.V.R.I., Izatnagar, U.P.
18. Davies, F.G. 1976. J. Hyg. Camb. 76: 163-166.
19. Deqiu, S., Donglou, X. and Jiming, Y. 2002. Vet. Microbiol. 90:165-182.
20. Dhand, N.K., Sharma, C.S., Sandhu, K.S., Sharma, D.R. and Singh, J. 2002. Indian J. Anim. Sci. 72 (10) : 853-855.
21. Dhar, P., Sreenivasa, B.P., Barret, T., Corteyn, M., Singh, R.P. and Bandyopadhyay, S.K. 2002. Vet. Microbiol. 88:153-158.
22. Diallo, A., Libeau, G., Couacy-Hymann, E. and Barbron, M. 1995. Vet. Microbiol. 44:307-317.
23. Diallo, A., Taylor, W.P., Lefevre, P.C. and Provost, A. 1989. J. Virol. Meth. 23: 127-131.

24. Falade, S. 1978. Res. Vet. Sci. 24 : 376-379.

25. Forsyth, M.A. and Barrett, T. 1995. Virus Res. 39: 151-155.

26. Fraser, C.M. 1986. The Merck Manual. 6th ed. Merck & Co. Inc., U.S.A.

27. Furley, C.W., Taylor, W.P. and Obi, T.U. 1987. Vet. Rec. 121: 443-445.

28. Garin- Bastuji, B., Gerbier, G., Douzal, Y., Vaucel, D., Hummel, N., Thiebaud, M., Grayon, M. and Verger, J.M. 1994. Epidemiol Sante Anim. 26: 103-130.

29. Hamdy, F.M. , Dardiri, H.A. and Breese, S.S. 1975. Annual Meeting of American Society for Microbiology. 75:265-268.

30. Howerth, E.W., Greene, C.E. and Prestwood, A.K. 1988. Am. J. Vet. Res. 49: 1906-1913.

31. Ilango, K. 2006. Current Science. 90 (2): 163-167.

32. Jain, N.C., Prasad, G., Gupta, Y. and Mahajan, N.K. 1988. Rev. Sci. Tech. Off. Int. Epiz. 7:375-378.

33. Jayaraman, M.S. and Mallick, B.B. 1961. Indian Vet. J. 38: 226-227.

34. Kitching, P. (1983). Vaccine 1 : 4-9.

35. Kitching, R.P. and Taylor, W.P. 1985. Trop. Anim. Hlth. Prod. 17 : 64-74.

36. Knight, P.A., Queminet, J., Blancharrd, J.H. and Tilleray, J.H. 1990. Biologicals. 18:263-266.

37. Libeau, G. 1995. Res. Vet. Sci. 58:50.

38. Luedke, A.J., Jochim, M.M. and Bowne, J.G. 1964. Can J. Comp. Med. 29:229-231.

39. MacLachlan, N.J. 1994. Comp. Immunol. Microbiol. Infect. Dis. 17:197-206.

40. Malik, Y., Chand, P. and Batra, S.K. 1998. Indian J. Anim. Sci. 68 : 1207-1208.

41. Mellor, P.S. 2000. J. Comp. Path. 123:231-247.

42. Mrunalini, N., Rama Sastry, P., Pandarinath, G.N. and Ramakrishna Rao, M 2000. Indian Vet. J. 77:932-935.

43. Mullens, B.A., Tabachnick, W.J., Holbrook, F.R. (1995). Med. Vet. Entomol. 9:71-76.

44. Murphy, F.A., Gibbs, E.P.J., Horzinek, M.C. and Studdert, M.J. 1999. Veterinary Virology. 3rd edn. Academic Press.

45. Nanda, Y.P., Chatterjee, A. and Purohit, A.K. 1996. Vet. Microbiol. 51 : 207-211.

46. Nandi, S., Pandey, A.B., Suresh, I. and Audarya, S.D. 2005. Indian J. Comp. Microbiol. Immunol. Infect. Dis. 26: 1: 44-46.

47. Obi, T.U. 1980. Proc. Int. Workshop on PPR. Ibadan, p. 71.

48. Obi, T.U. and Patrick, D.1984. J. Hyg. 93:579-586.

49. OIE 2004. Caprine and Ovine brucellosis. In: Manual of Diagnostic Tests and Vaccines for Terrestrial Animals. 5th edition. Office International of Epizooties.

50. Parhi, N.K., Kar, B.C., Rao, A.T. and Ray, P.K. 1993. Indian J. Comp. Microbiol. Immunol. Infect. Dis. 14:46.

51. Parsonsosn, I.P. 1990. Curr. Top. Microbiol. Immunol. 162: 119-141.

52. Pini, A. 1976. Onderstepoort J. Vet. Res. 43: 159-164.

53. Prasad, G., Jain, N.C. and Gupta, Y. 1992. Rev. Sci. Tech. 11: 699-711.

54. Radostits, O.M., Gay, C.C., Blood, D.C. and Hinchcliff, K.W. 2000. Veterinary Medicine, 9th edn. W.B. Saunders Company Ltd.

55. Rao, T.B., Das, J.H., Sarma, D.R. and Singh, S.S.S. 1994. Livestock Advisor. 19 : 3-6.

56. Rao, T.V.S. and Bandyopadhyay, S.K. 2000. Anim. Hlth Res. Rev. 1(2) : 127-136.

57. Rao, T.V.S. and Negi, B.S. 1997. Trop. Anim. Hlth. Prod. 29: 235-239.

58. Rao, T.V.S., Malik, P., Nandi, S. and Negi, B.S. 1998. Acta Virol. 41:345-348.

59. Rao, T.V.S., Negi, B.S. and Bansal, M.P. (1997). Indian J. Anim. Sci. 67: 642-645.

60. Roger, F., Yesus, M.G., Libeau, G., Diallo, A., Yigezu, L.M. and Yilma, T. 2001. Vet. Microbiol. 152(3) :265-268.

61. Saha, G.R., Nayak, N.C. and Bhowmik, M.K. 1991. Indian J. Anim. Res. 25 : 1-4.

62. Saha, G.R., Roy, S.S., Mukherjee, N., Mitra, K. 1985. Indian J. Anim. Hlth. 24 : 29-34.

63. Sapre, S.N. 1964. Indian Vet. Rev. 15:69-71.

64. Sarkar, J. Sreenivasa, B.P., Singh, R.P., Dhar,P., and Bandyopadhyay, S.K. 2004. Vaccine. 21 :4728-4735.

65. Shaila, M.S., Puroshottam, V., Venugopal, K. and Venkatasan, R.A. 1989. Vet. Rec. 9: 602-604.

66. Shaila, M.S., Puroshottam, V., Venugopal, K. and Venkatasan, R.A. 1990. Indian Veterinary J. 67: 385-386.

67. Sharma, M.M., Uppal, P.K., Lonkar, P.S. and Mathur, P.B. 1986. Indian J. Anim. Sci. 56: 1183-1186.

68. Sharma, S.N. and Dhanda, M.R. 1971. Indian J. Anim. Sci. 41 : 864-867.

69. Soman, J.P., Singh, R.P. and Jha, G.J. 1985. Indian Vet. J. 62:907-909.

70. Srinivas, R.P. and Gopal, T. 1996. Livestock Advisor. XXI: 22-26.

71. Thompson, J.J. and Liardet, D.M. 1966. New Zealand Vet. J. 14 : 126-129.

72. Voller, A. and Bidwell, D. 1986. ELISA. In : Manual of Clinical Laboratory Immunology. Rose, N.R., Friedman, H. and Fahey, J.L. (Eds.), 3rd edn., American Soc. Microbiol, Washington, D.C. pp 99-109.

73. Wilson, A.A., Chitra, S. and Palaniswamy, K.S. 2000. Intas Polivet. 1:209-212.

74. Zymlyakova, V.P. 1974. Veterinariya. 11:71-73.

–ΔΔΔ–

13

Mycobacterium Subspecies Paratuberculosis Infection in Ruminants in India Diagnosis and Measures

S.V. SINGH
Division of Animal Health
Central Institute for Research on Goats, Makhdoom
PO - FARAH, Mathura (U.P.)

Johne's disease (Paratuberculosis) is a chronic granulomatous enteritis primarily infecting ruminants, caused by *Mycobacterium avium* subspecies *paratuberculosis* (Map), emerged as number one killer disease and is endemic in most of the countries. Status of disease with respect to national level prevalence, production losses, geno-types present, diagnostic tests, control measures, prevalence in other animals and human beings is not known in most of the developing and less developed countries. In India information is limited in the huge ruminant population of nearly 500 million domestic ruminants, except some isolated studies, in buffaloes[75,61],cattle[43], sheep[20] and goats[70, 50]. Though majority of Indian livestock are poor in health and production. Much talked claim of being number one in milk production is number driven and per animals productivity (milk or meat) is far below world and Asian averages. This is mainly due to in-attention and low priority JD received in last 50 years for control and eradication. Presently country lacks diagnostic kits and reagents to screen large domestic and wild ruminant population. Johnin used for years to screen livestock in India is also not available at present. JD is in the list B of OIE and in the wake of WTO regime, it invites trade related barriers and certification of breeding animals is mandatory. Map escapes pasteurization temperature[42], therefore, continuously enters into human food chain and potentially infects human population (Crohn's disease).

The aim of the present paper is to review the status of information on different aspects of Johne's disease in caprines (CJD) and ovines (OJD) with especial reference to diagnosis and control measures in India..

ECONOMIC LOSSES IN INDIA

Though JD is main cause of low productivity disease has not been investigated properly in India and economic losses never been estimated or realized. Precious genetic material is wasted and large numbers of unproductive cattle roam on roads and low producing buffaloes find their way to slaughterhouses, prematurely. In heavily infected herds, this leads to a high rate of wastage in cattle in the prime productive life at 2 to 4 years of age. It has been recognized as serious health problem of goats and sheep located at central and state government farms[35, 34, 18, 14, 64, 44, 22, 30, 66, 48, 61]. It causes high mortality and morbidity in young and adult goats[1, 70,49, 50]. Economic losses are significant through lost productivity, infertility and susceptibility to other disease, lost export markets and direct expenditure on diagnosis and control [50,41,19]. Prevalence of disease was significantly higher in females as compared to male goats. Peri-natal and lactation stresses were the predisposing factors and in 3-4 months after parturition there was 3 fold increased incidence of JD in young goats.

JD affected goats showed marked reproductive in-efficiency. Of 193 JD positive goats, 8 breedable does (57 Barbari and 30 Jamunapari) suffered from reproductive in-efficiency, but only 38 females became sub-fertile or infertile, in 2-4 breeding season[52].

HOST RANGE

Map infects wide variety of ruminant species, like domestic livestock (cattle, sheep, goats, buffalo, camel), free ranging wildlife (deer, elk, antelopes, blue bulls, bison, llamas, bighorn sheep and alpacas) [19,24] and non-ruminants (rabbits, fox, weasels, horses, pigs, chickens and mice) [9,6]. Both natural and experimental infections have been documented. Wild life also act as reservoir for Map[9].

CLINICAL DISEASE

Small Ruminants

JD has a protracted incubation period and clinical disease is the terminal stage of this process. Onset of clinical disease is unpredictable, but may be precipitated by parturition, lactation or by stressful events (Nutritional, transport, pregnancy, lactation, concurrent infections etc.). In all ruminants, clinical disease is characterized by a progressive, a-febrile weight loss that leads to emaciation and poor coat quality (roughness, loss of pigmentation, alopecia, 'wool-slip'), despite maintaining a good appetite. Drop in milk yield has been reported in the lactation preceding that in which clinical disease became apparent. A major feature of illness is chronic, intractable diarrhea, although in some animals this may be intermittent. In contrast to cattle, diarrhea is not a feature in small ruminants. Decreased serum concentrations of calcium, total protein and albumin have been reported in both cattle and sheep with clinical JD[32, 13]. Though JD is always chronic in nature but in some instances, disease progresses relatively rapidly with the interval between the appearance of wasting and death measured in months. In other cases, after the initial loss of condition, there may be no clinical deterioration for long periods. First signs of clinical disease are progressive weight loss and drop in milk production and most farmers may cull the animal without further diagnosis. For this reason the actual number of clinical cases of JD is greatly underestimated[2].

Transmission

It occurs by fecal oral route, but milk is an important source of transmission of disease in suckling animals. Maps were isolated from milk of clinically infected goat with Johne's disease[54]. This is the first report of live cultivable Map from goat milk in India. Milk and colostrums are important source of infection to young animals.

Susceptibility to Johne's Disease with Respect to Species, Breed, Sex, Age, Management, Climate

Significantly higher incidence of JD ($P<0.01$) in Barbari (13.15%) than in Jamunapari (5.58%) goat breed was reported[53]. Incidence of disease was significantly different ($P>0.01$) in both the breeds. Few goats (0.37%) were infected with JD below 1 year of age. High percent of animals were infected with the disease in the age group of 1-3 years (26.7 - 29.5%) and above 5 years (37.5%) in Barbari herds. High percentage (10.9-11.3%) of disease cases in Jamunapari goats were in 2-6 years of age group and very low percentage in age group of 0-1 years. Female had significantly higher incidence ($P> 0.01$) of disease in compared to males, showing appreciable potential to become clinical shedders. Muzaffarnagri sheep flock had quit a few cases of JD as compared to goat herds, despite sharing common grazing area, watering sources and closely situated to one of the goatherds. Significant effect of parturition and lactation was reported on the precipitation of JD and showed that 80.0% of parturient animals precipitated in disease and became shedders within 6 months[53] (optimum 1-2 month). In Jamunapari 53.0% of parturited goats became clinical shedder within 6 months, when they were in lactation stress, which might have been accentuated due to nutritional inadequacy. Incidence of JD in 30 days to 10 months age was very low. Precipitation of disease was inhibited during pregnancy. Incidence of JD was directly proportional to the herd size. Small herd sizes with extensive system of management were apparently useful to reduce the incidence of JD in goats. Timely culling of weak and unthrifty kids by farmer's reduced the burden of JD in herds under village conditions. Low incidence of JD has been reported in goatherds located in arid regions as compared to semi-arid regions[53].

PREVALENCE

Goatherds

Map has been recorded in all sheep and goats rearing states of the country [31, 65, 64,70,22]. Significantly high prevalence in goatherds has been reported [29, 48]. Organized farms continue to suffer heavy losses due to high morbidity, mortality and premature culling caused by paratuberculosis[48]. High incidence has been reported from farmer's herds. Under intensive management system, control of JD is critical to the economic viability of these farms. 13.5% sero-prevalence reported in Pashmina goats[8]. 11.3% prevalence of JD reported from the northern belt of country by screening 30.0% fecal samples (459) from 12 breeding farms located in UP, HP, Orissa and west Bengal on comparison of 1536 goats[52]. The incidence of JD was ranging from 4.1-44.4% on the basis of fecal examination in different goat breed in different climatic and rearing practices [52,70]. It was reported 18.9% and 2.5% JD incidence in goat's herds located in semi-arid and arid regions respectively and isolated Map from 139 of 1005 (37.4%) by culture[50]. Post mortem based study on the status of paratuberculosis in goats, showed overall incidence of 3.5% in goats which was lower than the earlier reports based on the retrospective study on the necropsy data (8.6%) and fecal cultures (18.9% and 2.5%) at two separate farms[50].

Goat Milk

Goat milk is convenient clinical material for the screening of lactating goats. 69.8%, 54.7% and 48.6% incidence by milk culture, ELISA and IS 900 PCR, respectively in 53 lactating goats of different breeds from the organized herds was reported[56]. Overall incidence of apparently weak goats (Jamunapari) was 69.2% and 61.5% in the culture of milk samples and milk ELISA test respectively. In organized goatherds breed-wise incidence of Map was 66.6% 75.0% 33.3% 80.0% in milk culture and 66.6% 50.0%, 33.3% and 20.0% in milk ELISA in Sirohi, Marwari, Jakhrana and Barbari goats breeds respectively. Incidence of Map was 71.1% in milk samples of lactating goats in organized herd, whereas in farmers lactating goat herds 58.2%[17].

Young Kids

On the basis of screening of fecal couture and direct microscopy, 56.0% and 38.0% kids, respectively were infected with Map. Screening of slaughtered kids (farmer's herds in Agra region) by culture of tissues (MLN and intestine), 59.3% kids positive (18.1% from organized herds and 67.3% from farmer's herds). Independently 38.3% and 43.6% kids were positive in MLN and intestines culture, respectively. Using ELISA kit 58.6% and 35.4% kids were found positive from organized and farmer's herds, respectively[59].

Sheep Flocks

On the basis of the pathology 15.0% incidence of the paratuberculosis and suggested endemic nature of the disease was recorded[36]. Some workers outside India have reported the breed susceptibility. High incidence of the disease in sheep above 2-3 years of age was reported[36]. Typical Map colonies were isolated from 3.7% sheep subjected to screening by fecal culture from organized sheep flock in Mathura[50].

Cultural Characteristics

Map isolates were cultivated and maintained on solid HEY medium. All the typical colonies (goat or sheep origin), characterized as that of Map were slow growing and strictly Mycobactin J dependent (studied up to 25th passage levels). Isolated 37.4% typical Map colonies and 62.6% colonies were of fast growing Mycobacteria from culture of 1005 fecal samples of goats[50] and adoption of Map bacilli in the liquid medium[10]. Map isolates V15, U 46,V29, and V113 obtained from fecal samples of terminally sick goats from organized herds due to Johne's disease on HEY medium with mycobactin J and 0.9% HPC as decontaminant While X 15 isolate was from goat tissues[52]. Colonies were small, pinhead sized, translucent and glistening with entire margin and central nipple. Mycobactin J dependency was total. The isolates did not grow in liquid medium.

Large number of orange, round, dome shaped, hemispherical and non-mycobactin J dependent colonies obtained from fecal samples sheep and goats belonging to organized herds in Mathura district. These atypical Mycobacteria have since been characterized as *M. asiaticum* from Spain[20].

Biochemical Characterization of Map

Though Map is extremely slow growing and visible typical Map colonies did not appear before 8 weeks in primary cultures. Size of colonies depended on number of colonies per culture. Biologically Map is almost non-reactive. Characterized Map isolates biochemically were (catalase, slow positive at RT and 60^0 C; Nitrate negative; Tween 80, negative; Aryl sulphate, negative; and P, Nitrobenzoic acid test, negative) [50] and close to *M. avium group* in biochemical characteristics[20].

Lipids of Map

Multiple tests are required to classify a single mycobacterium at sub-species level. Mycobacterial cell wall is rich in lipids and has made them attractive molecules for chemo-systematic studies. High performance liquid chromatography was used to identify specific lipid marker for diagnostic and taxonomic purposes. Among various lipids, mycolic acid patterns are of stable taxonomic characters It was reported the lipid patterns of 18 wild type clinical Map isolates of goat and sheep origin, and compared with Map (Tepes) strain used for commercial Johnin production. Fourteen of these Map isolates had typical pattern of Alpha and Keto mycolic acids in the ratio of 30:70, which resembled the lipid patterns of *Mycobacterium avium*, therefore, providing evidence of their identity as Map. Additional band of Methoxy mycolic acid was present in the Map (TEPES) strain and 3 other isolates, suggested major differences in their identity as Map. PDIM pattern of these isolates was also determined[53].

RFLP Studies

RFLP analysis using *M. avium* probes and dot blot hybridization using IS 900 based probes the Map strains were characterized. Map strains were related to *M.avium* as shown by RFLP analysis and of the 18 isolates only 4 could hybridized with IS 900 probes using Dot Blot hybridization[52]. First time wild type Map isolates were characterized at molecular level in India.

Mycobacterial Antigens

Using counter immunoelectrophoresis CIE[16] reported 35 and 15 precipitin bands respectively with rabbit and cattle hyperimmune sera against *M. paratuberculosis.* On crossed immunoelectrophoresis sonicate antigen of *M. paratuberculosis* with hyperimmune goat sera showed more than 18 immunogenic precipitin lines suggestive of their immunodominant nature[8].

Antigen analysis of 8 well characterized Map strains by SDS-PAGE and western blot[52]. The 86 and 193 kDa protein bands were common in all the 8 strains. While characteristic 42 and 45 kDa proteins were immuno-dominant and were present only in V15 Map isolate. Western blot analysis showed that 193-kDa band was highly cross-reactive. Besides, few minor bands (14.2 and 29 kDa) were also revealed by V 15 hyper-immune serum. The 45-kDa proteins were seen as immuno-dominant antigen in V15 and were not shared by any other Map strain.

Genotyping of Map by PCR-RE Analysis

The Map isolates of goat and sheep origin[20], were first time genotype as 'Bison Type' by using IS 1311 PCR-REA analysis[74]. Genotypic differences have been used to characterized isolates of *M. paratuberculosis.* The IS-1311 polymerase chain reaction-restriction endonuclease analysis (PCR-REA) is a genetic typing technique that distinguishes between C (Cattle), S (Sheep) and B (Bison type strain of paratuberculosis) Map strain S 5 isolated from clinical cases of paratuberculosis in organized goatherds and sheep flocks[40,74] and in cattle and buffalo population[57] of North India were genotyped as 'Bison type'. This is the first report on the incidence of B type strains in species other than Bison. Multiplex pulsed field gel electrophoresis (PFGE) and restricted fragment length polymorphism analysis results indicated that they are not the same strains. These results suggest that there are genetic differences between Map isolates related to geographic and host factors that can help in potential for the epidemiological tracing of new paratuberculosis isolates. Map 'Bison type' isolates were shown to be shared by goats, cattle, sheep, buffalo and blue bull population of North India[57]. This is the first report of 'Bison type' Map bacilli infecting many livestock species and causing frank JD lesions in these animals. The strain similarity also suggested that cattle and goats should not be allow to share grazing pasture as they infect each other[67].

Diagnosis

Diagnosis of asymptomatic shedder and sub-clinically infected animals is difficult. Country lacks diagnostic kits and reagents for cattle, buffaloes, goats and sheep and is dependent on imported diagnostic kits. Lack of expertise and specialized laboratories on this disease is big constraint. Sub-clinically infected animals and asymptomatic shedder animals are more difficult to identify. No single test can identify all JD infected animals in a herd. Culture is considered the 'Gold standard' test

Clinical History

Clinical history is important to suspect the problem of JD in herds. It was reported that clinical shedders were emaciated usually accompanied by sub-mandibular edema, which had a tendency to disappear as diarrhea developed[53]. Diarrhea was either intermittent or continuous with marked tendency to improve in the pregnant goats and reappeared in severe form soon after parturition. Rapid loss in body weight (3.30 -5.39 kg) was seen in later stages of disease or before the disease was clinically confirmed. Appetite, body temperature, breathing and pulse appeared normal till the end when goats was unable to stand or was in recumbence. In 24.5% cases, there was no appearance of diarrhea and only goats suffered from general weakness and loss in body condition. Diarrhea was not always present in goats and also not severe although loose and soft feces were generally present.

Direct Tests

Symptomatic animals may shed the Map in their feces and can be diagnosed by smears and culture, (to detect AFB), PCR tests and gross and histo-pathological lesions. Serology and tests of CMI are useful in some animals, but immunity can diminish in the later stages of disease.

Direct Microscopy (DM)/Fecal/Tissue Examination

Microscopic examination of smears of fecal samples (clinical case) and tissues or sections from ileum and mesenteric lymph nodes at necropsy is oldest procedure for diagnosis of JD [34, 23, 33, 46]. The test is still useful and relevant under the field conditions and in the absence of other diagnostic tests. DM of fecal samples using sediments of decontaminated samples, which was superior to centrifugation methods[52]. DM examination of fecal samples of goats in organized herds revealed 46.4% goats positive for acid-fast bacilli (AFB) [10].

Fecal Culture

Traditional culture of Map from feces as the disease progresses is the most sensitive test available and has 100 % specificity. Though used since 1989[70] for the screening of fecal samples in goats, It was standardized culture of Map from feces, tissues, milk, blood etc., by centrifugation, decontamination (0.9% HPC), and inoculation on HEYM with and without mycobactin J[52] and reported high sensitivity using milk[50], Human[45]. Of the 4 different tests (DM, culture, IS 900 PCR, ELISA), used on regular basis, culture was the most sensitive method under any stage of disease. Though, the sensitivity of culture method depends on the phase of infection and type of clinical sample used[61]. Isolated Map colonies from 18.31% fecal samples of young goats (142) from organized herds and Of the 74 tissue samples, Map colonies were isolated from 36.99% goats[11]. Incidence of Map in necropsied animals (goat and sheep) of organized herds/flocks was 90.9% (40/44), whereas, in farmer's herds/flocks was 76.4% (13/17) [72].

Small amount of moisture was detrimental to the growth of Map colonies on HEY medium. Problems are encountered in the examination of faecal samples as the detection limit is around 10^4 organisms/gm of clinical sample and because inhibitors of the PCR naturally occur in feces further limit test sensitivity[73]. Test is time consuming and therefore expensive. Stool / fecal culture is usually diagnostic in clinical cases, but is less reliable for asymptomatic animals. The typical Map colonies were identified on the basis of mycobactin J dependency, acid fastness, slow growth and cultural characteristics.

Nucleic Acid Probes

DNA probe test are also standardized for the specific diagnosis of infected animals. Nucleic acid probes used to characterize the Map isolated of goat and sheep origin[52]. Sensitivity and specificity of DNA probe test was 33.5% and 100% respectively for sub clinical case of paratuberculosis was reported[63].

Polymerase Chain Reaction

Specific IS 900 PCR had been employed to quickly identified Map DNA in the clinical samples like milk tissues, feces etc. Insertion element IS 900 is considered unique to Map and IS 900 based PCR is used for the unique[69]. Map diagnosis. Test can be completed within a day as opposite to several months for culture. High specificity can be achieved in PCR since no sample would be lost through fungal overgrowth. The sensitivity depends on the quality of reagents, stage of infection, level of infection etc. IS 900 PCR has been standardized in India, to screen fecal, tissue, milk, and Map cultures, from goats, sheep, cattle, buffaloes and blue bulls [17,75,43,19].

The procedure adopted for isolating RNA and DNA from cultures of Map (feces and tissues) was after passing these cultures through French press[52, 5]. 60.0% DNA from organized herds and 100.0% of the farmer's herd were amplified from the MLN [72], whereas 20% DNA from the organized herds and 100% of the farmer's herds were amplified from the intestine from the decontaminated tissue pellet. Tripathi et al, (2005) [68] compared tissue PCR to other test and reported that ELISA was most sensitive serological test among clinical test and PCR was most sensitive among postmortem test. PCR tests require skilled technical input and specialized equipment and are therefore, expensive to perform.

HISTO-PATHOLOGICAL ANALYSIS

Gross and Histo-Pathological Lesions

Chronic productive (focal or diffused) pathological changes have been described in goats. In early infections, no visible lesions except enlargement and hyperplasia of lymphoid follicles in mesenteric lymph nodes. In less severe cases intestinal mucosa was granular and congested. In advance stage among goats, emaciation set in and are devoid of fat deposits with enlargement of mesenteric lymph nodes [21, 52, 11, 48, 34, 33, 18, 25, 29]. The presence of the multiple granulomas in liver as observed[36], suggests the hematogenous spread of the disease. Histo-pathological details of the ileum and the corresponding MLN from 10 naturally infected goats was studied[27]. They found ileum lost epithelial cells and the infiltration of epitheloid cells, macrophages and low number of lymphocytes, plasma cells and occasional eosinophils were seen. Macrophages were fixed in the process of engulfing bacteria or contained bacteria in phagosomes or phagolysosomes. Larger phagolysosomes were common. The MLNs were infiltrated with epitheloid cells and macrophages. Macroscopic changes due to JD were confined to the intestine near ileo-caecal valve region and mesenteric lymph nodes in goats. Calcified foci in MLN of infected goats were demonstrated occasionally. While, changes in small intestinal mucosa were few and were limited to denudation of mucosal lining. The intestinal wall was thickened. However, in sections of macroscopically normal intestine marked epithelial cell infiltration and abundant acid-fast bacilli were not un-common. In some cases productive inflammation with tubercle formation was seen in lymph nodes in goats. In general, the lesions of paratuberculosis remained confined to the intestine near ileo-caecal vale and its regional lymph nodes, as reported by other workers[34]. Productive inflammation in the form of tubercles formation was also demonstrated in lymph nodes of infected goats. Unlike cattle intestinal changes in goats were few. Only moderate thickening of mucosa was seen, while corrugations were rare.

INDIRECT TESTS (IMMUNOLOGICAL TEST)

Cutaneous Skin Testing

Cellular immunity tests can be performed in vivo or in vitro, using respectively the intradermal test or the gamma-interferon assay. Both tests are performed using avian purified protein derivative (PPD) or Johnin (the paratuberculosis equivalent) to stimulate the response. In the intradermal test, skin thickness is measured before and 72 hours after injection with PPD. An increase in thickness of more than 2 mm is regarded as a positive reaction. Johnin production has been discontinued in the country and is not available for use. Skin test (DHR) though not reliable has been used frequently in the screening of animals[23]. Therefore, it cannot be used as sole test for culling [29,30]. Use of intravenous Johnin causes febrile reaction in sensitized goats [64,47].

Stimulating aliquots of heparinised blood with PPD performs the gamma-interferon assay and the subsequent release of gamma-interferon is measured in an ELISA.

Humoral Immune Response

Antibodies usually develop 10 to 17 months after infection; titers tend to increase as the intestinal lesions become more extensive but can decrease in late stage disease[37]. Serology is unreliable for asymptomatic cases. Serologic tests for diagnosis of JD such as agar gel immuno-diffusion, enzyme-linked immunosorbent assay (ELISA), and complement fixation are relatively easy to perform but suffer from a lack of specificity. Additional serological tests used in the diagnosis JD in goats being FAT, Agar gel precipitation test (AGP) and counter immuno-electrophoresis [29,28]. Serological methods are more sensitive and specific for the detection of paratuberculosis infection in comparison to methods used for detection of bacteria such as fecal PCR and fecal culture. Absorbed ELISA is most widely used and accepted by OIE.

Complement Fixation Test

CF test works well on clinically suspect animals, but does not have sufficient specificity to enable its use in the general population for control purposes. Despite the fact that the outcome of the test gives no guarantee about the status of the individual animal, the CFT test is often the only test demanded for import / export of cattle within Europe.

Agar-gel-immuno-diffusion Test (AGID)

The AGID test is useful for the confirmation of the disease in clinically suspect cattle, sheep and goats. The relatively low cost of the AGID make this test a very attractive test for use in small ruminants. Its sensitivity is lower than ELISA test and, as in the former, is closely related to the immuno-pathological forms of the infection. The relatively low cost of the AGID make this test a very attractive test for use in small ruminants. 4 protoplasmic soluble antigens were compared (S3 - typical Map of goat origin, V95 - typical Map of sheep origin, V30 - atypical *M. asiaticum* strain and Map 'Bovine' origin purified antigen from USA). US antigen has slight edge to V95 antigen in AGID test in goats and V95 antigen was slightly superior to US antigen in sheep[20].

ELISA

The convenience of sample collection, rapid laboratory turnaround time and low cost has make an ELISA the preferred test for herd analysis [3, 38]. In general, the detection of serum antibodies against *Map* is achieved using different crude fractions of the bacterium; lipo-arabinomannan (LAM) or PPDs. The specificity of the test is increased in the so-called absorbed ELISA by absorption with sonicates from other (environmental) Mycobacteria, like *M. phlei* or *M. vaccae*. Since the commercial absorbed ELISA has become available, it has become the most widely used serological test for paratuberculosis in cattle. Another approach to avoid the occurrence of ELISA false positive results would be the use of *Map* specific recombinant antigens.

Indigenous plate ELISA kit has been standardized for the screening of serum and milk samples of ruminants against JD[49]. Soluble protoplasmic antigen from the novel 'Bison Type' MAP strain of goat origin has been used in the kit. Present ELISA kit has 40.75% sensitivity and 78.6%, specificity[1]. ELISA test was standardized for Map infection in goats using sonicated soluble protoplasmic antigen @ 1 micro gram / ml and the serum dilution was 1:50[52]. Plate ELISA test detected all the experimentally infected kids as positive. Earliest kids were detected in ELISA at 60 DPI. In clinically positive and naturally infected goats the ELISA using species-specific antigen was 91.3% sensitive. While 15.62% goats from slaughterhouse were found positive by ELISA and showed 76.5%, 90.0%, 69.89% and 77.47% similarity in Map-GGH1, TEPES *M. avium* and *M. tuberculosis* strains, respectively using anti-sera against Map V15 strain[47]. At present country lacks the diagnostic kits (plate ELISA kit), and reagents (Johnin), for the detection of Johne's disease in livestock and human beings. This ELISA kit developed using Map 'Bison type' of goat origin at the rate of 0.1 micro gm per well, for the screening of goats, sheep, cattle, buffaloes and human beings. These kits have been extensively evaluated in animals and human population at CIRG, and has been standardized and released for use in the country.

Control

Optimizing the overall economic benefit from a paratuberculosis control program depends on many factors including, but not limited to: within-herd prevalence, cost of the disease, cost of the diagnostic test, accuracy of the diagnostic test, actions taken by the producer based on test results, level of sanitation (kid rearing hygiene in particular) on the farm, the costs of attaining a more stringent level of herd hygiene, and the economic state of the goat (i.e. milk price, cull goat price, cost of herd replacements, etc.). Integration of these factors to define the economically optimal paratuberculosis control program for producers is complex, especially given the diversity in production, animal health and management styles among herds. Decision analysis models clarify the decision making process by comparing alternative courses of action in a structured and efficient manner while providing a means of measuring the net economic outcome of each decision. Control of paratuberculosis requires first and foremost effective herd biosecurity management protocols and possibly the use of diagnostic tests[12]. Control strategies that require herd-wide testing is a significant financial burden to dairy producers: culling (slaughter of test-positive animals) and herd management changes to limit infection transmission add even more costs. In fact, the efficacy of test-and-cull strategies to control paratuberculosis has been questioned. Data from case histories, and simulation models have led experts to conclude that changing herd management is more effective than test-and-cull programs for paratuberculosis control[3]. However, modeling efforts to date have focused solely on the cost–benefit of testing without rigorously scrutinizing the cost benefit of herd management changes. Optimizing the overall economic benefit from a paratuberculosis control. Although owners of heavily Map-infected herds recognize increased cull rates as one of the major economic burdens associated with the disease, it is difficult to estimate the cost. Increased cull rate is one of the unapparent costs associated with premature involuntary culling (IC). Lost opportunity cost due to premature involuntary culling results in lost future income that decreases profit by: increasing replacement costs (more replacements must be raised); increasing herd health costs (more goats with health problems) due to the inability of the producer to selectively cull for other diseases; decreasing the average age of goat in the herd, which decreases the proportion of goat that are at higher milk production; and, decreasing opportunities for culling low producing goat, which also decreases average production per goat in the herd. Some have estimated that the decreased lifetime production of goats due to premature culling and slowed genetic improvement of a herd due to involuntary culling of genetically valuable animals is high per clinical case of paratuberculosis due to unrealized future income[15]. The disease cost (DC) incurred while the infected cow is in the herd included decreased milk production, decreased fertility, increased mortality (replacement cost plus lost slaughter value), and increased premature involuntary culling as defined by the following equation

Test and cull policy has been employed to reduce the Map infection in the goatherds located at CIRG and other parts of country for past 25 years, leading to huge loss of valuable genetic material, has not yielded the desired results. At best this could help to keep the infection within control, if applied rigorously, which is not usually possible since screening of herds involves lot of inputs[1].

Vaccine

Johne's disease being chronic in nature and bacilli being very resistant to environment, countries like US could not reduce or control the JD infection in their cattle, despite large-scale screening and elimination of the positive reactors. Workers have injected living cultures of Map suspended in mineral oil into cattle and have observed that allergy is produced, and in some cases, at least, a certain resistance to infection. Subsequently Countries like Spain, Norway, Australia, New Zealand, have already adopted vaccination to control and eradicate disease from their countries [26, 39, 7]. Using Map 'Bison type' (S 5), isolate an indigenous inactivated vaccine has been developed, standardized and used in sheep and goats and yielded very good protection from development of JD and also re-sorption of the lesions in the already infected or clinical cases of paratuberculosis. Vaccine yielded good CMI and humoral response under controlled experiment and subsequent challenge by highly pathogenic strain of Map. Vaccine also gave very good response in infected and advance cases of naturally infected animals.

References

1. Anon. 1980 -1995. Paratuberculosis, (Johne's disease). In: Annual Reports, Indian Veterinary Research Institute, Bareilly, India.
2. Chiodini R J, Van Kruiningen HJ, Merkal RS, Thayer W and Coutu JA. 1984. Characteristics of an unclassified *Mycobacterium* species isolated from patients with Crohn's disease. *Journal of Clinical Microbiology* 20: 966-71.
3. Collins M and Morgan IR. 1991. Economic decision analysis model of a paratuberculosis test and cull program. *Journal of the American Veterinary Medical Association* 199: 1724–29.
4. Collins M T, Wells S J, Petrini K R, Collins J E, Schultz R D and Whitlock R H. 2005. Evaluation of five-antibody detection tests for diagnosis of bovine paratuberculosis. *Clinical and Diagnostic Laboratory Immunology* 12: 685-92.
5. Cox R A and Katoch VM. 1986. Evidence for genetic diversity in ribosomal RNA genes in Mycobacteria. *FEBS letters* 95: 1-2.
6. Daniels M J, Hutchings M R, Beard P M, Henderson D, Greig A, Stevenson K and Sharp J M. 2003. Do non-ruminant wildlife pose a risk of paratuberculosis to domestic livestock and vice versa in Scotland? *Journal of Wildlife Disease* 39: 10–15.
7. Eppleston J, Reddacliff L A, Windsor P A, Whittington R J and Jones S.2005. Efficacy of a killed Mycobacterium paratuberculosis vaccine for the control of OJD in Australian sheep flocks. In: Neilsen, S.S. (eds) 8Th ICP. Copenhagen, Denmark. 2005. p 187-194.
8. Goswami T K, Joardar S N and Ram G C. 2000. Antigenic analysis of *M. paratuberculosis* strain TEPS by SDS-PAGE and two-dimensional immuno-electrophoresis. *Indian Journal of Animal Science* 70: 789-91.
9. Greig A, Stevenson K, Henderson D, Perez V, Hughes V, Pavlik I, Hines M E, McKendrick I and Sharp J M. 1999. Epidemiological study of paratuberculosis in wild rabbits in Scotland. *Journal of Clinical Microbiology* 37:1746-51.

10. Gupta VK, Singh N, Singh SV and Shankar H. 1997. Adoption of M. *paratuberculosis* (isolated from goat and sheep) on to liquid medium from solid medium. Indian *Veterinary. Medicine Journal* 21: 14-17.

11. Hajra S. 2003. Pathobiology of spontaneous and experimental paratuberculosis in goats with special reference to early lesions. M.V.Sc., Thesis submitted to CSA. Univ.of Agric. & Tech., Kanpur, UP.

12. Harris N and Barletta RG. 2001. Mycobacterium avium subsp. paratuberculosis in veterinary medicine. *Clinical Microbiology Reviews* 14 :489–512

13. Jones D and Kay J.M. 1996. Serum biochemistry and the diagnosis of Johne's disease (paratuberculosis) in sheep. *Veterinary Record* 139, 498-99.

14. Kaul G L, Somvanshi S and Biswas J C. 1989. Mortality pattern in Pashmina goats. *Indian Veterinary Journal* 65: 847-49.

15. Kennedy D J and Benedictus G. 2001. Control of *Mycobacterium avium* subsp. *paratuberculosis* infection in agricultural species. *Revue scientifique et technique (International Office of Epizootics)* 20: 151 – 79.

16. Krishnappa G, Jaganath C and Rao B U. 1989. The specificity of antibody response in experimental and natural bovine paratuberculosis studies by C. I. E. with intermediate gel. *Veterinary Microbiology* 21: 67-78.

17. Kumar P.2004. Prevalence and molecular characterization of *Mycobacterium avium subsp. paratuberculosis,* the cause of juvenile paratuberculosis (Johne's disease) in goat kids in Agra region. M.V.Sc., Thesis submitted to UP Pt.DDU Pashu Chikitsa Vigyan Vishwavidyalay evum Go Anusandhan Sansthan. Mathura.

18. Kumar R, Prasad M C and Paliwal O P. 1998. Paratuberculosis in goats: a retrospective study. *Indian Veterinary Journal* 65: 582-84.

19. Kumar S. 2005: Isolation and characterization of *Mycobacterium avium* subsp. *paratuberculosis,*(cause of Johne's disease in ruminants), in Blue Bull (*Boselaphus tragocamelus*) population by culture and molecular methods (IS 900 PCR). M.Sc., dissertation submitted to Bundelkhand Univ., Jhansi, UP.

20. Kumar V. 2002. Identification of *Mycobacterium avium* subspecies *paratuberculosis* in sheep and goats and prospective evaluation of an ELISA and AGID test for the diagnosis of Johne's disease. M.V.Sc., thesis submitted to CSA University, Kanpur. UP.

21. Kurade N P, Tripathi B N, Rajkumar K and Parihar N S. 2004. Sequential development of histological lesion and their relationship with bacterial isolation, fecal shedding and immune responses during progressive stages of experimental infection lambs with *Mycobacterium avium* subsp *paratuberculosis. Veterinary Pathology* 41: 378-87.

22. Lal Krishna V J, Katoch R C and Gupta V K. 1989. Paratuberculosis (Johne's disease) in goats in Himanchal Pradesh. *Indian Journal of Comparative immunology, microbiology and infectious diseases* 10: 43-45.

23. Lall, J M. 1966. Johne's disease in cattle, sheep and goats. Monograph, ICAR Research Series No.-19, New Delhi.

24. Manning, E JB and Collins M T. 2001. *Mycobacterium johnei*: pathogen, pathogenesis and diagnosis. *Revue scientifique et technique (International Office of Epizootics)* 20 : 133-50.

25. Mohiyuddin S and Malaki A. 1987. Histopathological study of Johne's disease lesions in a natural infected goat. *Indian Veterinary Journal* 44: 743-45.

26. Munoz M, Garcia Marin J F, Garcia-Pariente C, Reyes L E, Verna A, Moreno O, Furetes M, Doce J, Puentes E, Garrido J and Perez V. 2005.Efficacy of a killed vaccine (SILIRUM) in claves challenged with MAP. In: Neilsen, S.S. (eds) 8[Th] ICP. Copenhagen, Denmark. 2005. p 208-217.

27. Paliwal O P and Rehbinder C .1981. Ultrastructural studies of paratuberculosis (Johne's disease) in goats. *Acta veterinaria Scandinavica* 22: 180-88.

28. Paliwal O P, Kumar R and Somvanshi R. 1985. The immune spectrum of *Mycobacterium johnei* infection in goats. *Indian Veterinary Journal* 69: 743-47.

29. Paliwal O P, Rajya BS and Krishna SG .1983. Fluorescent antibody technique (FAT) in the diagnosis of *Mycobacterium johnei* infection in goats. *Indian Veterinary Pathology Journal* 7: 6-14.

30. Paliwal O P. 1984. Pathology of Johne's disease in goats. Pathoanatomical and immunofluoresscent studies. Ph.D. Thesis, Rohilkhand Univ., UP.

31. Patel S M, Jhala V M and Jaiswal T N. 1996. Serum Immunocongaglutinin level in relation to Johne's disease in sheep. Indian *Veterinary Journal* 63: 349-52.

32. Patterson DS, Allen WM, Berrett S, Ivins LN and Sweasey D. 1968. Some biochemical aspects of clinical Johne's disease in cattle. *Research in Veterinary Science* 9: 117.

33. Rajan A, Valsala KV, Mariamma K and Krishnan. 1980. Pathology of endocrinology in Johne's disease in goats. *Kerala Journal of Veterinary Science* 11: 166-73.

34. Rajya B S and Singh C M. 1961. Studies on the pathology of Johne's disease in sheep. III Pathological changes in sheep with naturally occurring infection. *American Journal of Veterinary Respondents* 22: 189-203.

35. Rao KR. 1950. Some observation on investigation of Johne's disease in Mysore. *Indian Journal of Animal Science* 20: 17-26

36. Reddy K P, Sriraman P K, Naidu N R G and Rao P R. 1984. Pathology of Johne's disease in sheep. *Indian Veterinary Journal* 61: 179-184.

37. Rideout BA, Brown ST, Davis WC, Gay J.M, Giannella RA, et al. 2003. Diagnosis and Control of Johne's Disease, Committee on Diagnosis and Control of Johne's Disease, National Research Council, 244 pages.

38. Salgado M, Manning E J B and Collins MT. 2005. Performance of a Johne's disease enzyme linked immunosorbent assay adapted for milk samples from goats. *Journal of Veterinary Diagnostic Investigation* 17: 350-54.

39. Saxegaard F and Fodstad FH .1985. Control of paratuberculosis (Johne's disease) in goats by vaccination. *The Veterinary Record* 116, 439-41

40. Sevilla I, Singh SV, Garrido JM, Aduriz G, Rodriguez S, Geijo MV, Whittington RJ, Saunders V, Whitlock RH and Juste RA. 2005. Molecular typing of *Mycobacterium avium* subspecies *paratuberculosis* strains from different hosts and regions. *Revue scientifique et technique (International Office of Epizootics)* 24: 1061 - 66

41. Sevilla I, Singh SV, Juste RA, Kumar V and Bhatia A K. 2004. Paratuberculosis 'Bison' genotype strains isolated from sheep and goats in India. *Journal of Immunology and immunopathology* 6: 137.

42. Shankar H. 2006. Isolation and characterization of *Mycobacterium avium subsp. paratuberculosis,* in the pasteurized and un-pasteurized milk and dairy products in North India. M.Sc., dissertation submitted to Dr. B.R.Ambedkar Univ., Agra.

43. Sharma G. 2004. Prevalence and molecular characterization of *Mycobacterium avium subsp. paratuberculosis,* the cause of Johne's disease (Paratuberculosis) in cow milk in Agra region. M.Sc., dissertation submitted to Jiwaji Univ., Gwalior, MP.

44. Sharma M M, Lonkar P S, Srivastava C P, Maru A and Dubey S C. 1987. Prevalence and Johne's disease (JD) in sheep and goats. *Indian Journal of Comparative Microbiology Immunology and Infectious Disease* 8: 48-49.

45. Singh AV, Singh SV, Singh PK, Gupta VK and Vihan VS. 2005d. Association of *Mycobacterium avium subsp. paratuberculosis* in the cases of chronic colitis (Crohn's Disease) in human beings. XXIX Ann. Conf.IAMM, Oct, 2005. SRM Coll. & Res. Institute, Chennai. (Abstract MBO – 16), p. 86.

46. Singh N, Singh SV and Vihan VS. 1986.Prevalence of clinical form of Johne's disease in goats. National symposium on immune deficiencies and emerging diseases of animals and birds. 5-7, March, 1986, Bombay

47. Singh N, Vihan VS, and Singh SV.1988.infertility in female goats associated with paratuberculosis. 2nd National Seminar on Sheep and Goat Disease. Sept 1988, CSWRI, Avikanagar Rajasthan, India.

48. Singh N, Vihan VS, Singh SV, and Gupta V K 1992. Prevalence of Johne's disease in organized goat flocks. V International Conference on Goats. 2-8, March 1992, New Delhi. pp. 513.

49. Singh N, Kala S N, Vihan VS and Singh SV. 1990. Genetic study on the susceptibility to Johne's disease in goats. *Indian Journal of Animal Sciences* 60: 1163-65.

50. Singh N, Singh SV, Gupta VK, Sharma VD, Sharma R K and Katoch VM. 1996. Isolation and identification of *Mycobacterium paratuberculosis* from naturally infected goatherds in India. *Indian Journal of Veterinary Pathology* 20 :104-08.

51. Singh N, Vihan VS, Singh SV and Gupta VK. 1992. Prevalence of Johne's disease in organized goat flocks. In: Recent advances in goat production. Post conference proceedings of V International conference on goats, 2-8, March, New Delhi. p. 1813-17.

52. Singh N, Vihan VS, Singh SV and Gupta VK. 1998. Prevalence of Johne's disease in organized goatherds. *Indian Journal of Animal Science* 68: 41-42.

53. Singh SV and Singh N. 2000. Effect of climate on Johne's disease incidence in goatherds. *Indian Journal of Animal Science* 70: 492 - 93.

54. Singh SV and Vihan VS. 2004. Incidence of *Mycobacterium avium* subsp. *paratuberculosis* in clinically suspected small ruminants. *The Indian Journal of Small Ruminants* 9: 56-59.

55. Singh SV, Kumar P, Singh AV, Srivastava AK, Sevilla I, Juste R and Gupta V.K, 2005c. Prevalence and molecular characterization of juvenile form of paratuberculosis in Indian goatherds. In: Neilsen, S.S. (eds) 8Th ICP. Copenhagen, Denmark. 2005. p 61.

56. Singh SV, Kumar S, Singh AV, Sevilla I, Juste RA and Gupta VK. 2005a. Comparative evaluation of culture ELISA and PCR in screening of milk samples for the detection *of Mycobacterium. avium* subsp. *paratuberculosis* infection in lactating goats. In: Neilsen, S.S. (eds) 8Th ICP. Copenhagen, Denmark. 2005. p. 119.

57. Singh SV, Sevilla I, Juste RA, Whittington RJ, Kumar V, Gupta VK, Bhatiya AK and Singh A.V. 2005c. *Mycobacterium avium* subspecies *paratuberculosis* ('Bison type') genotype infecting goat population in India. In: Neilsen, S.S. (eds) 8Th ICP. Copenhagen, Denmark. 2005. p 88.

58. Singh SV, Srivastava AK, Singh AV, Sharma G, Kumar P, Yadav D and Kumar S. 2004. Isolation of *Mycobacterium avium subsp. paratuberculosis* from the tissues samples of hog dear and its molecular characterization. XXII Annual Con. of IAVMI. Oct, 2004, Palampur, HP. P. 66

59. Singh SV, Singh AV, Misra S, Shukla N, Vihan VS and Gupta VK. 2005 b. Incidence of paratuberculosis in ruminant population of India using antigen from novel Map 'Bison type' of goat origin by indigenous plate ELISA kit. (Communicated)

60. Singh SV. 1998. Diagnosis of paratuberculosis in goats. Ph.D. thesis, submitted to CSA. Univ.of Agric. & Tech., Kanpur, UP.

61. Sivakumar P, Singh N and Tripathi BN. 2005. Pathology and diagnosis of paratuberculosis in water buffaloes (*Bubalus bubalis*). In: Neilsen, S.S. (eds) 8Th ICP. Copenhagen, Denmark. 2005. p. 38.

62. Sivakumar P. 2003. Pathobiology of bovine paratuberculosis. M.V.Sc. Thesis. IVRI. Izatnagar.

63. Sockett DC, Conrad T A, Thomas C B and Collins M T. 1992. Evaluation of four serological tests for bovine paratuberculosis. *Journal of Clinical Microbiology* 30: 1134-1139

64. Somvanshi R, Kaul G L and Biswas J C. 1987. Assessment of intravenous Johnin test in diagnosis of paratuberculosis in goats. *Indian Journal of Animal Science* 57: 974-976.

65. Sreenivasulu D, Krishnaswamy S and Jaankarah-Sharma B. 1986. Study on in vitro technique to assess the cell mediated immune response to vaccination against Johne's disease in sheep. *Indian Veterinary Journal* 63: 619-22

66. Srivastava A K and More B K. 1987. Prevalence of Johne's disease in local, angora and crossbreed goat in Maharashtra. *Livestock Advisor* 12: 33-37.

67. Tripathi B N and Stevenson K. 2005. Detection of *Mycobacterium. avium* subsp. *paratuberculosis* in formalin fixed paraffin embedded tissues of goats by IS 900 PCR. In: Neilsen, S.S. (eds) 8Th ICP. Copenhagen, Denmark. 2005. p. 122.

68. Tripathi B N, Sivakumar P, Paliwal O and Singh N. 2005. Tissue PCR and its comparison with other test for the diagnosis of natural paratuberculosis in goats. In: Neilsen, S.S. (eds) 8[Th] ICP. Copenhagen, Denmark. 2005.p. 123.

69. Vary P H, Andersen PR, Green E, Hermon-Talyor J and Mc Fadden JJ. 1990. Use of highly specific DNA probes and the polymerase chain reaction to detect *Mycobacterium paratuberculosis* in Johne's disease. . *Journal of Clinical Microbiology* 28: 933-37.

70. Vihan VS, Singh N and Singh SV. 1989. Preliminary observations on faecal culture test for the detection of *Mycobacterium paratuberculosis* in goats. *Indian Journal of Animal Science* 59: 1522-24

71. Vihan VS, Singh N and Singh SV.1996. Evaluation of combined therapeutic regimens in treatment of natural form of paratuberculosis in Goats. *Indian Journal of Animal science* 59, 1522-24.

72. Vohra J. 2005. Studies on spectrum of *Mycobacterium avium subsp. paratuberculosis,* infection in digestive, reproductive and udder tissues by culture and molecular methods (IS 900 PCR) vis a vis serum ELISA in goats and sheep. M.Sc., dissertation submitted to Allahabad Agricultural Institute Deemed Univ., Allahabad.

73. Whipple D L, Kapke P A and Andersen P R. 1992. Comparisons of a commercial DNA probe test and three cultivation procedures for detection of *Mycobacterium paratuberculosis* in bovine faeces. J. Vet. Deign. Invest. 4: 23-27.

74. Whittington RJ. 2002. Personal communication.

75. Yadav D. 2004. Prevalence and molecular characterization of *Mycobacterium avium subsp. paratuberculosis,* the cause of Johne's disease (Paratuberculosis) in buffaloes in Agra region. M.Sc., dissertation submitted to Dr. B.R.Ambedkar Univ., Agra, UP.

-ΔΔΔ-

14

Caprine and Ovine Abortions of Infectious Nature - An overview

S V S MALIK, SAMIR DAS AND V M VAIDYA

Division of Veterinary Public Health, Indian Veterinary Research Institute, Izatnagar

Abortions in small ruminants, especially when assuming alarming proportions, significantly cut into profit potentials of sheep and goat industry. Moreover, abortions on account of infectious etiologies may also lead to nagging and life-threatening health consequences in these animals and human beings in their contact. Normally, about 1.5 to 2.0% (up to 5%) of the ewes in a flock may abort, however, few such isolated cases can also quickly escalate into an abortion "storm," resulting in 20-30% percent abortions or as high as 80% lamb mortality.

Factors for abortions include following infectious and non-infectious causes

(A) Non-infectious causes

i) Physical : Heat and cold stress, fight, vaccination, fall from height, narrow passage, hunger stress, dissolution of corpus luteum, mating during pregnancy etc.

ii) Toxic : Ingestion of pesticides, castor oil, paints etc.

iii) Psychological : Fear, sudden explosion, sudden lighting etc.

iv) Hormonal : Excess of estrogen, low level of progesterone, excess of oxytocin etc.

v) Nutritional : Mineral and vitamin insufficiency etc.
vi) Genetic : Inbreeding, spontaneous mutations etc.

(B) Infectious causes

There are several infectious agents including bacterial, viral, fungal, protozoal agents that are known to cause abortions in small ruminants. Of the bacteria, *Chlamydophila abortus* (formerly called as *Chlamydia psittaci* serovar 1); also known as agent for Enzootic Abortion in Ewes (EAE), is the major pathogen of pregnancy in sheep and goats world over including the parts of continental Europe. Moreover, *Campylobacter fetus* sub sp. *fetus* (formerly called as Vibrio), *Salmonella* spp., *L. monocytogenes* and *B. melitensis* can also cause severe losses, although the last does not occur in many developed countries. Of the protozoa, *Toxoplasma gondii* is a major cause of abortion in many sheep-producing countries, while *Sarcocystis* species and *Neospora caninum* also have the potential to cause fetal loss. The rickettsia *Coxiella burnetii* probably infects the ovine placenta more often than is recognized, though fetal death is relatively uncommon, while abortion due to tickborne fever is secondary to the febrile response. The most important viral infections associated with ovine and caprine abortions are border disease, bluetongue (BT), rift valley fever (RVF), Wesselsbron disease, and Akabane virus disease[9].

In case of ewes, the most common abortifacient agents are *C. fetus, Ch. abortus, T. gondii, Listeria* spp., *Leptospira* spp., *Brucella ovis, C. burnetii*, border disease virus (BDV), and, RVF virus (RVFV), bluetongue virus (BTV). The major conditions associated with infectious abortion in goats are chlamydophilosis, toxoplasmosis, leptospirosis, brucellosis, Q fever and listeriosis.

An investigation of infectious abortion in sheep and goat for abortifacients including *T. gondii, N. caninum, Ch. abortus* and Border disease virus infections revealed *Ch. abortus* as the most commonly involved agent (sheep 39%, goats 23%), followed by *T. gondii* (sheep 19%, goats 15%) and *C. burnetti* (sheep 1%, goats 10%)[11]. The most important infectious agents identified in goats in North America include *Ch. abortus, C. burnetii, T. gondii* and *L. monocytogenes*[46]. A recent study of infectious causes of ovine abortion on 148 farms in northern Spain during 1999-2003 revealed Border disease in 16% of the flocks, toxoplasmosis in 15%, chlamydiosis in 12%, salmonellosis in 10%, Q fever in 3% and, miscellaneous infections in 7% (*Yersinia* spp., *Listeria* spp., *Brucella* spp.). The inflammatory lesions compatible with an infectious cause were seen in 7%, the non-infectious causes were identified in 1%, while diagnosis could not be ascertained in 38% of the flocks[33].

Of the protozoa, *Toxoplasma gondii* is a major cause of abortion in many sheep-producing countries, while *Sarcocystis* species and *Neospora caninum* also have the potential to cause fetal loss. Of the bacteria, *Chlamydia psittaci* is the major pathogen of pregnancy in the UK, as in parts of continental Europe; *Campylobacter fetus* sub sp. *fetus*, *Salmonella* spp., *L. monocytogenes* and *B. melitensis* can also cause severe losses, although the last does not occur in the UK. The rickettsia *Coxiella burnetii* probably infects the ovine placenta more often than is recognized, though fetal death is relatively uncommon, while abortion due to tickborne fever is secondary to the febrile response. The most important viral infection in the UK is border disease; Wesselsbron disease, bluetongue and Akabane virus disease occur outside the UK[9].

Chlamydial Abortions

1. Chlamydophilosis (Chlamydial Abortion) or Enzootic Abortion of Ewes (EAE)

Etiology: A chlamydial organism called-*Chlamydophila abortus* (formerly called as *Chlamydia psittaci* serovar 1), also known as agent for Enzootic Abortion in Ewes (EAE), which causes abortion during the last month of pregnancy, stillbirths, and weak lambs that soon die.

Global scenario: Ch. abortus is the major pathogen of pregnancy in sheep and goats world over. In many areas, chlamydial abortion is the second cause of infectious abortions after brucellosis. Most of the work on chlamydial infections of small ruminants pertains to ewes, however, chlamydiosis has also an economic and public health impact in numerous goat farms throughout the world. Chlamydial abortions were first identified in Germany in 1959, and subsequently reported from Bulgaria, Spain, USA, France, India, Japan, United Kingdom, Chad, Greece, and Tunisia.

Indian Scenario: In our country, chlamydiosis in sheep has been identified as one of the important emerging diseases. *C. psittaci* has been isolated from endometritis cases (27.3%) among sheep and goats (Katoch *et al.*, 1997). Seropositivity of *C. psittaci* infection has been detected in apparently healthy exotic (7.7%) and crossbred (8.8%) service rams in Hisar, Hariana by ELISA. Chlamydial antibodies had been detected in voluntary blood donors (55.69%) in Delhi by microimmunofluorescence (MIF) assay using specific antigens of *C. trachomatis*, *C. psittaci* and *C. pneumoniae* detected; with a positivity of 12.5%, 0.3%, 42.77%, respectively[43].

Transmission: The agent is transmitted from aborting sheep to susceptible individuals as well as to uninfected sheep, with ewe lambs being most susceptible to abortions. Infected does excreted large numbers of chlamydiae in placenta and fetal fluids at the time of kidding and at the time of abortion. Some goats may shed chlamydiae in vaginal fluids from more than two weeks before abortion to more than two weeks after abortion. This may explain the higher incidence of abortion in newly infected herds of goats, since the susceptibility to infection varies in relation to the physiological status of the animal. Goats that are less than 100 days pregnant are more susceptible than those at the end of gestation or those that are barren. Smaller amounts of chlamydiae can also be shed in urine, milk and feces during several days after abortion. Young goats born from infected mothers may retain the infection in the herd or transmit it to other herds. The survey of a group of 27 yearlings in an infected herd during their first year of life has demonstrated that they can spread the disease by not being detected by their serological response.

Clinical Symptoms: Chlamydiosis is clinically characterized by abortion during the last months of pregnancy, stillbirths or premature births of weak kids with low birth weight. Abortions occur without previous clinical-specific signs even if some goats may develop persistent cough without breathlessness, or arthritis and keratoconjunctivitis. In experimental infections, slight vaginal discharge may be observed the day before abortion in some goats. Retained placentas and metritis are not usual, even if they are more frequent than in ewes[28]. After the abortion, goats may recover rapidly[5] or may present brown discharge from the vagina. In experimental infections[14] or in some natural infections with a high rate of abortions [41], only 50% or less of goats that aborted have been found to recover quickly whereas post abortive sickness in ewes is unusual. This could be due to virulence differences between strains since very little is known on virulence mechanisms of strains. No difference in virulence could be shown in mouse models between ovine and caprine strains[8], but amplified fragment length polymorphism (AFLP) revealed genomic differences between caprine strain AC1 and ovine serotype-1 *C. psittaci* strains[40]. In a newly infected flock the rate of abortion is severe.

Frequently 30% or more, sometimes 90% of pregnant does may abort and milk production may decrease. The high rate of abortion is observed for 2 or 3 years after which the disease takes on a cyclic nature. Every year, about 10% of pregnant females will abort for several years until a new outbreak occurs and then all the yearlings will abort. The high level of immunity produced after abortion is responsible of the cyclic evolution of the disease in the herd. It is exceptional for a goat to abort twice. Some of the ewes that aborted can become chronically infected[35]. No research has been done to determine the incidence of chronic infections in goat herds. Chlamydial antigens and DNA can be detected in the vagina, uterus and uterine tubes during the peri-ovulatory period of ewes that aborted. The fetus does not display specific macroscopic lesion. Kids delivered close to term may be covered by brown material. Clear or blood-stained diffuse edema, blood-stained fluids in abdominal and pleural cavities and petechiae on the tongue, in the buccal cavity and on the hooves are often observed.

Public health consequences: The notable zoonotic diseases caused by *C. psittaci* are ornithosis/ psittacosis associated with avian strains and, abortions resulting from infection of pregnant women with the strain that causes enzootic abortion of ewes (EAE). Sheep are a recognized source of human infection, the most common route of transmission being direct contact with aborted fetuses or infected ewes during lambing time. Human disease occurs usually in the form of an acute generalized infection. In pregnant women the infection can be life threatening. Infection in non-pregnant individuals is less severe and may not always be diagnosed. There is no evidence to suggest that the transmission of EAE strain of *C. psittaci* occurs from human-to-human[27].

Diagnosis: It is usually performed by the detection of bacteria in smears or impression of the placenta combined with serological analysis of at least ten sera samples. Identification of the organism is attempted in the aborted fetus, tissue, or uterine discharge either by staining with Gimenez or Giemsa staining, or molecular tools. The confirmatory diagnosis is done either by isolation of the agent in developing chick embryo or cell-line followed by its molecular fingerprinting.

Prevention: The disease can be prevented by maintaining flock isolation and rigid sanitation during gestation. Abortion epidemics due to EAE can usually be stopped or dramatically reduced by treatment of the entire flock with tetracycline. Vaccination with killed vaccines can reduce the incidence of abortion but not the excretion of Chlamydiae at kidding[39].

Treatment: Tetracyclines, which affect the replication of chlamydiae, can be effective in preventing abortions. Oxytetracycline given @20 mg/kg by intramuscular route at 105 and 120 days of pregnancy can prevent abortion but not the chlamydial shedding at kidding.

Bacterial Abortions

1. Brucellosis

Etiology: The organism belonging to genus-*Brucella* are gram-negative coccobacilli or short rods. Out of six known species of the brucellae (*B. abortus, B. melitensis, B. suis, B. ovis, B. canis,* and *B. neotomae*), *Brucella melitensis* mainly infects sheep and goats, but has also been seen in cattle and dogs. *Brucella abortus* is found in cattle, bison, and water buffalo and occasionally in sheep, goats, and dogs. *B. ovis* can cause epididymitis and infertility in rams. Four of the species namely *B. abortus, B. melitensis, B. suis* and, rarely, *B. canis,* are zoonotic in nature.

Global scenario: Brucellosis occurs worldwide barring most of the developed countries having effective measures for its control. It is a major livestock health problem in Africa, the Middle East, Central and South-East Asia, South America, and some Mediterranean countries.

Indian scenario: A national survey of brucellosis has revealed the overall seroprevalence of the disease in sheep and goats as 7.9 % and 2.2 %, respectively, with varying prevalence of ovine brucellosis in different states, *viz.,* Karnataka (8.5%), Maharashtra (2.4%), Andhra Pradesh (2.3%), Himachal Pradesh (22%), Orissa (6.6%) and Gujarat (6%). The prevalence of brucellosis in goats in different states, *viz.,* Karnataka (4.6%), Maharashtra (2.4%), Andhra Pradesh (2.1%) and Orissa (0.8%) was found to be lower than that seen in case of sheep[17]. Goat samples (foetus, vaginal discharge and serum) have shown predominance positivity for *B. melitensis* biotype[54].

Transmission: Brucella in animals is usually transmitted by contact with the placenta, fetus, fetal fluids, and vaginal discharges from infected animals. Bacteria can also be found in the blood, urine, milk, and semen; and shedding in milk and semen can be prolonged or lifelong. Infection occurs by ingestion and through mucous membranes, broken skin, and possibly intact skin, and even through fomites. The mammary gland can be infected by direct contact. Brucellae are destroyed by several hours of exposure to direct sunlight but can remain viable for several months in water, aborted fetuses, manure, wool, hay, equipment, and clothes, especially under high humidity, low temperatures, and no sunlight.

Clinical signs: Ruminants usually abort only during their first gestation. In sheep and goats, *B. melitensis* can cause abortion, retained placenta, orchitis, and epididymitis. Abortions usually occur late in gestation in sheep and during the fourth month of gestation in goats. In goats, mastitis and lameness may be seen but this arthritis is rare in sheep. *B. ovis* affects sheep but not goats. This organism can cause epididymitis, orchitis, and impaired fertility in rams. Initially, only poor quality semen may be seen; later on, the lesions may be palpable in the epididymis and scrotum leading to permanent atrophy of testes. Systemic signs are rare; abortions, placentitis, and perinatal mortality can be seen but are uncommon; and deaths are rare in adult animals.

Post mortem lesions: At slaughter, signs of infection including bacteria, purulent lesions, and granulomas may be present in the reproductive tract, mammary tissues, or articular tissues, as well as spleen, liver, lymph nodes, kidneys, and lungs. The ruminant fetus may be autolyzed, normal, or have evidence of bronchopneumonia. The cotyledons may be red, yellow, normal, or necrotic. The intercotyledonary region is typically leathery, with a wet appearance and focal thickening. Placentitis, with edema and necrosis of the cotyledons and a thickened and leathery intercotyledonary region, can also be seen in sheep infected with *B. melitensis*. The placenta is usually normal in goats.

Diagnosis: Stained slides of infected materials may be indicative of brucellosis, but definitive diagnosis is made by culture and identification of the organism. In sheep and goats, *B. melitensis* can be diagnosed with buffered Brucella antigen tests (BBAT, also known as the card and plate agglutination tests), complement fixation, ELISAs. The bulk milk ring test is not used in small ruminants. *B. ovis* infections can be diagnosed by ELISA, complement fixation, hemagglutination inhibition, indirect agglutination, and gel diffusion tests. A brucellin allergic skin test is sometimes used to test pigs for *B. suis* or unvaccinated sheep and goats for *B. melitensis*.

Zoonotic aspects: The prevalence of brucellosis in man is higher in countries experiencing high occurrence of disease in sheep and goats.

The disease in humans is mainly caused by *B. melitesis,* which runs an acute or chronic course. *B. melitensis* infection is primarily foodborne and associated with consumption of unpasteurized dairy products. In India, human brucellosis cases have been reported from almost all the states. A high proportion of occupationally exposed personnel in Delhi such as those engaged in veterinary hospitals and slaughter houses (27.7%); and the abattoir personnels (63%) have shown seropositivity for brucellosis. Majority of human cases have been found to be infected with *B. melitensis* though *B. abortus* and *B. canis* have also been isolated.

Prevention and control: Quarantine and slaughter of infected animals is necessary to eradicate the disease. The programmes to slaughter infected cattle and small ruminants have successfully eliminated brucellosis from domestic animals in many countries, however, to maintain the disease-free status seems to be problematic in these countries because of occasional re-introduction of the infection through imported infected livestock and latent carriers of brucellae that have eluded serological testing. Any area exposed to infected animals and their discharges should be thoroughly cleaned and disinfected. The *Brucella* organism can survive for months in the environment, but is destroyed by heat and sun exposure.

Treatment: Heavy doses (1-2g/day) of tetracycline or streptomycin are administered for several weeks to obtain favorable results[53].

2. Campylobacteriosis (formerly known as Vibriosis)

Etiology: caused by *Campylobacter fetus,* subsp. *intestinalis,* is a contagious disease of sheep, which causes abortion in the last month of pregnancy, stillbirths, and birth of weak lambs that die soon. The disease usually starts with one or two abortions in a flock, followed by a rapid increase in abortion rate. *C. fetus* (sub-species *venerealis*) causes a sexually transmitted disease of sheep, cattle, and goats. Its role as an agent of abortion in these animals has considerable economic impact on the livestock industry.

Transmission: Vibrio abortions are usually introduced into a flock via a carrier animal previously has aborted, but carries the infectious bacteria. Ewes infected with typically abort during the last 6 to 8 weeks of pregnancy or give birth to weak or dead lambs.

Clinical symptom: Campylobacters produce varied symptoms in different species of animals. Sheep may suffer with enteritis, or there may be still births (in case of sheep mainly) or abortions (mainly by *C. fetus*).

Diagnosis: Diagnosis is by identification and cultural examination of the organism in the freshly aborted fetus and placental membranes at a diagnostic laboratory.

Prevention and control: A campylobacter vaccine is available that will effectively prevent the disease in most flocks. Ewes should be vaccinated twice the first year in early gestation and again in mid-gestation and once yearly thereafter, usually immediately after the breeding season.

Public health consequences: Abortion in females by *C. fetus* subsp. fetus has been documented. Approximately 99% of all human Campylobacter isolates reported to the Centers for Disease Control from 1982 to 1986 were *C. jejuni,* while less than 1% were *C. fetus* subsp. *fetus*[51].

Treatment: can be done by supplementing the feed with oxytetracycline or chlortetracycline @5mg/lb body weight for 5 to 7 days.

3. Listeriosis

Etiology: Listeriosis is caused by *Listeria* spp. which is widely found in soil, feedstuffs and feaces of healthy animals. Listeriosis can occur sporadically, but it is most commonly associated with feeding of moldy silage or spoiled hay. The causative organism, *Listeria monocytogenes,* can grow and survive in moldy or poorly fermented corn silage. Moldy or spoiled corn silage should never be fed to sheep.

Clinical symptom: This disease can cause abortion in ewes as well as problems of the brain and central nervous system, although these latter two forms of the disease are usually not found in a sheep flock at the same time; mortality can be high; Sheep may also become depressed and disoriented and they may walk in circles with a head tilt and facial paralysis. Listeric infections and abortions usually develop in the late winter or early spring. Abortions are most commonly recognized in the last trimester of pregnancy, frequently in the absence of other clinical signs[21].

Global Scenario: World over, abortions in sheep and goat are usually associated with *L. monocytogenes* & *L. ivanovii,* the latter being implicated as the most frequent cause in case of sheep (McLaughlin, 1987) with abortion rates ranging from 1.65%[3] to 12%[45]. *L. monocytogenes* infection in these species is normally low but may reach as high as 15%[7]. In U.K., listeric abortion is most common in sheep[4] and in exceptional cases approximately 50% of the flock have been affected[25].

Indian Scenario: L. monocytogenes serotype 4 under experimental conditions has been shown to cause abortion and mortality upto 33% in sheep and 3.33% in goats[15]. In Himachal Pradesh, migratory flocks of 216 sheep and 331 goats with history of abortions revealed *L. monocytogenes* from vaginal swabs of 8 sheep and 11 goats whereas *L. ivanovii* could only be isolated from 2 sheep but not from goats, respectively[47].

Clinical Diagnosis: Diagnosis can be made through isolation of the causal agent. The cold enrichment method is a tool in epidemiology investigations.

Public health consequences: About one-third of listeriosis cases happen during pregnancy, which result in spontaneous abortion in the 2nd/3rd trimester (CDC, 2005). It is estimated that 800 cases occur every year in pregnant women and fetuses in the United States alone.

Treatment: When abortions occur, flock treatment with high levels of one of the tetracyclines may help reduce the incidence of problems. A high dose of penicillin and tetracycline is effective.

4. Leptospirosis

Etiology: The disease in caused by *Leptospira interrogans.* Sheep are generally more resistant to Leptospirosis than cattle, swine, and most other domestic animals. Abortion due to this disease may occur during the last month of pregnancy.

Transmission: ***Leptospira*** is shed in the urine of infected animal and rodents play important role in the transmission.

Global scenario: In one of the study from Southern Spain (1970-1985), with 973 outbreaks of abortion in sheep and 262 outbreaks of abortion in goats. Of these, 1.7 and 2.6% respectively were caused by *Leptospira* spp., chiefly by the serovar pomona (11 flocks of sheep, 64.7%, and six herds of goats, 75.0%) and in a lower proportion the serovars sejroe, icterohaemorrhagiae and grippotyphosa[23].

Indian scenario: In India leptospirosis had shown an increased trend being involved in cases of abortions, repeat breeding and other reproductive problems in livestock resulting in huge economic losses. In a nationwide study of animal Leptospirosis of 4462 sera by microscopic agglutination test (MAT) from cattle, buffalo, sheep (551) and goats (271) which were mainly from the cases of reproductive disorders *viz.* abortion, repeat breeding, sterility showed 15.2% seropositivity in sheep and 14.3% in goats[50].

Diagnosis: Microscopic agglutination test (MAT) is the gold standard test. The aborted sample can be tested for *Leptospira* organism by dark field microscopy. A blood test of aborting sheep will confirm the diagnosis.

Prevention and control: Avoid close contact with infected swine and cattle. Effective rodent control is also important because some types of Leptospirosis may be transmitted to sheep by infected rats and other vermin.

Treatment: Prevention can be done with annual vaccination of a five-strain leptospirosis vaccine. A combination of vaccination and treatment with oxytetracycline or chlortetracycline can control leptospirosis epidemics.

5. Salmonella

In the U.S., salmonella abortion is a distant fourth in frequency as a cause of abortion, but probably occurs more often than recognized. The two major factors determining whether a pregnant ewe will abort from Salmonella are stress on the ewe and the number of Salmonella bacteria the ewe ingests. Abortions may occur earlier in gestation but are most common in the last month of gestation. Most of the ewes show diarrhea and some will die from metritis, peritonitis and septicemia. Healthy lambs may also contract the disease and die[44]. *Salmonella enterica* serovar Abortusovis is one of the common pathogens responsible for abortion in sheep in Iran[31].

Indian Scenario: In one of the study from Himachal Pradesh, in 115 does and 163 ewes with reproductive disorder *viz.* abortions (15 does and 28 ewes), endometritis (93 does and 133 ewes) and cervicitis and/or vaginitis (7 does and 2 ewes). Six isolations of salmonellae were made with two isolates of *S. dublin* from abortions in does, w one isolate of *S. typhimurium* was identified from an abortion in a ewe. Three isolates of *S. rostock* were recorded from endometritis, i.e., two isolates from does and one from a ewe[55].

Parasitic

1. Toxoplasmosis

A relatively common cause of abortion, toxoplasma generally causes ewes to abort during the last month of pregnancy or give birth to dead or weak lambs.

Etiology: Toxoplasmosis, caused by parasitic protozoan, *Toxoplasma gondii,* causes abortions in the last month of pregnancy, stillbirths, and weak, uncoordinated lambs. The organism is a coccidium of cats spread to sheep when feed, water, and the environment are contaminated with cat feces.

Global scenario: With an overall global incidence of 33% in human beings, the disease has been reported from over 180 countries including more than 60 countries from Asia-Pacific region alone where it is also causing great economic losses.

Indian Scenario: Sero-prevalence study of *Toxoplasma* infection in sheep and goats from Hissar region was studied using the indirect haemagglutination test and an incidence 6·6 % in sheep and 11·5 % in goats was seen. Seropositivity was higher in the female than in the male in both species[48].

Mode of Transmission: The intermediate hosts of *T. gondii* include sheep, goats, rodents, swine, cattle, chickens and birds. These hosts may carry an infective stage (cystozoite or bradyzoite) of *T.gondii* encysted in tissue, especially muscle and brain. Tissue cysts remain viable for long periods, perhaps for the life of the animal. *T. gondii* is frequently implicated in ovine reproductive failure. The organism, an obligate intracellular protozoal parasite with a 2-stage life cycle, is capable of producing infection in humans[1]. Ovine infection with *T. gondii* produces variable clinical disease, depending on the ewe's immune and reproductive status and stage of gestation[2]. Sheep may ingest sporulated oocysts in contaminated feed or pasture. The oocysts excyst in the small intestine, invade the intestinal mucosa, and multiply as tachyzoites in regional lymph nodes. Hematogenous spread of the tachyzoites establishes a generalized infection. Host immune response is mounted, and the tachyzoites are cleared from the system or transformed into bradyzoites. Bradyzoites, sequestered in tissue cysts, maintain the infection and a lifelong immunity[2]. Incubation period is form 10 to 23 days in one common-source outbreak from ingestion of undercoked meat, five to 20 days in outbreak associated with cats.

Prevention and control: Control of disease can be done by keeping farm cats away from pregnant sheep. Any ewe that aborts should be immediately isolated. The area in which she aborted should be carefully cleaned and disinfected. The oocyst is destroyed by freezing (0°F, -14°C), heating (150 °F, 65°C), boiling water, iodine, and ammonia. There are no vaccines or treatment available for toxoplasmosis.

Diagnosis: Several diagnostic tools exist for diagnosis of fetal loss induced by *T. gondii.* Histologic examination of the cotyledon may reveal multifocal, mineralized, necrotic lesions and are highly suggestive of *T. gondii* infection[56]. Laboratory examination of aborted lambs and placental membranes can confirm the diagnosis. Ewes are asymptomatic, the agent causes abortion in ewes and perinatal mortality in lambs. Depending on the stage of pregnancy at which the ewe is infected, the fetus may be aborted, retained, mummified, or stillborn. If fetus survives the infection, it may be born alive but weak. There is a characteristic inflammation of the placenta and fetus can be tested for antibodies to *Toxoplasma*. Ewes that abort due to Toxoplasma one year typically have a normal lamb the next (immunity develops).

Public health consequences: Humans can be exposed to *Toxoplasma* infected sheep when assisting with lambing, handling placental tissues from infected ewes or caring for infected lambs. No vaccine is as yet available to prevent either Toxoplasma infection or toxoplasmosis in cats, humans, or other species[56].

Table 1. Sheep and goat sera found positive for *Leptospira antibodies* during 1990-2000 from India (Srivastava and Kumar , 2003)

State	Goat sample (positive)	Sheep sample (positive)
A.P	54 (12.9)	390 (20.2)
U.P	33 (30.3)	19 (5.2)
M.S.	15 (13.3)	18 (0)
M.P.	19 (47.3)	0
H.P.	110 (5.4)	124 (3.2)
Delhi	40 (12.5)	0
Total (positive)	271 (14.3)	551 (15.2)

Table 2. Year wise Seroprevalance of Leptospiral antibodies in sheep and goat in Andhra Pradesh state, India (Srivastava and Kumar, 2003)

Year	Goat sample (positive)/%	Sheep sample (positive)/%
1990-91	16	30
1991-92	-	-
1992-93	5(1)/20%	-
1993-94	8	-
1994-95	-	9
1995-96	-	68(21)/30%
1996-97	-	22(1)/4.5%
1997-98	25(6)/24%	214(45)/21%
1998-99	-	38(10)/26%
1999-2000	-	9(2)/22.2%
Total	54(7)/12.9%	390(79)/20.25%

Rickettsial

1. Q-fever

Etiology: Caused by *Coxiella burnetii* - a gram-negative obligate intracellular, highly pleomorphic coccobacillus (0.2 to 0.4 μm wide, 0.4 to 1.0 μm long bacterium. The causal agent is widely found in nature and responsible for infection in humans, domestic and wild animals, pets, birds, poikilotherms and arthropods. The economic losses due to *C. burnetii* infection in farm animals may be associated with abortions and the birth of dead or weak offspring.

Indian Scenario: The isolation of *C. burnetii* from aborted human placenta by specific seroconversion in guinea pigs was first reported in India, who also detected antibodies against the agent in 14.37% of 153 human milk samples by CAT.[27]

Global Scenario: In a study to determine the causes of caprine abortion in California, *C. burnetii* was thought to be the cause in 9 percent (19 cases) of the abortions.

Source of Infection and Mode of Transmission: The symptoms associated with chronic Q fever in sheep, goats, and cattle are infertility, abortion, and the birth of full-term dead or weak, lower weight offspring. Abortions typically occur over a 2 to 4 week period and may affect 5 to 50% of the flock. Most of the abortions occurred in the last trimester of pregnancy or near term. Goats naturally infected with *C. burnetii* may abort without apparent clinical signs, but anorexia and depression for 1 to 2 days before abortions have been observed. In some affected does there is placental retention for 2 to 5 days, and agalactia may occur within 1 week of abortion.

Public Health consequences: In human infections, the most commonly identified sources are farm animals, especially cattle, goats and sheep[37] shedding large quantities of infectious agent into the environment during an abortion or normal delivery, not only in the birth fluids, placenta and fetal membranes but also in the urine and faeces[26]. Therefore, in man, the most important route of infection is thought to be through the inhalation of aerosols that have been contaminated with the products of parturition of infected animals. In abattoir, aerosols are generated by the handling fetuses, placentas, uteruses, hide and wool.

Clinical Diagnosis: Aborted samples can be stained with Gimenez's stain, a very large number of *C. burnetii* is seen as characteristic thin, pink-stained coccobacillary bodies against a blue or green background[32]. *C. burnetii* being an obligate intracellular organism can be grown only in a live system such as embryonated eggs, cell cultures and laboratory animals.

Treatment: Chloramphenicol, tetracycline, doxycycline

Viral

1. Bluetongue

Bluetongue disease (also called catarrhal fever) - caused by a reovirus is a non-contagious, arthropod-borne viral disease of ruminants, mainly sheep and less frequently of cattle, goats, buffalo, deer, dromedaries and antelope. It is transmitted by the insect vector *Culicoides* spp. Diagnostic efforts and epidemiological studies indicate that bluetongue virus (BTV) appears to be associated with some of the reproductive losses. A wave of abortions, stillbirths and deformities in sheep was seen in Al-Ahsa oasis in the eastern region of Saudi Arabia in year 1999 and the virus was identified as bluetongue virus[16]. Sero surveillance study of sheep in Tamil Nadu, India showed 29.4% morbidity and 45.5% mortality with clinical symptoms as abortion, stillbirth and stomatitis. In another study of 1209 sera samples around Ahmedabad district, Gujrat around 638 (52.77%) sero-positivity was seen by cELISA.

2. Border disease

Most often seen in newborn lambs, the border disease is also known as hairy-shaker disease or fuzzy lamb syndrome as affected lambs generally have hairy coats and tremble uncontrollably. Other signs include weak and frail lambs, abortions, mummified lambs and an unusual number of open ewes. Border disease virus (BDV) is related to bovine viral diarrhea (BVD) in cattle. The disease is generally brought into a flock through new additions, or when sheep are mixed with cattle shedding BVD virus.

3. Wesselsbron disease

Wesselsbron disease is an acute, arthropod borne flavivirus infection of sheep, cattle, and goats. Infection is common in sheep and goat but clinical disease is infrequent. Mortality in newborn animals may reach 27% in lambs and 18% in goat kids. Infection in adults is usually sub clinical, but disease may be severe in the presence of pre-existing liver pathology. Occasional abortion in ewes, together with congenital malformation of the CNS with arthrogryposis of the ovine fetus and hydrops amnii in ewes, is also seen.

4. Cache Valley Virus

Cache Valley Virus- A bunyavirus is an occasional cause of abortion outbreaks in sheep. It is spread by mosquitoes to pregnant ewes. The timing of infection in relation to gestation has a major effect on clinical effects. If a ewe is infected at less than 28 days of gestation, the embryos usually die and are reabsorbed. If infected after 45 days, there are usually no adverse effects. However, if infection occurs between 28 and 45 days of gestation, the fetus(es) usually develop the "arthrogryposis-hydranencephaly (A-H) syndrome", resulting in various congenital abnormities affecting the central nervous system. Ewes that are infected usually show no signs of disease and develop a good immunity that lasts for several years. Cache Valley virus is similar to Akabane Disease except that in only affects sheep[44].

5. Rift Valley Fever (RVF)

RVF also called as Rift valley disease and Infectious enzootic hepatitis, is a viral disease of sub-Saharan Africa. The RVF virus attacks the liver and causes symptoms ranging from fevers and listlessness to hemorrhage and abortion rates approaching 100% in pregnant sheep. It is transmitted by mosquitoes. In India, the occurrence of antibodies to RVF in sheep and goats was first reported from Rajasthan in 1995. Subsequently, an outbreak of RVF-like disease among sheep was reported from Tamil Nadu. Although, no virus was isolated from sheep during the epizootic, detection of HI and anti - RVF IgG antibodies in sheep sera (convalescent and survey) was suggestive of an RVF-like illness[19]. There is a need to carefully monitor this newly emerging virus infection in our country. There is no specific therapy for infected animals. Vaccination of animals against RVF has been used to prevent disease in endemic areas and to control epizootics[44].

6. Caprine herpesvirus-1 abortion in goat

Caprine herpesvirus-1 (CapHV-1) has rarely been reported to cause abortion in goats. This virus is closely related to other ruminant á-herpesviruses, especially -BHV-1. Over the last 3 decades, infection by CapHV-1 has been reported in many countries, including the USA[12]. Natural outbreaks of the disease are apparently rare even though seropositivity has been detected in many countries, showing that the virus is widespread[13]. Infections with CapHV-1 have been associated with late term abortions and gastroenteric and respiratory disease in 1- to 2-week-old kids. Infection in adult goats is generally subclinical, although vulvovaginitis or a balanoposthitis may develop. Rarely, pulmonary lesions have been reported in does[13].

7. Contagious caprine pleuropneumonia (CCPP)

CCPP caused by *Mycoplasma* results in abortions and also cause a terrible type of mastitis (and sick babies). The distinguishing feature of Contagious Caprine Pleuropneumonia is respiratory distress and fever. Infectious abortion due to Mycoplasma is around 6% of abortion in sheep and goats which have been isolated from genital secretions and identified as *M.mycoides* subsp.mycoides, *M.capricoium, M.arginini, Acholeplasma iaidlawii.* Caprine abortion has also been reported through *Mycoplasma capricolum* subsp. capricolum[38].

Major zoonotic abortions in small ruminants

PATHOGEN	TRANSMISSION	ANIMAL DISEASE	HUMAN DISEASE
Brucella spp.	direct contact urine, semen or by handling fetal membranes	abortions, epididymitis in sheep abortions, arthritis, spondylitis, mastitis, orchitis in goats	fever, chills, sweating, anorexia, constipation, insomnia, headache neurologic signs
Campylobacterios is (*C. jejuni, C. fetus)*	direct contact, contaminated water, or fecal-oral route	late-term abortions or very weak neonates in sheep, metritis, placentitis leading to septicemia and death	acute enteritis, bloody/mucoid diarrhea, abdominal pain, vomiting, headache, muscle and joint pain
Leptospirosis (*Leptospira interrogans)*	direct contact with urine, contaminated water, aerosol	fever, anorexia, jaundice, hemoglobinuria, anemia abortion, infertility	infection may resolve asymptomatically or develop severe icterus, fever, headache, conjunctivitis, gastrointestinal signs, gastro intestinal hemorrhage
Listeria	fecal-oral	encephalitis, septicemia, depression, paralysis of facial muscles, profuse salivation, strabismus	meningitis, septicemia, abortions very severe in immunocompromised individuals
Q-Fever *(Coxiella burnetii)*	aerosol	usually asymptomatic abortions	fever, chills, anorexia, ocular pain, pneumonitis, gastroenteritis
Salmonellosis *(S. typhimurium, dublin, newport* etc)	fecal-oral, direct contact, fomites	Abortion, Acute and chronic enteritis Septicemia in young animals	Severe Diarrhea and debilitation. Increased severity in the very old, the very young, and the immunocompromised.
Toxoplasmosis (*Toxoplsma* spp.)	Fecal-oral, vertical transmission	Abortion	Abortion in pregnant women

References

1. Acha, P.N and Szyfres, B 2003. Zoonoses and Communicable Diseases Common to Man and Animals. Pan Amer. Hlth. Org., Washington, D.C., 76-85.

2. Aitken I.D. 1993–94. Chlamydial abortion and toxoplasmosis: where are we now? Proc Sheep Vet Soc 18:89–93.

3. Alexander, A.V., Walker, R.L., Johnson, B.J., Charlton, B.R. and Woods, L.W. 1992. *J. Amer. Vet. Med. Asso.*, 200 : 711-714.

4. Anon. 1992. Veterinary Investigation Diagnosis Analysis III. Ministry of Agriculture, Fisheries and Food. Central Veterinary Laboratory, Weybridge (cited and Low and Donachie, 1997).

5. Appleyard, W.T; Aitken, I.D. and Anderson, I. 1983.Outbreak of chlamydial abortion of sheep and goats. *Vet. Rec.,* 113:63.

6. Aruni, A.W; Babu, M *et al.,* 1999. A report on blue tongue in Tamil Nadu. *Indian Vet J.,* 76(11): 953-935.

7. Blood, D.C; Henderson, I.A. and Radostits, O.M. 1979. *Vet. Medicine,* 5th edn. ELBS and Bailliere, London.

8. Boumedine, S.K and Rodolakis, A. 1998. AFLP allows the identification of genomic markers of ruminant *Chlamydia psittaci* strains useful for typing and epidemiological studies. *Res Microbiol.,* 49:735-744.

9. Buxton, D and Henderson, D. 1999. Infectious abortion in sheep. *In Practice.,* 21: 360-368.

10. CDC 2005. How great is the risk for listeriosis? National center for infectious diseases/Division of bacterial and mycotic diseases. www.cdc.gov/ncidod/dbmd/diseaseinfo/listeriosis_r.htm.

11. Chanton-Greutmann, H; Thoma, R; Corboz, L; Borel, N and Pospischil A. 2002. Abortion in small ruminants in Switzerland: investigations during two lambing seasons (1996-1998) with special regard to chlamydial abortions. *Schweiz Arch Tierheilkd.,* 144(9):483-92.

12. Chénier, S ; Montpetit, C and Hélie, P. 2004. Caprine herpesvirus-1 abortion storm in a goat herd in Quebec. *Can Vet J.,* 45(3): 241–243.

13. Engels M, Thiry E. 2000. L'infection de la chèvre par l'herpesvirus caprin de type 1. *Point Vét.,* 31:37–42.

14. Eugster, A.K.; Jones, L.P. and Gayle, L.G. 1977. Epizootics of chlamydial abortions. In: Proceedings of the Ann. Med. Amer. Assoc. *Vet Lab Diagn.,* 20:69-78.

15. Gupta, B.L; Sharma, S.N. and Tanwani, S.K. 1980. Studies on experimental infection of sheep, goat, rabbits and mice and serological response with *Listeria monocytogenes* serotype 4. *Ind. J. Anim. Sci.,* 50(9): 739-743.

16. Housawi, F.M; Abu Elzein, E.M; Ramadan, R.O; Gameel AA; Al-Afaleq AI and Al-Mousa, J. 2004. Abortions, stillbirths and deformities in sheep at the Al-Ahsa oasis in eastern Saudi Arabia: isolation of a bluetongue serogroup virus from the affected lambs. *Rev Sci Tech.,* 23(3):913-20.

17. Isoor, S; Renukaradhya, G.J. and Rajasekhar, M. 1998. A serological survey of bovine brucellosis in india. *Rev. Sci. tech. Int. Epizoot.,* 17: 781-785.

18. Ivanov, I. 1962. Untrsuchungen uber die Listeriose der schafe in Bulgarien. *Monatshete fur Veterinarmedizin,* 17: 729-736 (cited by Low and Donachie, 1997).

19. Joshi, M.V. 2002. Re-emerging and newly emerging viral zoonoses with special reference to India: An Update. In: IX Annual Conference of IAAVR and National Symposium on "Concepts in sustainable livestock production and health in new millennium", Feb. 4-5, Vety. College, Nagpur, pp. 9.

20. Jubb, K.V.F. and Huxtable, C.R. 1992. Pathology of domestic animals. Vol. I, 4th ed., V.F. Jubb, F.C. Kennedy and N. Palmer (eds.), Academic Press, London, pp 393-397 (cited by Low and Donachie, 1997).

21. Jubb, K.V.F; Kennedy, P.E. and Palmer, N. 1985. Pathology of domestic animals. Orlando, Fla : Academic Press, 3: 353.

22. Kotoch, R.C; Mandeep,S et al.1997. Investigations on endometritis among sheep and goats. *Indian vet J.*, 74(5):365-67.

23. Leon-Vizcaino L; Hermoso de Mendoza M and Garrido F. 1987. Incidence of abortions caused by leptospirosis in sheep and goats in Spain. *Comp Immunol Microbiol Infect Dis.*, 10(2):149-53.

24. Low, J.C. and Donachie, W. 1997. A review of *Listeria monocytogenes* and listeriosis. *The Vet J* 153 : 9-29.

25. Low, J.C. and Renton, C.P. 1985. Septicaemia, encepyhalitis and abortions in a housed flock of sheep caused by *Listeria monocytogenes* type 1/2. *Vet. Rec.*, 116(6): 147-150.

26. McQuiston, J.H. and Childs, J.E. 2002. Q fever in humans and animals in the United States. *Vector borne zoonotic dis.* 2(3): 179-191.

27. Malik, S.V.S. and Vaidya, V.M. 2005. Q fever: A neglected zoonosis in India? In: III Annual Conference of Indian Association Veterinary Public Health Specialists and National Symposium on "New Approaches in Food Safety and Quality control with special reference to emerging foodborne diseases and intoxications., Feb 9-10, College of Veterinary Sciences, Punjab Agricultural University, Ludhiana, pp.44-54.

28. Mc Cauley, E.H. and Tieken, E.L. 1968. Psittacosis-lymphogranuloma venereum agent isolated during abortion epizootic in goats. *J. Am. Vet. Med. Assoc.*, 152:1758-1765.

29. McLauchlin, J. 1987. *Listeria monocytogenes*, recent advances in the taxonomy and epidemiology of listeriosis in humans. *J. Appl. Bacteriol.*, 63: 1-11.

30. Nandi, S; Paney, A.B; Suresh, I and Audarya, S.D. 2005. Seroprevalance of Bluetongue disease in sheep by cELISA in Ahmedabad district of Guarat state (India). *Indian J Comp. Microbiol. Immunol Inf. dis.*, 26(1):44-46.

31. Nikbakht, M; Raffatellu, S; Uzzau ,H; Tadjbakhsh and Rubino, S. 2002. IS200 fingerprinting of Salmonella enterica serotype Abortusovis strains isolated in Iran. *Epi and Infect.*, 128: 333-336.

32. Office International des Epizootics (OIE) 2004. Q fever. In: Manual of Diagnostic Tests and Vaccines for Terrestrial Animals. Part 2, Section 2.2.2, Chapter 2.2.10.

33. Oporto, B; Barandika, J.F; Hurtado, A.; Aduriz, G; Moreno, B. and Garcia-Perez. A.L. 2006. Incidence of Ovine Abortion by *Coxiella burnetii* in Northern Spain. *Annal. New York Acad. Sci..* 1078 : 498.

34. Pand,H.K; Singh,P.P; Pathak,R.C. 1997. Isolation and identification of genital Mycoplasma/Acholeplasma spp. from sheep and goats with clinical conditions. 74(2):101-103.

35. Papp, J.R and Shewen, P.E. 1996 Localization of chronic *Chlamydia psittaci* infection in the reproductive tract of sheep. *J Infect Dis.*, 174:1296-1302.

36. Raborokgwe M; Binta MG and Mushi EZ, 1998. Wesselsbron disease virus associated with abortions in goats in Botswana. *J Vet Diagn Invest.,* 10(2):191.

37. Raoult, D. and Marrie, T. 1995. Q fever. *Clin. Infect. Dis.* 20: 489-496.

38. Rodriguez, J.L; DaMassa, A.J and Brooks, D.L. 1996. Caprine abortion following exposure to Mycoplasma capricolum subsp. capricolum. *J Vet Diagn Invest.* 8(4):492-4.

39. Rodolakis 2001. Chlamydiosis in Goats In: Recent Advances in Goat Diseases, Tempesta M. (Ed.) Publisher: International Veterinary Information Service (www.ivis.org).

40. Rodolakis, A and Souriau, A. 1986. Response of goats to vaccination with temperature-sensitive mutants of *Chlamydia psittaci* obtained by nitrosoguanidine mutagenesis. *Amer J Vet Res.,* 47:2627-2631.

41. Rodolakis, A; Bernard, F and Lantier, F. 1989. Mouse models for evaluation of virulence of *Chlamydia psittaci* isolated from ruminants. *Res Vet Science.,* 46:34-39.

42. Rodolakis, A; Boullet, C and Souriau, A. 1984. *Chlamydia psittaci* experimental abortion in goats. *Amer. J. Vet. Res* ., 45:2086-2089.

43. Satpathy, G; Sharma, A and Vishalakshi, P. 2001. Species specific chlamydiae antibodies in voluntary blood donors of Delhi. *Indian J Med Res.,* 114:164-8.

44. Schoenian, S. 2006. Sheep Diseases A-Z Sheep & Goat Specialist,Western Maryland Research & Education Center, Maryland Cooperative .Extensionhttp://www.sheepandgoat.com/articles/SheepDiseasesA-Z2006.pdf

45. Sergeant, E.S.G; Love, S.C. J. and McInnes, A. 1991. Abortions in sheep due to *Listeria ivanovii. Aust. Vet. J.,* 68 : 39.

46. Seyedmehdi M; Heath AM and Pugh DG. 2002. Theriogenology of sheep and goats. In: Pugh DG, ed. Sheep and Goat Medicine. 1st ed. Philadelphia: WB Saunders, pp129–186.

47. Sharda, R; Moghe, M.N. and Tanwani, S.K. 1991. Antibiotic sensitivity pattern of bacteria isolated from repeat breeding animals. *Indian Vet. J.,* 68 : 197-200.

48. Sharma, S. P. and Gautam, O. P. 1972. Prevalence of *Toxoplasma* antibodies in sheep and goats in the area of Hissar, Haryana, *India Trop Ani Hlth and Prod Bio and Lif Sci., 4(4)*: 339-343.

49. Smith, M.C and Sherman, D.M. 1994. Goat Medicine. Lea & Febiger Publishing, Malvern, PA.

50. Srivastava, S.K and Kumar, A.A. 2003. Seroprevalance of Leptospirosis in animals nd human beings in various regions of the country. *Indian J. Comp. Microbiol. Immunol. Infect. Dis.,* 24 (2): 155-159.

51. Steinkraus, G.E and Wright, B. D. 1994. Septic Abortion with Intact Fetal Membranes Caused by *Campylobacter fetus* subsp. *fetus. J Clin Microbiol.*, 32(6): pp. 1608-1609.

52. The Merck Veterinary Manual. Wesselsbron Disease. http://www.merckvetmanual.com/mvm/index.jsp?cfile=htm/bc/56600.htm

53. Thapliyal, 2001. *Brucellosis.* Diseases of animals transmissible to man. International book distributing. pp111-120.

54. Verma, S; Katoch, R.C. and Gupta, V.K. 2000. Micrological and histopahtological studies on Gaddi goats infected with *Brucella melitensis* biotype 1. *Intas polivet.,* 1:175-177.

55. Verma, S., R. C. Katoch, J. C. Verma, P. Nigam, M. K. Batta. 1998. Reproductive failures in does and ewes due to salmonellosis in Himachal Pradesh. *India. Vet. Arhiv.*,68: 177-181.

56. Weissmann, J. 2003. Presumptive *Toxoplasma gondii* abortion in a sheep *Can Vet J.* 44(4): 322–324.

-ΔΔΔ-

15

Mycoplasma Infections in Goat and Sheep in Indian Perspective

NEM SINGH, V P SINGH AND RAJNEESH RANA

National Referral Lab on
Mycoplasma I.V.R.I., Izatnagar and
Animal Health Division, C.I.R.G., Makhdoom, Mathura

Goat and sheep farming is one of the enterprises, which has been practiced by a large section of human population especially in developing world and plays an important role in employment generation and improving the household nutrition. In spite of the various developmental efforts in livestock improvement sector in the country, the status of sheep and goat has largely remained as resource for poor farmers. According to the livestock census, India possessed 123.5 million goats and 58.2 million sheep[1]. Over 94% of the goat population of the world is distributed in two continents of Asia and Africa.

It is a potential source of meat, milk, fibre, hide, manure for landless rural poor people and provides dependable source of income to 40% of rural population below the poverty line of India. The goat industry has contributed about 3.95% of total milk and 10.0% of total meat productions while; sheep industry contributed around 4.8% of total milk production[1].

Number of factors are responsible for economic losses to the goat and sheep industry, among them the health is utmost important. The goat population in our country is frequently exposed to ravages of infectious diseases. Pneumonia has been noticed as one of the most frequently encountered condition and is responsible for around 28.7%

mortality[2]. Amongst various pneumonic conditions, contagious caprine pleuropneumonia (CCPP) is one of the most dreaded diseases of goats and belongs to OIE list B.

In 1994, FAO established an Emergency Prevention System (EMPRES) for Transboundary Animal and Plant Pests and Disease in order to minimize the risk of such emergencies developing. Initial priority was given to two transboundary pest and diseases problems. Contagious caprine pleuropneumonia is one of the important major transboundary livestock diseases of goats. Such types of diseases are among the most contagious and place a serious burden on the economics of the countries in which they occur.

Contagious caprine pleuropneumonia is a disease of major economic importance of nearly 40 countries worldwide including Africa, Middle East, Mediterranean Littoral and Asia. The direct losses of the disease result from its high mortality, reduced milk production and meat yield, cost of treatment, control, disease diagnosis and surveillance. In addition to these, there are indirect losses due to the imposition of trade restrictions[1].

In India, first record of the existence of CCPP was made by Steel in the year 1889 from Khandesh, Bombay followed by other workers in number of states[3]. The disease still exists in most of the states of the country including Andhra Pradesh, Assam, Gujarat, Himachal Pradesh, Kerala, Maharastra, Manipur, Rajasthan, Tamil Nadu and Meghalaya[4,5,6,7]. CCPP has been reported to affect only goats and does not cause disease in sheep, neither spontaneously nor experimentally[8,9]. In natural infections, the organisms are acquired by susceptible goats through inhalation of contaminated droplets from infected goats[10]. The goats of all ages and either sex are susceptible to infection however, mortality was found higher among the younger animals than adults[1].

Different members of *Mycoplasma mycoides* cluster group viz. *M.mycoides* subsp. *capri*, *M.mycoides* subsp. *mycoides* type LC, *M.capricolum* subsp. *capricolum* and *M.capricolum* subsp. *capripneumoniae* are play an important role in pleuropneumonia infections of goats. In India, *M.mycoides* subsp. *capri (M.capri)* has been found to be mainly responsible for caprine pleuropneumonia on the basis of isolation and serosurveillence studies[14,15,16,17].

Mycoplasma mycoides subsp. *mycoides* (LC) has also been described with respiratory tract infection and pleuropneumonia in goats[6,16,18].

Apart from these species, *M.capricolum* has been reported from goats with respiratory affections[19,20]. Although, there is no report of any isolation of *M.capricolum* subsp. *capripneumoniae* (F-38) from goats however, serological evidence has been found in our country[7].

Another important disease, contagious agalactia (CA) is predominantly a disease of milking sheep and goats appearing in a herd in the spring soon after lactation and begins as a result of the activation of latent infection. The young ruminants become infected directly at sucking while the adults are contaminated via the milker's hands, milking machines or by bedding which often provides a rich source of mycoplasmas. The CA infection frequently occurs as an enzootic. In lactating female animals, it is usually manifested by mastitis. Males, young animals and non-lactating females suffer from arthritis, keratoconjunctivitis and respiratory problems.

The major causal agent of *Contagious agalactia* disease in both sheep and goats is *Mycoplasma agalactiae* and reported by number of workers in our country[6,22,23,24,25,48]. *M.mycoides* subsp. *mycoides* LC is also one of the causative mycoplasma of contagious agalactia, which is characterized by mastitis, arthritis, keratoconjunctivitis and occasionally abortion. The joints are also affected and painful with accumulation of synovial fluid. In chronic cases, ankylosis may develop which makes the animal limp or prevents it from standing.

Mycoplasma ovipneumoniae has been isolated from goats[6,49] and affects sheep and goats. It is one of many agents associated with pneumonia complex and also causes an atypical pneumonia. It is often associated with *Pasteurella haemolytica*.

A number of mycoplasmas have been implicated in the pathogenesis of various inflammatory ocular diseases of sheep and goats and it is mainly due to the *Mycoplasma conjunctivae* which responsible for epidemic keratoconjunctivitis[25].

The role of *Mycoplasma* serogroup 11 was also observed in case of chronic mastitis and agalactia and play an important role in granular vulvovaginitis and other genital tract infections of goats[26,27].

Unusual mycoplasmas were also reported from the sheep and goats viz. *M.fermentans, M.putrifaciens* and *M.ovine/caprine* serogroup 11 and identified and characterized on the basis of 16sRNA sequencing[24].

Therapy against the incidence of the mycoplasmal diseases is also tried which based solely on antibiotics namely tetracycline, macrolide, florfenicol, tiamulin and fluoroquinolones[28,29,30]. However, many workers feel that bacteriological treatment is an illusory objective. If the therapeutic dose is not exactly defined and the relevant antibiotic is not administered for a sufficiently long period, the resulting effect may be very poor as the causal agents continue to be shed into the environment and there is a possibility that resistant strains will develop[31]. The recent guide lines of OIE suggested that antibiotic therapy is generally ineffective for control of CCPP, CA and most of the mycoplasmal infections as such therapy tends to reduce clinical signs but promotes the carrier state.

The diagnosis of mycoplasmal infection is utmost essential for its successful control therefore, many workers have described various serodiagnostic tests viz. SAT, IHA, CFT and ELISA for qualitative and quantitative detection of antibodies in abroad as well as in India[7,35,36,37,38,39]. In field conditions, the slide agglutination test (SAT) is working very well for the detection of antibodies while, CFT and ELISA can be used in the laboratory for confirmatory diagnosis.

Recent development in the molecular techniques viz; DNA probe hybridization, pulse field gel electrophoresis, restriction endonuclease analysis, fingerprinting and polymerase chain reaction (PCR) have overcome the problems of isolation and sero cross reactivity to a great extent. Amongst them, PCR is the most specific and takes very short time for detection of infections including chronic and sub clinical cases, which may not be detected by routine diagnostic procedures[42,43,44].

To control the mycoplasmal diseases, various vaccines have been developed for protecting the sheep and goats in India as well as abroad. Number of vaccines against caprine pleuropneumonia viz. formalized killed vaccine, alum precipitated killed vaccine and fractionated *M.mycoides* subsp. *capri* vaccine were developed in our country which gave protection in goats for 3 to 7 months of duration[3,20,45,46,47,50]. Recently, Singh *et al.* (unpublished report) have also tried sonicate saponified *M.capri* vaccine in goats and observed very encouraging results upto a year. As concern the vaccine against contagious agalactia infection, the live attenuated vaccines are prepared using *M. agalactia* in Turkey, which produced better protection than dead vaccines in controlled experiments. This vaccine has been used successfully in the field in Turkey but it has not been adopted elsewhere in the world.

It is the need of time, to develop the reliable vaccines against contagious agalactia and other mycoplasmal infections, as they are responsible for a great loss to goat and sheep industry of our country.

References

1. FAO (Food and Agriculture Organization of the United Nations) 2001 and 2002 Production Year Book, FAO, Rome, Italy.

2. Kumar R and Paliwal O P. 1995. Pathological conditions in mortality of goats – A retrospective study, compendium. National Symposium on impact of disease on livestock and poultry production and wildlife conservation, pp. 92, held during 9-10 Nov. at Dept. of Pathology, Orissa Veterinary College, Bhubaneswar.

3. Longley E O. 1940. Contagious pleuropneumonia of goats. Indian Journal of Veterinary Sciences 10: 127-97.

4. Iyer P R K, Sharma G and Bhalla N P. 1968. Contagious caprine pleuropneumonia, ICAR Technical Bulletin (Animal Husbandry) No.18: 1-13.

5. Ojo M O. 1976. Caprine pleuropneumonia IV: Pathogenicity of *Mycoplasma mycoides* subsp. *capri* and caprine strains of *Mycoplasma mycoides* subsp. *mycoides* for goats. Journal of Comparative Pathology 86: 519-29.

6. Gupta M M and Verma B B. 1984. Isolation and characterization of mycoplasma from the respiratory tract of kids in Bihar. Indian Journal of Animal Sciences 54: 58-62.

7. Singh R K, Srivastava N C, Singh V P and Singh V P. 1999. Serosurveillence of contagious caprine pleuropneumonia using enzyme linked immunosorbant assay. Indian Journal of Animal Sciences 69: 999-1001.

8. McMartin D A, MacOwan K J and Swift L L. 1980. A contrary of classical contagious pleuropneumonia : from original description of etiology. Br. Veterinary Journal 136: 507-15.

9. Thiaucourt F and Bolske G. 1996. Contagious caprine plueopneumonia and other pulmonary mycoplasmoses of sheep and goats. Revue Scientifique et Technique Office International des epizooties 15: 1377-1414.

10. MacOwan K J. 1976. A mycoplasma from chronic caprine pleuropneumonia in Kenya. Tropical Animal Health Production 8: 28-56.

11. Singh V P, Srivastava N C, Tripathi B N and Rana R. 1997. Isolation of pathogenic *Mycoplasma mycoides* subsp. *capri* from pneumonia kid. Indian Journal of Veterinary Research 6: 43-45.

12. Mohanty B C, Kar B C, Mishra P R, Roy P K, Mishra R and Bal M K. 2002. Kid mortality in Orissa. Indian Journal of Animal Sciences 72: 546-48.

13. Singh V P, Srivastava N C, Kumar Manoj M, Jai Sunder and Varshney J P. 2004. Isolation and characterization of an Indian strain of *M.mycoides* subsp. *mycoides* type LC from a case of caprine arthiritis. Comparative Immunology Microbiology Infectious Diseases 27: 273-84.

14. Mittal V, Kumar M and Singh V P. 2006. Sero-epidemiological study of CCPP in goats in Uttaranchal. Indian Veterinary Journal 83: 29-30.

15. Nayak N C and Bhaumik M K. 1991. Pathogenicity of *Mycoplasma mycoides* subsp. *mycoides* for goat kids. Small Ruminant Research 5: 155-67.

16. Sikdar A, Srivastava N C and Uppal P K. 1993. Development of a killed vaccine against caprine pleuropneumonia : preliminary investigation in goats. Indian Veterinary Journal 70:600-04.

17. Singh M P and Prasad R L. 1994. Pathogenicity of mycoplasmas isolated from goats. Indian Journal of Animal Sciences 64: 1126-31.

18. Srivastava N C, Chattopadhyaya S K and Singh V P. 1996. Isolation of *Mycoplasma agalactia* from pneumonic sheep. Indian Journal of Animal Sciences 66: 1000-02.

19. Nicholus R A J. 2002. Improvements in the diagnosis and control of diseases of small ruminants caused by mycoplasmas. Small Ruminant Research 45: 145-49.

20. Mondal D, Ghosh S K and Pramanik A K. 2003. Caprine mycoplasmal keratoconjunctivitis. Indian Journal of Animal Health 42: 98-100.

21. Rana J S, Gupta P P and Banga H S.1993. Pathology of genital tract of goats experimentally infected with Mycoplasma serogroup 11. Indian Journal of Animal Scinces 63: 706-09.

22. Kumar D, Gupta P P, Rana J S and Banga H S. 1993. Pathology of Mycoplasma serogroup 11 induced mastitis in sheep. Australian Veterinary Journal 70: 70-71.

23. Bhaumik A, Verma B B, Thakur D K, Pandey S N and Benerjee N C. 1990. Effect of oral administration of tylosin tartrate in the treatment of experimental and natural cases of caprine mycoplasmosis. Indian Veterinary Journal 67: 948-51.

24. Bagherwal R K and Sisodia R S. 1991. Long acting oxytetracycline as chemotherapeutic agent against pneumonia in kids due to mycoplasma infection. Indian Journal of Veterinary Medicine 15: 140-41.

25. Fadda A, Cannas E A, Cubeddu G M, De Palmas S and Petracca G. 1995. Susceptibility of Mycoplasma agalactiae to tilmicosin. In Proc. 3rd International Mastitis Seminar. Tel Aviv, 28 May – 1 June, Lachmann Printers, Ltd., Bet Dagan, 126-27.

26. Madanat A, Zendulkova D and Pospisil Z. 2001. Contagious agalactia of sheep and goats. Acta Veterinary Brno 70: 403-12.

27. Singh V P and Srivastava N C. 1994. Agglutination and precipitating antibodies against different in immunogenic fractions of *Mycoplasma mycoides* subsp. *capri* in goats. Indian Journal of Comparative Microbiology Immunology and Infectious Diseases 15: 15-18.

28. Singh V P and Srivastava N C. 1995. Indirect haemagglutination and complement fixing immune responses in goats against various fractions of *Mycoplasma mycoides* subsp *capri*. Indian Journal of Animal Sciences 65: 845-48.

29. Rana R and Srivastava N C. 1996. Monitoring of serological response in goats and rabbits immunized with *Mycoplasma mycoides* subsp. *capri.* Indian Veterinary Journal 73: 379-83.

30. Manimaran K, Singh V P, Ltu K, Das S, Kumar A A and Srivastava S K. 2006. Immune response against different inactivated vaccines of *Mycoplasma mycoides* subsp. *capri* in goats. Journal of Applied Animal Research 29: 117-20.

31. Kumar M M, Singh V P, Srivastava N C, Singh V P, Sharma B, Jaisunder and Kumar A. 2001. Rapid and specific detection of *Mycoplasma mycoides* cluster and differentiation of *mycoides* group from *capricolum* group by polymerase chain reaction. Indian Journal of Comparative Microbiology Immunology Infectious Diseases 22: 118-21.

32. Kumar M M, Singh V P, Srivastava N C, Das S and Singh V P. 2003. Identification of *Mycoplasma capricolum* subsp. *capricolum* causing MAKePS syndrome in small ruminants by PCR. Indian Journal of Animal Sciences 73: 55-57.

33. Das S and SinghV P. 2003.Multiplex PCR for differentiation of *Mycoplasma mycoides* cluster from other mycoplasma. Journal of Applied Animal Research 24: 95-100.

34. Srivastava N C, Singh V P and Rana R. 1998. Immunogenicity of *Mycoplasma mycoides* subsp. *capri* saponin vaccine in goats. Indian Journal of Animal Sciences 68: 595-596.

35. Rajneesh Rana, Srivastava N.C and Singh V.P. 1999. Passive mouse protection assay of killed mycoplasma vaccines. Ind. J. Ani. Sc. 69 (9): 681-682.

36. Jai Sunder, Srivastava N C and Singh V P. 2002. Preliminary trial on development of vaccine against *M.mycoides* subsp. *mycoides* type LC infection in goats. Journal of Applied Animal Research 21: 75-80.

37. Singh N, Rajya B S and Mohanty G C. 1975. Pathology of *Mycoplasma agalactiae* induced granular vulvovaginitis in goats. Cornell Veterinary 65: 353-73.

38. Mondal D and Srivastava M K. 2004. Ovine pneumonia due to *Mycoplasma agalactiae.* Indian Journal of Animal Sciences 74: 253-55.

39. Singh V P and Srivastava N C. 1993. Immunogenicity of *Mycoplasma mycoides* subsp. *capri* in goats. Indian Veterinary Journal 70: 989-92.

-ΔΔΔ-

16

Infectious Agents Transmitted through Semen and Embryo

K P AGRAWAL

National Agricultural Innovation Project
Indian Council of Agricultural Research, KAB – II
Pusa Campus, New Delhi

For breed improvement in livestock through out the world, the exotic germplasm is introduced from one country to another either as live animal, frozen semen or embryo. There has been considerable concern about the semen and embryo quality and their impact on disease transmission. Advent of cryopreservation of semen and embryo as an adjunct to AI and ET and import/export of germplasm had made the users to re-think about disease transmission problems. Survival of many pathogens under cryopreservation process and after the addition of antibiotics to the media used for semen and embryo, reawakened interest in this area and led to health regulations being developed for global movement of semen and embryos. Infectious agents transmissible through semen and embryos have been discussed in present communication.

Semen is a potential source of several agents, viz : infectious bovine rhinotracheitis virus, infectious pustular vulvo-vaginitis virus, bovine viral diarrhoea virus, mucosal disease complex virus, foot and mouth disease virus, bluetongue virus, African swine fever virus, porcine parvovirus, Brucella abortus, Chlamydia, Trichomonas, Toxoplasma spp, Leptospira spp., Mycoplasma spp, Campylobacter spp, Ureaplasma spp., Mycobacterium spp. There is great potential for the spread of infectious disease to large number of females through artificial insemination because of several reasons : 1) using single infected ejaculate after dilution in large number of females, 2) allowing source of infection for a long time if infected semen is frozen for future use and 3) bypassing the natural defences of the female vaginal and cervical mucosa against invading micro-organisms. There is also possibility of semen contamination by micro-organisms present in the atmosphere, on the body of teaser and donor animal, in unsterilized equipments, in the semen extender, and in case of frozen semen, from liquid nitrogen or its containers.On the other hand, Al if carried out under strict sanitary and controlled conditions, provides an excellent means for disease control because : 1) when semen is diluted, the number of organisms per ml of inseminating dose is decreased approximately ten fold which makes the number below the minimum infective dose, 2) the technique avoids animal to animal contact, 3) collection of semen from clinically healthy and disease free donor ensures earliest and safest approach for obtaining semen that is free from infectious disease causing organisms, 4) possibility of testing the ejaculate for the presence of micro-organisms prior to use.

The disease transmission potential of embryos is much less then that of semen. A few characteristics of embryo, viz : presence of zona pellucida and secondary resistance factors such as young age, small size and limited mobility serve to reduce the potential exposure of embryos. It is difficult for bacterial and/or fungal agents to penetrate the zona pellucida to gain access to the embryonic cells. Thus infectious agents that are of concern in embryo transfer are mostly viral rather than bacterial or fungal.Pathogens through embryo transfer are transmitted either within or on the embryo or in fluids transferred with the embryo. Transmission through embryo association occurs under the following circumstances : 1) If the pathogen is in the ovum at the time of fertilization , 2)if the pathogen is carried into the ovum at fertilization either within or attached to the spermatozoa, 3) if the pathogen passes through, becomes embedded in, or adheres to the surface of the zona pellucida following fertilization and 4) if the pathogen reaches the embryo as a result of damage to the Z. P. during filtration, freezing or micro-manipulation.Other factors which help to reduce the potential of early embryos to transmit disease are inherent in the techniques that are used in the collection and processing of embryos. For instance, embryos are flushed from the uterus of the donor with a considerable volume of fluid which helps to dilute the pathogens that might be present in the uterus. In addition, freezing has been found very effective in inactivating low levels of many viruses. Washing and enzyme treatment also help to control the spread of diseases.

Diseases of livestock can cause major economic losses, both to livestock owners and the country as a whole. Inadequate health management, lack of control efforts and lapses during transport of germplasm from one country to another have resulted in large number of deaths through out the world.

For breed improvement in livestock through out the world, the exotic germplasm is introduced from one country to another either as live animal, frozen semen or embryo. There has been considerable concern about the semen and embryo quality and their impact on disease transmission. Advent of cryopreservation of semen and embryo as an adjunct to AI and ET and import/export of germplasm had made the users to re-think about disease transmission problems. Survival of many pathogens under cryopreservation process and after the addition of antibiotics to the media used for semen and embryo, reawakened interest in this area and led to health regulations being developed for global movement of semen and embryos. Infectious agents transmissible through semen and embryos have been discussed in present communication.

Infectious Agents Transmitted Through Semen

There are two aspects of disease transmission through semen. One is transmission of pathogens which are present in the semen under natural conditions. The other is when animals are bred artificially (Al) using extended semen. Agents transmissible by semen in cattle, sheep, goat and pig are listed below.

Cattle

1. FMD virus
2. Bluetongue virus
3. Bovine ephemeral fever virus
4. BVD/MD virus
5. *B. abortus*
6. *Campylobacter* spp.
7. *M. paratuberculosis*
8. *Trichomonas*
9. *Leptospira* spp
10. *Mycoplasma* spp
11. *Ureaplasma* spp.

Sheep & Goats

1. FMD virus
2. Rinderpest virus

8. *M. paratubcrculosis*
9. *Leptospira* spp.

3. PPR virus
4. Border disease virus
5. *B. ovis*
6. *B. melitensis*
7. *Campylobacter* spp.
10. *Ureaplasma* spp
11. *Chlamydia* strains
12. *Toxoplasma gondii*
13. *Actinobacillus seminis*

Pig

FMD virus
SVD virus
Hog cholera virus
Pseudorabies virus
Porcine parvovirus
B. suis
Leptospira spp.
Ureaplasma spp.

Control Measures

Health status of donor animals is most important when Al is practiced. Emphasis should be to use disease free donor animals as there is very likelihood of semen being healthy from such animals. Health status of donor animals prior to their entry to the Al centre is determined at different levels : 1) potential donors free from infectious diseases should be moved to an Al centre and kept under quarantine, 2) donors during quarantine are subjected to further examinations and tests, prior to being moved in with the main male herd/flock, 3) regular testing of animals following their entry into the herd/flock. The countries having national disease control programme and those interested in the import/export of semen adopt norms recommended by the O.I.E. Policies adopted by most of the countries for the import/export of semen are based on disease-free categories as listed below. The diseases included in each category vary from one country to another, but in general for diseases included in O.I. E. List "A", the territory has to be free; for diseases included in O. I. E. list "B", the herd has to be free, and for the diseases in O. I. E. list "Others", the donor has to be free.

O. I. E. List "A" (territory has to be free)

1. FMD
2. Rinderpest
3. Vesicular stomatitis
4. CBPP
8. PPR
9. Sheep and Goat pox
10. Swine vesicular disease
11. African swine fever

5. Lumpy skin disease
6. Bluetongue
7. Rift Valley fever
12. Hog cholera
13. Teschen disease

O. I. E. List "B" (herd has to be free)

1. Anaplasma marginale
2. Babesia spp.
3. B. abortus
4. B. ovis
5. B. melitensis
6. B. suis
7. Campylobacter spp.
8. Mycobacterium spp.
9. P. multocida (HS)
10. Bovine leukemia virus
11. IBR/IPV
12. M. paratuberculosis
13. Trichomonas
14. R. ruminantium
15. Leptospira spp.
16. Coxiella burnettii
17. Mycoplasma agalactiae
18. Mycoplasma spp. (CCPI)
19. Nairobi sheep disease
20. Wesselsbron disease
21. S. abortus ovis
22. Scrapie agent
23. Pseudorabies
24. Transmissible gastroenteritis
25. Theileria spp
26. Trypanosoma spp
27. Chlamydia Psittaci (EAE)
28. Caprine arthritis-encephalitis (CAE)

O. I. E. List "Others" (donor has to be free)

1. Listeria monocytogenes
2. Bovine ephemeral fever
3. BVD/MD
4. Border disease
5. Malignant catarrhal fever
6. Bovine parainfluenza
7. EHD
8. H. somnus
9. Mycoplasma spp.
10. Ureaplasma app
15. E. rhusiopathiae
16. Mycobacterium spp
17. Vesicular exanthema
18. Porcine parvovirus
19. Enteroviruses
20. Japanese encephalitis
21. Parapox virus
22. Adeno virus
23. Reovirus
24. Cytomegalo virus

11. Chlamydia strains
12. Toxoplasma gondii
13. Actinobacillus seminis
14. Ulcerative dermititis
25. Swine influenza virus
26. Transmissible genital papilloma virus
27. Haemagglutinating encephalomyelitis virus (HEV)
28. Haemophilus spp

The micro-organisms which are most likely to contaminate the semen may be present in the preputial cavity, in the hair of preputial orifice and its surrounding, on the body of donor/teaser animal and in the atmosphere, improperly sterilized collection equipments, semen extenders, media and liquid nitrogen etc. These micro-organisms are Pseudomonas aeruginosa, Corynebacterium pyogenes, Staphylococcus spp., Proteus spp., Streptococcus spp., Escherichia coli etc.

The number of micro-organisms can be reduced to a minimum by adherence to strict aseptic procedures and sanitation of AI centre, Regular douching of the preputial cavity and clipping of the hair surrounding the preputial orifice help to reduce the level of contaminants. When antibiotics (penicillin/streptomycin) are added to semen extenders to reduce or eliminate infections in the semen, the care has to be taken to ensure that the antibiotic : 1) does not have a deleterious effect on fertilizing capability of spermatozoa and 2) is effective in the semen extender in which it is to be used. The following facts should also be kept in mind when antibiotics are added to semen: 1) that the antibiotics be added to semen before freezing, 2) that the action of antibiotics against some organisms being bacteriostatic, its concentration in the semen be kept optimum to avoid recommence of growth of the organisms and 3) that the use of antibiotics should not become an excuse for relaxing aseptic conditions in the preparation of donor/teaser animals for collection, during collection of semen and handling of semen in the laboratory.

Infectious Agents Transmitted Through Embryo

With the advent of successful embryo transfer technology, global transport of germplasm in the form of embryo provided a new alternative and safe method of genetic improvement. The possibility of transmission of infectious agents through embryo is minimized due to specific embryo characteristics. These are presence of zona pellucida and secondary resistance features such as young age, small size and limited mobility that serve to reduce the potential exposure of embryos. Many of the bacterial, fungal and viral pathogens being too large to penetrate the zona pellucida and incapable of surviving in media reduce the possibility of transmitting an infectious agent through embryo. Only small viruses might penetrate the zona pellucida. Transmission through embryo is possible under the following circumstances.

1. If the pathogen is in the ovum at the time of fertilization and is capable of expressing itself in the embryo, foetus or in the newborn animal. A few viruses including RNA viruses have been implicated in infection of gametes.
2. If the pathogen either within or attached to the spermatozoa is carried into the ovum at fertilization and subsequently capable of infecting the embryo or foetus.
3. The other mode of infection of pre-implantation embryos is from the uterine environment of the donor or the recipient.
4. Pathogen may reach the embryo as a result of damage to the zona pellucida during embryo filter, freeze-thawing or micro-manipulation.
5. The other source of pathogen transmission is via fluids transferred with the embryo. Transmission of pathogens via fluids can be avoided if embryos are washed several times in sterile medium prior to transfer. Exposure of the embryo to trypsin or inactivation with antiserum helps in this process. Serum used in the washing and transfer fluid should also be sterile to avoid transfer of pathogens.

Control Measures

The success of the embryo transfer programme depends on the strict quality control at different steps involved in the embryo transfer programme and any compromise in the quality control may prove futile. Two main approaches considered to control pathogen transmission through embryo transfer are : one is based on the health status of the donor animals (male and female), the other is based on the health status of the embryo.

1. Control based on the health status of the donor is indirect and traditional one. It is based on the fact that if the donor animals are free from infectious disease, the embryos received from them are most likely to be healthy, provided that sterile materials and aseptic procedures are used in collection, handling and transfer

of embryo. Approaches used by different countries vary. Regulations requiring donor females to be quarantined also vary (e.g. territorial, herd/flock and individual donor) from country to country. Routine health testing and vaccination of donors (both male and female) are absolutely necessary. Any animal arriving to ET centre should be tested for infectious diseases (e.g. FMD, Brucella and Tuberculosis).

2. Control based on the health status of the embryo depends either on scientific knowledge obtained through experiments and/or field trials designed to test the likelihood of disease transmission by embryos under real life situations, or through diagnostic tests on individual embryos. Different aspects of embryo transfer that require sanitation and quality control measures to keep embryos free of pathogens are embryo collection, embryo handling and processing, embryo transfer, laboratory hygiene, sterilization and embryo transport.

-ΔΔΔ-

17

Status of Mycotic Diseases of Small Ruminants and its Zoonotic Importance

MAHENDRA PAL

Ex-Professor and Head
Department of Veterinary Public Health
College of Veterinary Science
Anand Agricultural University, Anand 388001, Gujarat

INTRODUCTION

India is primarily an agricultural based country where over 70 per cent of the population resides in villages. Small ruminants play an important role in supporting the livelihoods of poor farmers. The economically weaker section of our society rear these animals and earns their livelihood by the sale of animals and their products viz. milk, meat, skin, hair and wool. These animals suffer from many infectious diseases resulting into high morbidity and mortality. The close association of man with animals has given great opportunity to many infectious agents to produce serious and life threatening diseases in human beings. Such diseases which are naturally transmitted between vertebrate animals and man are called 'zoonoses'. Today, there are about 300 zoonotic infections of varied etiologies that may occur in sporadic and epidemic form[31].

Mycoses are the infectious diseases of man and animals that are caused by fungal pathogens. There are about 2,50,000 species of fungi in the world, of which only 100 are considered as pathogens of man and animals including birds. Most of the mycotic agents grow, proliferate and survive in diverse types of environmental materials such as the soil, wood, vegetables, fruits, pigeon droppings, parrot excreta, bat guano, litter, sewage, munia bird faeces, plant debris, rice husk, hay, grass, straw, tree, shrub, bagasse, etc. [19, 35, 21, 27] The infection is mainly acquired through the respiratory tract following inhalation of large number of infectious fungal cells from the saprobic environment. Transmission can also occur through the traumatic implantation of the fungal contaminated objects into the skin or mucous membranes[23, 31].

Mycotic infections are global in distribution, and maximum cases have been reported from tropical and subtropical countries of the world. As mycoses are not included in the category of notifiable diseases, the exact information on their prevalence and incidence is lacking[26]. It is presumed that at least 800 million people in the world suffer or have suffered from one or other type of mycotic infections. The incidence of mycoses seem to be higher in the patients who have underlying diseases such as tuberculosis, diabetes mellitus, acidosis, malnutrition, cancer and AIDS or have received prolonged treatment of broad-spectrum antibacterial antibiotics, steroids or other immunosuppressant drugs. These conditions predispose the host to fungal infection by impairing the natural defense system of the body. The disease can be fatal in immunocompromised patients. Many occupational groups such as animal attendants, veterinarians, pet owners, dairy farmers, zoo employees, etc. are at a greater risk of acquiring the fungal disease. In India, the work on mycotic diseases is very meager that appears as the 'tip of iceberg'. The present paper is, therefore, an attempt to delineates the status of mycotic disease of small ruminants and its zoonotic importance.

Aspergillosis

Aspergillosis is a highly infectious, global mycotic disease of avians and mammals including human beings. It is primarily a respiratory mycosis that may become generalized. The first case of aspergillosis was reported by Virchow in 1856. There are over 185 species of the genus *Aspergillus* and only 20 have been implicated in the disease. *Aspergillus fumigatus* is the principal species that is most commonly found in man and animal infections throughout the world. These fungi are ubiquitous in nature and are present in the soil, plant material, litter etc. The main portal of entry for *Asperpgillus* spores is the respiratory tract. The lesions may occur in any part of the body including the brain, skin, eye, mammary gland, bone and heart. The nosocomial outbreaks of invasive aspergillosis is attributed to the environmental reservoirs. The disease is considered as the second most common mycosis that requires hospitalization in USA. *Aspergillus* infection accounts for about 5 per cent of the CNS fungal infections. The case fatality is very high, often 95 per cent in acute invasive pulmonary aspergillosis. The disease has been recorded in small ruminants.[6,44,10,11,37] In India, cases of pulmonary and systemic aspergillosis are diagnosed in goats and sheep.[10,11,9,37] The association of *Aspergillus* spp. is also established with caprine mastitis.[20] The species isolated from goats and sheep include *A. flavus, A. fumigatus, A. oryzae and A. terreus.*

Candidosis

Candidosis is a commonly encountered, infectious, global fungal mycosis; and is also known as candidiasis, moniliasis and thrush[23]. The disease affects many organs of the body including the nail, brain and skin. There are 163 species of genus *Candida* of which 20 are considered to be significant producing infections in man and animals. The chief species *C. albicans* is associated in approximately 90 per cent of candidal infections. The infection can occur by direct contact of macerated, abraded skin or mucous membrane with yeast contaminated objects. The yeast *Candida* is the fourth most common primary blood stream organism. Nosocomial infection due to *C. albicans* are frequently reported. Vaginal candidosis is one of the most common infection seen in general practice. About 90 per cent of AIDS patients suffer from oropharyngeal or esophageal candidosis. Candiura is a marker for haematogenous seeding to the kidneys. Alimentary, pulmonary and systemic candidosis are reported in small ruminants[8,34]. In India, *C. parakrusei* is isolated from the goat[43]. The role of *C. albicans* has been established in mastitis[20] and pneumonia in goats[34]. Experimentally, the infection is produced in sheep by *C. albicans.*

Coccidioidomycosis

Coccidioidomycosis is a non-contagious, dust-borne, severe and fatal systemic fungal disease. The disease is known by various names such as coccidioidal granuloma, desert fever, desert rheumatism, Posada-Wernicke's disease, San Joaquin fever and Valley fever[29]. The disease is caused by *Coccidioides immitis,* a dimorphic fungus, which prefers to grow in desert soil as saprobe. The yeast phase occurs in host tissue and the mould form in the soil. The infection is limited to arid and semi-arid regions of USA and South America. In USA, about 1,00,000 cases of coccidioidomycosis occur each year. Inhalation of the dust particles, containing the infectious arthroconidia of *C. immitis* may result in primary focus of infection in the lung. The disease in cattle was first reported by Giltner in 1918 from USA. Subclinical infections have been described in animals particularly in ruminants and pigs. There is no record of coccidioidomycosis in goat, however, the disease is encountered in sheep[13]. Mycotic pneumonia in a sheep due to *C. immitis* has been reported from India[9]. The diagnosis was confirmed on histopathological examination of the lung.

Cryptococcosis

Cryptococcosis also known as Busse-Buschke's disease, European blastomycosis and torulopsis, is an acute, subacute or chronic mycotic disease of global importance[23]. The disease occurs both in immunocompetent as well as immunocompromised hosts. The genus *Cryptococcus* contains 34 species of which only three namely *C. neoformans, C. albidus* and *C. laurentii* are incriminated in the etiology of disease. The chief causative agent is *C. neoformans* which is an encapsulated basidiomycetous. Yeast has two varieties – *C. neoformans* var. *neoformans* (Telemorph *Fillobasidiella neoformans* var. *neoformans*) and *C. neoformans* var. *gattii* (Telemorph *F. neoformans* var. *bacillispora*). Both varieties exist in the environment as saprophyte and has been frequently recovered from the avian excreta and less commonly from the soil, wood, fruits, vegetables, etc. The respiratory tract is considered the prime portal of entry of *C. neoformans*. It is believed that man and animals may acquire the infection from the environment where the fungus grows abundantly and produce million of infectious cells. The agent can infect the bone, lymph node, brain, skin, eye, heart, and other visceral organs. Cryptococcosis once described as the 'sleeping disease' in 1970 has emerged as the 'awakening giant' due to advent of HIV/AIDS which is considered the most important predisposing factor. The global incidence of cryptococcal meningoencephalitis in AIDS patients is recorded 2-33 per cent. In USA, 85 per cent of cases occur in HIV-infected persons. The first report of cryptococcosis in animal (horse) was published by Frothingham in 1902. Since then, the natural infection is recorded in a wide variety of animals including sheep and goats[4,23]. The outbreaks of cutaneous cryptococcosis in sheep and pulmonary cryptococcosis in goat are described[45, 3]. Pal (1975) is credited to elucidate the etiologic role of *C. neoformans* in caprine mastitis[18]. The udder of the goat showed induration and diffuse thickening and infected teat yielded a mucilaginous, grayish section with floccules. In India, cryptococcosis has been diagnosed in small ruminants[36, 22]. The pathogen *C. neoformans* was isolated from mastitic milk of goat and the pneumonic lungs of sheep. The first report of ovine pulmonary cryptococcosis in India is published by Pal[22]. The genetic crossing experiment on Pal's medium indicated that the strains of *C. neoformans* were of 'alpha' mating type.

Dermatophytosis

Dermatophytosis (ringworm, tinea) is a cosmopolitan, superficial and highly contagious mycosis of man and animals[23]. The disease occurs in sporadic and epidemic forms and is caused by dermatophytes which attack the keratinized tissues of the skin, hair and nail. Presently, 40 different species of dermatophytes are identified. *Microsporum canis* is emerged as a global zoophilic dermatophyte. About 80 per cent of ring worm cases in man is contracted from animals. Many species of animals such as buffalo, camel, cat, cattle, goat, monkey, etc. may serve as reservoir of infection to human beings. It is an important occupational zoonoses of veterinarians, animal handlers, dairy workers, agricultural workers, pet owner, butchers and livestock raisers. The disease has been diagnosed in small ruminants[41,1,17,39]. The dermatophytes isolated from the ringworm cases of small ruminants are summarized in tabular form.

Goat	Sheep
1. *Microsporum canis*	*Microsporum canis*
2. *M. fulvum*	*T. mentagrophytes*
3. *M. gypseum*	*T. quinckeanum*
4. *T. mentagrophytes*	*T. rubrum*
5. *T. verrucosum*	*T. verrucosum*

The goat and sheep can get infection from anthrophilic, geophilic and zoophilic dermatophytes. In India, the zoonotic implications of *M. canis* in dermatitis of goat handlers is established by Pal[25] and Pal and Dave[33]. In both cases of human ringworm, the diseased goats were the principal reservoir of infection. The zoophilic dermatophyte *M.canis* was demonstrated by culture as well as direct microscopy in the cutaneous lesions of man and goats.

Geotrichosis

Geotrichosis is a sporadic, usually non-fatal, rare opportunistic fungal disease of man and animals. The disease is caused by *Geotrichum candidum*, a saprophytic fungus that has been recovered from air, water, soil, milk, fruits, vegetable, cereal and decaying organic matter. The infection was reported for the first time by Link in 1809 by isolating *G. candidum* from a patient. The disease has been diagnosed in HIV patients. It is believed that human and animal infection result from saprobic reservoirs. In India, *G. candidum* is isolated from the mastitic milk of a goat on Sabouraud medium and sunflower seed agar[20]. The goat had a history of chronic mastitis and did not respond to conventional treatment of mastitis. There is no report of geotrichosis in sheep from India.

Histoplasmosis

Histoplasmosis, also known as caveran disease, Darling's disease, reticuloendothelial cytomycosis and Tingo Maria fever, is a highly infectious, chronic, noncontagious, inhalation acquired systemic mycosis[23]. Dr. Samuel T. Darling (1905-1906) is credited to record the disease for the first time in a human patient from USA. The disease is caused by the dimorphic fungus *Histoplasma capsulatum* that grows as a saprophyte in the bat guano, soil and avian excreta. The infection is commonly acquired through respiratory tract. The disease is endemic in USA and maximum cases are reported from Ohio, Missouri and Missisippi. It is estimated that more than 40 million people have been infected and 50,000 new cases occur each year in the United States. The disease has been reported in many species of animals including sheep[13, 1, 23]. Menges[14] recorded 33 per cent histoplasmin sensitivity in sheep in USA. Hitherto, there is no record of disease in goat. Histoplasmosis has not been reported so far from small ruminants in India.

Paecilomycosis

Paecilomycosis is a rare opportunistic mycotic disease of animals and man. Several species of *Paecilomyces* such as *P. lilacinus, P. variotii* and *P. virides* are associated with the disease. These fungi are common inhabitant of the soil. *P. variotii* has been isolated from endocarditis mainly involving prosthetic valves. The etiologic significance of *P. variotii* is established in caprine mastitis by Pepin and Pritchard[40]. There is no record of disease in goat and sheep in India.

Paracoccidioidomycosis

Paracoccidioidomycosis (Brazilian blastomycosis, Lutz's mycosis, South American blastomycosis) is a chronic, granulomatous, noncontagious, systemic mycosis primarily of man. The disease is caused by *Paracoccidioides brasiliensis,* a dimorphic fungus that is isolated from humid and protein rich soil. Recently, the pathogen has been isolated from the Armadillos in Brazil. In India, the pulmonary form of disease in goat and sheep is reported on the basis of histopathological findings[42]. However, *P. brasiliensis,* the principal etiological agent of disease, was not isolated from the infected lungs.

Rhinosporidiosis

Rhinosporidiosis is a benign, chronic, nonfatal, mycotic disease particularly of the nasal mucosa and rarely of other organs. The disease is characterized by polypoid growths and is caused by *Rhinosporidium seeberi*. The organism has not been cultured on laboratory media [(38)]. The ecological habitat of *R. seeberi* is still unknown. The traumatic injury to the mucosa or skin may predispose the host to infection. The first report of animal rhinosporidiosis is published by Zschokke in 1913. The disease is reported from many countries of the world. The prevalence rate in human is higher in India particularly in the states of Kerala and Tamil Nadu. The outbreak of disease is reported from India which involved cattle, buffaloes, goats and ponies. Pal and co-workers[38] reported nasal rhinosporidiosis in a she-goat. The diagnosis was made by demonstrating thick-walled sporangia and sporangiospores in the H and E stained smear of the nasal exudates of the animal. There is no record of the disease in sheep.

Rhodotorulosis

Rhodotorulosis is a rarely occurring opportunistic fungal disease of humans as well as animals. The disease is caused by *Rhodotorula glutinitis, R. mucilaginous* and *R. rubra*. The yeast is present in the environment. The source of infection is exogenous. The systemic infection is reported in immunocompromised patients. The pathogen has emerged as one of the most important causative agents of peritonitis. In India, pulmonary infection due to *R. rubra* in sheep is reported[15]. The author has isolated *Rhodotorula* spp. from a case of caprine mastitis.

Sporotrichosis

Sporotrichosis (Berumann's disease, Petmoss disease, Schenek's disease) is an infectious, sporadic, subcutaneous or chronic granulomatous mycotic disease of man and animals. The dimorphic fungus *Sporothrix schenckii* is the sole etiological agent of the disease. The organism is identified for the first time by Schenek in 1898. The infection usually occur following traumatic inoculation of *S. schenckii* into the skin by contaminated wooden splinter, thorn, plant, sphagnum, peat moss etc.[32] The disease is cosmopolitan in distribution, and cases have been recorded in man and animals[1,23,32]. Among the various animals, the cat is considered as an important species with greatest zoonotic potential. There are reports that the bite of infected cat and squirrel may produce disease in man. A solitary report of caprine sporotrichosis appeared from India[7]. A female goat with a history of diarrhoea and unresponsiveness to treatment showed congestion and oedema of the lungs. The pathogen was recovered from the trachea. There is no record of disease in sheep.

Zygomycosis

Zygomycosis, also known as mucormycosis, phycomycosis, is a sporadic, infectious, polymorphic, opportunistic fungal disease of man and animals. Many species of zygomycetes such as *Absidia corymbifera, Basidiobolus ranarum, Conidiobolus coronatus, Mucor racemosus, Rhizomucor pusillus, Rhizopus microsporus, R. oryzae* and *Saksenaea vasiformis* are incriminated as the etiological agent of the disease. These fungi contain broad, non-septate hyphae; and have the tendency to invade the blood vessel resulting into thrombosis and infarection. Man and animals acquire the infection from the saprobic environment. Disseminated form of the alimentary tract zygomycosis is diagnosed only at postmortem examination. The disease is reported from many countries of the world[16, 2, 1, 5]. In India, zygomycosis is encountered in goat and sheep[12, 5, 9]. The diagnosis in majority of cases is based on histopathology. The isolation of the fungus is seldom attempted.

DIAGNOSIS

Correct and early diagnosis is imperative in the management of mycotic diseases. Clinical observations may help in cases of ringworm, rhinosporidiosis and maduromycosis. Radiography, CT and MRI may be useful to locate the lesion in the organ. However, the demonstration of fungal organism in the clinical specimen by direct microscopy and its cultural isolation in pure growth is considered as 'Gold Standard' in the diagnosis of mycoses. Wet mount, KOH preparation (Fig. 1), India ink mount, stained smear with Gram, Giemsa, methylene blue, PAS (Fig. 2) are useful for direct detection of fungal elements in clinical specimens. The organism can be isolated from CSF, pus, blood, urine, skin scrapings (Fig.3), uterine discharge, corneal scraping, ear swab, milk (Fig. 4, 5), sputum, nasal exudates, biopsied and autopsied tissues on various mycological media such as Sabouraud dextrose agar with chloramphenicol, Pal's sunflower seed medium, brain heart infusion and determatophyte test medium (DTM). Pal's medium is an excellent selective medium for rapid identification of *Cryptococcus neoformans*[24]. DTM is employed for the recovery of dermatophytes from the skin scrapings[30]. A number of immunological tests such as AGID, CF, CIE, ELISA and LA are used to demonstrate antibodies and antigens in the sera of diseased patients. The skin test is of little diagnostic value but is found useful to identify the endemic areas of coccidioidomycosis and histoplasmosis. Laboratory animals particularly the mice are employed to verify the mycotic agent. Histopathological examination of tissues by Griedley stain, Gomori methanamine silver, periodic acid-Schiff, Mayer's mucicarmine, and haematoxylin eosin stain techniques are used to detect fungal agent. Very recently, molecular tools viz. PCR, RFLP, DNA probe etc. are employed for the diagnosis of mycoses. Routine application of molecular techniques due to high cost is restricted for the diagnosis of mycotic diseases. Genetic crossing of *Cryptococcus neoformans* and dermatophytes is done on 'Pal's sunflower seed medium' to know the mating types[22,25,31]. The detailed morphology of the fungi is studied by 'Narayan' staining method[28] (Fig. 6, 7).

CHEMOTHERAPY

Surgical excision of the affected tissue is advised in rhinosporidiosis, maduromycosis and sporotrichosis. Superficial mycosis can be treated with topical drugs such as clotrimazole, ketoconazole and miconazole[23]. Pulmonary and systemic mycoses are managed with amphotericin B (0.1 to 1.0 mg/kg IV), fluconazole (100 to 400 mg PO), flucytosine (15 to 150 mg/kg PO), itraconazole (50 to 400 mg PO) and ketoconazole (100 to 200 mg PO). Very recently, posaconazole and voriconazole are tried in coccidioidomycosis and aspergillosis, respectively. Despite toxicity, amphotericin B is still considered the main stay for the chemotherapy of mycozoonoses[31].

PREVENTION AND CONTROL

Since the fungi are widely distributed in our environment, complete eradication of mycotic diseases is very difficult. However, early diagnosis and chemotherapy, immediate attention to traumatic skin injury, care during handling of diseased patient or infected material, use of protective clothings, proper destruction of skin crusts and other pathological materials, thorough washing of hand with antiseptic solution, avoiding prolonged use of immunosuppresant drugs, good animal husbandry practice and decontamination of saprobic reservoirs may help in the control of mycoses. Greater awareness among the veterinarians and physicians about the growing role of fungi in various clinical disorders of man and animals, establishment of good mycology laboratories, appointment of well-qualified and experienced staff and introduction of course on diagnostic mycology are highly imperative in the better management of mycoses.

Conclusion

Mycoses are infectious diseases of man and animals that are global in distribution. Mycotic infections are recognized as a potential health hazard of considerable dimensions. Some of the diseases such as aspergillosis, dermatophytosis, sporotrichosis, histoplasmosis and cryptococcosis may occur in epidemic form. Most of the pulmonary and systemic mycoses are saprozoonoses as the environment serves as the common source of infection for humans as well as animals. Dermatophytosis, and blastomycosis are anthropozoonoses. Several, mycozoonoses are recorded in small ruminants. As veterinary mycology is still one of the neglected areas of medicine particularly in India, sincere attempts should be made to study the role of fungi in various clinical disorders of animals and also to elucidate their zoonotic significance. It is emphasized that further work on the diagnosis, epidemiology and chemotherapy should be undertaken.

References

1. Ainsworth, G.C. and Austwick, P.K.C. 1973. Fungal Diseases of Animals. 2nd edn. C.A.B. Farnham Royal, Slough, England.

2. Angus, K.W., Renwick, C.C. and Robinson, G.W. 1971. Segmental necrosis of the ileum in a lamb associated with a phycomycete. Veterinary Record 88: 654-656.

3. Baro, T., Torres-Rodriguez, J.M., De Mendoza, M.H., Morora, Y and Alia, C. 1998. First identification of autochthonous *Cryptococcus neoformans* var. *gattii* isolated from goats with predominantly severe pulmonary disease in Spain. Journal of Clinical Microbiology 36: 458-461.

4. Chapman, H.M., Rohinson, W.F., Bolton, J.R. and Robertson, J.P. 1990. *Cryptococcus neoformans* infection in goats. Australian Veterinary Journal 67: 263-265.

5. Chattopadhyay, S.K., Iyer, P.K.R. and Rao, M. 1986. Systemic phycomycosis in sheep: a case report. Indian Journal of Veterinary Pathology 9: 66-69.

6. Fragner, P., Vitovec, J., Vladik, P. and Tesarik, L. 1970. Bronchopulmonary aspergillosis in lambs. Mycopathologia 40: 337-340.

7. Gajendragad, M.R., Prabhudas, K. and Yadav, M.P. 1986. A case of sporotrichosis in a goat. Indian Veterinary Journal 63: 110.

8. Hajsig, M., Naglic, T., Hajsig, D. and Herceg, M. 1985. Zystemic mycoses in domestic and wild ruminants. I. Candidosis of forestomaches in the lamb, kid and newborn illama. Veterinary Archiv 55: 53-58.

9. Kamil, S.K. and Parihar, N.S. 1991. Pathology of mycotic pneumonia in sheep. Indian Journal of Animal Sciences 61: 13-18.

10. Krishna, L., Kulshrestha, S.B. and Paliwal, O.P. 1981. Systemic aspergillosis in lambs. Indian Veterinary Journal 5: 70-73.

11. Krishna, L., Kulshrestha, S.B. and Paliwal, O.P. 1984. Systemic *Aspergillus terreus* infection in a kid. Indian Journal of Animal Sciences 54: 308-310.

12. Krishna, L., Paliwal, O.P. and Jain, S.K. 1976. Phycomycotic abomasitis in a goat. Indian Journal of Veterinary Pathology 1: 39-40.

13. Maddy, K.T. 1954. Coccidioidomycosis in a sheep. Journal of American Veterinary Medical Association 124: 465.

14. Menges, R.W. 1954. Histoplasmin sensitivity in animals. Cornell Veterinarian 44: 21-31.

15. Monga, D.P. and Garg, D.N. 1979. Ovine pulmonary infection caused by *Rhodotrula rubra.* Mykosen, 23: 208-211.

16. Mostafa, I.E., Cerna, J. and Cerny, L. 1966. Mucormycosis in goats. Report of two cases. Ceylon Veterinary Journal 14: 79-82.

17. Nayak, N.C., Chakraborty, T. and Bhownik, M.K. 1989. Prevalence and pathology of caprine dermatophytosis due to *Trichophyton mentagrophytes*. Indian Journal of Animal Sciences 59: 1380-1383.

18. Pal, M., 1975. A study of cryptococcosis with special reference to its epidemiology. MVPH Dissertation, University of Calcutta, India.

19. Pal, M. 1978. Saprobic reservoirs of *Cryptococcus neoformans*. Indian Journal of Public Health 22: 327-328.

20. Pal, M. 1982. Association of fungi with caprine mastitis. In Proceedings of 3rd International Conference on Goat Production and Disease. Tuscon, Arizona, USA, January 10-15, 1982.

21. Pal, M. 1989. *Cryptococcus neoformans* var. *neoformans* and munia birds. Mycoses 32: 250-252.

22. Pal, M. 1989. Studies on the prevalence of cryptococcosis in respiratory disorders of domestic animals. Revista Iberica De Micologia 6: 29-33.

23. Pal, M. 1997. Zoonoses. Ist edn. R.M. Publishers and Distributors, Delhi, India.

24. Pal, M. 1997. Use of Pal's sunflower seed medium for an early diagnosis of cryptococcosis. Antiseptic 95: 175.

25. Pal, M. 2001. Dermatitis in a goat and its handler due to *Microsporum canis.* Indian Journal of Animal Sciences 71: 138-139.

26. Pal, M. 2002. A view on mycotic infections in India. Veterinarian 26: 3-5.

27. Pal, M. 2004. *Aspergillus fumigatus* isolated from ornamental plants in India. Revista Iberoamericana De Micologia 21: 218.

28. Pal, M. 2004. Efficacy of "Narayan stain" for morphological studies of moulds, yeasts and algae. Revista Iberoamericana De Micologia 21: 219.

29. Pal, M. 2005. Coccidioidomycosis: A mycotic zoonosis: Veterinary World 3: 10-13.

30. Pal, M. 2005. Diagnosis of dermatophytosis. Veterinary World 3: 304-305.

31. Pal, M. 2006. Mycotic zoonoses: Global health problem. Guest Lecture delivered at National Symposium on Emerging Zoonoses at Veterinary College, Jammu, April 18-19, 2006.

32. Pal, M. 2006. Sporotrichosis: A mycozoonosis. Veterinary World 4: 378-380.

33. Pal, M. and Dave, P.S. 2005. Tinea faciei in a goat handler due to *Microsporum canis*. Revista Iberoamericana De Micologia 22: 181-182.

34. Pal, M. and Lee, C.W. 1999. Spontaneous pneumonia in a goat due to *Candida albicans*. Korean Journal of Veterinary Clinical Medicine 16: 467-468.

35. Pal, M. and Mehrotra, B.S. 1985. Studies on the isolation of *Cryptococcus neoformans* from fruits and vegetables. Mykosen 28: 200-205.

36. Pal, M. and Randhawa, H.S. 1976. Caprine mastitis due to *Cryptococcus neoformans*. Sabouraudia 14: 261-263.

37. Pal, M., Sejra, S., Sejra, A.R. and Lee, C.W. 1995. Caprine pneumonia caused by *Aspergillus flavus*. Korean Journal of Veterinary Clinical Medicine 12: 203-206.

38. Pal, M., Singh, B., Padhi B. and Jain, H.S. 1984. Studies on the occurrence of nasal rhinosporidiosis in domestic ruminants. Indian Veterinary Medical Journal 8: 207-210.

39. Pal, M., Sukumaran, K., Sejra, A.R. and Lee, C.W. 1991. Caprine dermatitis caused by *Trichophyton mentagrophytes*. Korean Journal of Veterinary Clinical Medicine 8: 147-152.

40. Pepin, G.A. and Pritchard, G.C. 1984. Fungal mastitis in a goat due to infection with *Paecilomyces varioti*. Goat Veterinary Society Journal 5: 12.

41. Roberts, D.S. and Keep, J.M. 1965. Infection of the wool of sheep by *Microsporum canis*. Sabouraudia 4: 96-97.

42. Sharma, D.N. and Dwivedi, J.N. 1977. Pulmonary mycosis of sheep and goats in India. Indian Journal of Animal Science 47: 808-813.

43. Singh, M.P. and Singh, C.M. 1972. *Candida parakrusei* isolated from goats. Indian Journal of Animal Health 11: 111-112.

44. Young, N.E. 1970. Pulmonary aspergillosis in the lamb. Veterinary Record 86: 790.

45. Zaitoun, A.M. and Ali, H.S. 1992. An outbreak of cutaneous cryptococcosis in fed lot sheep at Assiut Governorate – Upper Egypt. Assiut Veterinary Medical Journal 28: 267-278.

–ΔΔΔ–

18

Alternative Methods of Control of Gastrointestinal Nematodosis in Small Ruminants

D SINGH AND C P SWARNKAR

Division of Animal Health
Central Sheep and Wool Research Institute
Avikanagar - 304501 (Rajasthan)

INTRODUCTION

In developing countries like India, small ruminant sector can make a sizable contribution to meet the rapidly increasing demand for livestock products. Expansion of small ruminant production would not necessarily mean numerical expansion of current breeding flocks but could be achieved through more efficient production by current stock. Parasitic infections, particularly, internal parasitism is one of the major disease syndrome of pastoral ruminants simply because environment which favour pastoralism also favour the development and survival of exogenous stages of helminth parasites. The effects of internal parasitism vary between wide extremes. At one end, death and total economic loss can occur and at the other, effects may be undetectable and control measures unwarranted. Economically major consequences includes increased mortality and reduction in live weight gain (35-40%), wool yield (35-40%) and quality, fertility and milk production (10-15%), rejection of carcasses or organs for human consumption, predisposing to other diseases and hamper the efforts for genetic selection[85,114]. Since long, the control of parasitic infection is primarily achieved by the frequent use of anthelmintics. The dependence on anthelmintics has several disadvantages including (i) development of anthelmintic resistant parasite strains (ii) consumer's concern regarding residues in food products and in the environment and (iii) the negative effects of preventive measures on the development of natural immunity against worms.

The emergence of anthelmintic resistant strains of parasites has put small ruminant production in jeopardy in some regions[118] and posses a major threat throughout other major production areas[99, 111]. This situation led to realization that unless the pharmaceutical industry can provide an endless supply of new molecules, intensive chemoprophylaxis, in the way that it is currently practiced, is not a sustainable option for the control of nematodes[46]. In addition, end of the anthelmintic spectrum, high cost and tendency of consumers to prefer products from livestock managed with minimal chemical intervention led to uniform agreement among parasitologists throughout the World that conventional paradigm of chemical-based worm control must be changed, from one of maximum, to one of the optimal worm management compatible with sustainability. Globally there are a number of researchers investigating alternatives to anthelmintics that may play a future role in worm control programme as there is increasing probability that chemical residue could be used as non-tariff trade barrier[43].

Alternative approaches can be used to manage the infection (i.e. assist in maintaining it below the economic or pathogenic threshold) with in a livestock production system. Unlike anthelmintics, these methods usually do not have a direct effect on parasitic infection in the host. Instead, the mode of action is based eg., on improving the immunity of the hosts or decreasing exposure of the hosts to the parasites. Several alternative technological solutions to worm control are currently being investigated and developed. A closer look on the control strategies alternative to chemotherapy reveals that some of them are in the technological state while others are still under research and development. Biological control by nematophagous fungi, worm vaccine, exploitation of breed / genetic resistance are still undergoing laboratory research, while grazing management and exploitation of nutrition - parasite interaction can immediately be applied as adjunct to chemotherapy[91]. The present communiqué elaborates the status and future of alternative approaches for the worm management in small ruminants.

Types of parasite populations

Throughout India, among gastrointestinal nematodes, the prevalent predominant species in small ruminants is *Haemonchus contortus* followed by *Trichostrongylus* spp., *Oesophagostomum* spp., *Strongyloides papillosus* and *Trichuris* spp. *Oestertagia, Teledorsagia, Bunostomum* and *Gaigeria* spp. were also on record from hilly parts of the country. In general, the life cycle of gastrointestinal nematode is consisting of free-living and parasitic phase. The stages in the free-living phase are known as "Supra-population" consisting of eggs and different stages of larvae (1-3) on pasture. Their development, survival and dissemination are mainly governed by the climatic factors and status of pasture. The stages in parasitic phase are known as "Infra-population" consisting of L_4, immature and mature worms residing in the host. Over 90% of the population of various stages of nematode parasites lives outside the animal as supra-population. Conventional control programme (based entirely on chemotherapeutics) are targeted only against infra-population (i.e. only against 10% of worm population).

For better understanding the different approaches used in worm management are categorized according to their target population as

A. Approaches targeted against infra-population

- Chemotherapeutic management (targeted selective treatment - TST)
- Nutritional management
- Immunological control (Worm vaccine)
- Genetic control (Breed resistance)
- Bioactive forages

B. Approaches targeted against supra-population

- Grazing and pasture management
- Biological control

In addition to these, simulation models are also being evolved as an aid to worm management programme. The successes of alternative approaches are highly dependent on understanding the epidemiology of the parasite in the context of a particular management system. Parasite epidemiology determines when the economic threshold is reached as well as the level of economic impact.

Targeted selective treatment (TST)

The main focus of anthelmintic research conducted outside the pharmaceutical industry is on the ways of ensuring efficacy of anthelmintic by reducing uses, targeting treatment to those stock that require them. TST seems to represent one of the most realistic method for restraining worm resistance and an opportunity for management by resource-poor farmers. TST offer the possibility at minimal cost to treat and utilize the relatively small proportion of individuals unable to withstand parasitic challenge[122]. As wit most parasites, burdens of nematodes infecting farm animals are grossly over-dispersed with vast majority of worms occurring in a small percentage of hosts[45]. The detailed epidemiological studies in sheep flocks of Rajasthan under All India Network Programme on Gastrointestinal Parasitism revealed that even during favourable monsoon season (July – September) around 20-30% and 50% of animals in field and farm conditions, respectively had faecal egg count (FEC) to the tune of >1000 epg. Other wise in rest of seasons, majority of animals had either < 200 epg or free of infection[101]. In recent times, it has been recognized that the phenomenon of parasite over-dispersion could be put to good use if those animals suffering from level of infection sufficient to cause considerable production loss or health effects, can be identified and treated individually. Following the same TST principle and based on strong correlation that exist between the coloration of the ocular mucous membrane of sheep and degree of aneamia caused by haematophagous parasite (*H. contortus*), the FAMACHA© (Faffa Malan Chart) system has been developed and validated by South African Scientists[72, 122]. With this approach not only cost of anthelmintic and labor will be reduced but it would help to maintain population in refugia which in turn should help to conserve the genes for susceptibility within parasite population.

Nutritional Management

As available feed resources are often low in quality, fibrous and deficient in components which promote efficient rumen function, the under-nutrition in energy, protein and minerals / trace elements has all been implicated directly or indirectly in predisposing the animals to parasitic infection[108]. In ruminants resistance to both infective larvae and adult worms has been associated with the development of acquired immunity[6] and expressed by increasing host ability (i) to reject incoming larvae, (ii) to depress worm fecundity and (iii) to expel adult worms. Nutritional interventions have a major influence on the severity of parasitism and some influence on the extent of acquired immune response in host. Research has shown that increased dietary intake of metabolizable protein and energy and high quality pasture can directly promote the host resistance (ability of animal to resist infection with worms and reject established worms) and host resilience (ability of animal to maintain a relatively unchanged level of production despite a worm infection) to worm infection[57, 120] and evident by observations as mentioned in table 1.

Table 1. Expression of resistance or resilience to GI nematodes through nutritional management

Response	Possible mechanism
Decreased establishment of larvae and reduction in worm count[18, 25]	Increased mast cell proteases activity
Rejection of established worms[121]	Antigenic in nature and / or mechanical or enzymatic damage to worms
Reduced faecal egg counts[95]	By reducing worm burden and fecundity of parasites
Increased body weight and growth rate[33]	Through enhancing host ability to mount protective immunity against worms
Suppression of periparturient rise in FEC[38]	Inhibition of relaxation of established immunity
Reduced severity of infection and drench frequency[60, 86]	By compensating the deficiencies/ damage caused by worms

Energy and proteins are vital component of nutrition for animal. Energy is used by the animal to move around and to perform body activities. Protein is necessary for production (wool, meat and milk), repair of various body structures and for immune response to pathogen. In grazing animals protein supply comes from either by conversion of plant protein to microbial protein in rumen and / or by incorporating by-pass protein as a supplementary feed.

Trichostrongylosis in sheep found to reduce surface area of small intestine. The tiny finger like projections that help nutrient absorption in the intestine disappear resulting in impaired absorption of nutrients. The adverse effect of parasitism on production could be masked by supplementing with protein/energy rich diet. Lambs on diet with additional protein found to produce significantly more antiparasitic immunoglobulin A (Ig A) leading to suppression of growth of *H.*

contortus and reduced fecundity of parasite[105]. Up to 30% reduction in FEC was recorded after offering cottonseed meal (@ 250 g/day) to GIN infected ewes during late pregnancy or early lactation[50]. Similarly, lower FEC and higher packed cell volume and growth rate was observed in *H. contortus* infected lambs supplemented with cotton seed cake as extra protein compared to those which were fed wheat bran[2]. Supplementation with high quality protein is often necessary to maintain the adequate productivity but high cost often limits their widespread use. In such situation low cost technologies based on incorporation of by-pass protein, provision of non-protein nitrogen (NPN) such as urea in the diet, feeding on plants having condensed tannins (CT) etc are useful.

Dietary deficiencies or variables inhibiting the utilization of minerals can also limit the ability of immune system to deal with parasites. Cells with shorter half-life (like lymphocytes) are particularly sensitive to trace element deficiency. Phosphorus (P) deficiency found to reduce immunity of host against worms. In grazing livestock P deficiency may occur due to feeding of low P diet, decreased absorption of dietary P due to worms[115] or grazing the animals on unfertilized pasture particularly during months when pasture growth is slow. In parasitized animals 40% increase in body weight gain was observed on maintaining them with a diet containing 0.28% P on DM basis compared to those maintained on diet containing 0.18% P on DM basis. Reduced worm burden and FEC and increased daily weight gain was recorded in lambs infected with *T. vitrinus* and offered diet fortified with P (@ 1.88 to 2.75 g/Kg DM)[24]. Copper has direct anthelmintic properties against abomasal parasites. It's supplementation (@ 2 – 5 g as copper oxide needles or copper oxide wire particles, COWP) found to reduce worm burden (75%), FEC (37-95%) and establishment of incoming larvae of *H. contortus* (95%)[7, 21, 58]. The another trace element having potential role in parasitism is Molybdenum (Mo) which may enhance the inflammatory response to nematodes by reducing the effectiveness of local Cu-dependent anti-inflammatory enzymes thus enhancing the rejection of abomasal worms. The studies on correlated responses in the intestinal antibody, jejunal mast cells and blood eosinophil counts suggested that Mo has an important enabling role in the cell-mediated immune response by enhancing the inflammatory response through increasing superoxide radical concentrations in the mucosa. Dietary intake of Mo (@ 0.15 to 0.30 mg /Kg body weight) results in significant reduction in FEC and worm burden in sheep infected with *H. contortus* and *Trichostrongylus* spp[106, 107].

In short term, the cost of nutritional interventions is larger than those using anthelmintics. However, the benefits of nutritional control of parasites in terms of productivity are often manifested over a longer period of times[51]. Cost-benefit analysis of such intervention in management of GI parasites has yet to be given the attention it deserves. Following points are of consideration while evaluating the gains of nutritional approach in worm management:

- Improved protein supplementation at times of scarcity will invariably improve resilience in the face of parasitism.
- Compared to their un-supplemented counter part, there will be better growth.
- Higher milk production in supplemented dams leads to higher pre-weaning weight gain of the lambs and kids.
- Improved nutrition can reduce worm burden and FEC. The former may contribute to reduce losses due to parasites *per se*, whilst the reduced FEC can contribute to lowered pasture infectivity.

Genetic control

Breeding or selecting lines of animals, either from within breed or between breed that are less reliant on anthelmintics for maintaining the health and productivity is one of another option to manage the supra-population of worms and associated of problem of anthelmintic resistance as well as to meet consumers demand for less drug usage in livestock. Worm resistant animals can be found in any flock, whether they are in a wormy, high rainfall environment or in a dry area. Studies have confirmed that at least part of the variation (20-30%) in an individual's ability to regulate worm is under genetic control[49] and is moderately heritable[98, 102] (Table 2). On comparison between a number of non-genetic and genetic strategies, it was observed that genetic strategy is most successful at reducing parasitism. Over a period of 224 days, FEC were reduced by 69% by genetic selection, 35% by protein supplementation, 28% by drenching and unaffected by vaccination in Merino flocks[39].

It has been recognized for decades that breeds of sheep differ in their resistance to nematode infection (Table 3). In India, at Nimbkar

Table 2. Heritability of FEC following natural/artificial infection of GIN in sheep

Breed	Age	Infection type	h^2
Merino[133]	Weaners	Artificial *T. colubriformis* after vaccination with irradiated *T. colubriformis*	0,41 ± 0.04 0.41 ± 0.09
Romney and its crosses[132]	Weaners	Natural *T. colubriformis* and *O. circumcincta*	0.34 ± 0.19
Merino[83]	18 months	Artificial *H. contortus*	0.23 ± 0.13
Merino[31]	Weaners	Natural *H. contortus*	0.42 ± 0.14
Romney[5]	Weaners	Natural *H. contortus*	0.34 ± 0.09
Romney[67]	Weaners	Natural *H. contortus*	0.13 ± 0.07
Merino[1]	Weaners	Artificial *H. contortus*	0.34 ± 0.10
B. Merino[110]	Hogget	Natural *H. contortus*	0.42 ± 0.27
Polish long wool[17]	Lambs Ewes	Natural *H. contortus* and *O. circumcincta* Natural *H. contortus* and *O. circumcincta*	0.22 to 0.33 0.18 to 0.25
Avikalin[98]	Hogget	Natural *H. contortus*	0.35 ± 0:18

Agricultural Research Institute (NARI), Phaltan (Maharastra), relatively high worm resistance was observed in Garole breed followed by Bannur, compared to the Deccani and Awassi breed[79]. An inverse relationship between FEC and inheritance of exotic blood in sheep was observed at Central Sheep and Wool Research Institute, Avikanagar, Rajasthan[109] and estimated moderate heritability (h^2) for FEC in artificially challenged Avikalin breed of sheep[98]. In various crosses of Garole, Deccani, Bannur, Awassi and Patanwdi breeds, the

Table 3. Resistance to parasites in different sheep breeds

Breed	Compared with	Type of infection	Results
Navajo[56]	Suffolk, Targhee, Rambouillet, Corriedale	Natural	Reduced number of worms
Barbados black belly[135]	Dorset and its crosses	Artificial /Natural	Decreased FEC and Worm count
Florida Native[66]	Rambouillet, Hampshire	Artificial /Natural	Decreased FEC, Hb and Worm count
Florida Native[47]	Rambouillet,	Artificial /Natural	Decreased FEC, Hb and PCV
Florida Native, St. Croix, Barbados[26]	Rambouillet Finn Dorset *x* Rambouillet,	Artificial /Natural	Reduced or no PPR
Florida Native, St. Croix[27, 28]	Finn Dorset x Rambouillet, Barbados blackbelly	Artificial /Natural	Decreased FEC, No difference in worm count
Florida Native[136]	Dorset *x* Rambouillet	Artificial /Natural	Decreased FEC, worm count
St. Croix[41]	Dorset	Artificial /Natural	Decreased FEC, worm count & globular leucocytes
Targhee x Barbados[117]	Targhee	Artificial	No difference in FEC, Worm count, PCY & Weight gain
Florida Native [19, 87]	Rambouillet	Artificial	Decreased FEC, worm count, PCV, pepsinogen, abomasal eosinophils
Florida Native, St. Croix[137]	Dorset X Rambouillet	Artificial	Decreased FEC, PCV, Total protein

h^2 for FEC was 0.12 ± 0.07 in artificially challenged lambs and 0.09 ± 0.02 in naturally challenged ewes[79].

Selection of resistant individual or breed is feasible using phenotypic markers such as FEC, packed cell volume, peripheral eosinophil counts etc. However, one of the major difficulties inherent in using phenotypic markers in naturally infected animals is the need for same uniformity of challenge. Since artificial challenge of large number of animals in different environment also possess problems, researchers are seeking to identify appropriate genetic markers[37]. Though, candidate genes and alleles have been identified however, to date not a single common gene has been identified [23, 29].

The genetic markers allow detection of specific region of DNA which is associated with worm resistance. The usefulness of these depends initially on the genetic control of the resistance trait and on the proximity of the markers on the chromosome to the gene of interest. If the trait is due to the combined action of several genes of small effect (polygenic) it is unlikely that genetic markers will be useful in identifying animals of superior resistance status. However, if the resistance is due to the effect of a single gene, or the action of one or a few genes of moderate or large effect (i.e. Quantitative Loci Trait, QTL) the probability of identifying useful markers is much greater[49]. The inclusion of resistance to GIN in sheep breeding programme might be facilitated by the identification of QTL or closely linked markers. However, the investigation of QTL for resistance to GIN in sheep is a difficult area of research, mainly due to a large number of problems associated with phenotypic and analytical complexity of this trait. A number of studies have investigated QTL for resistance in sheep with a diverse range of approach and have identified regions of interest on different chromosomes (viz., 1, 3, 6 and 20)[37]. A QTL conferring resistance to *H. contortus* and accounting for about 1/3rd of the total genetic variation in FEC was detected in the "Golden Ram" Merino flock in Australia[49]. Similarly evidence has been found for a putative QTL in mean FEC on primary challenge with *H. contortus* in Indonesian Thin Tailed sheep. The preliminary results with established experimental lines in New Zealand indicate that there may be several genes that have substantial effect on host resistance. Areas of the sheep genome containing one of these genes have already been patented as a marker[30] and work is underway on the other genes involved.

Further, the use of Marker Assisted Selection (MAS) techniques are considered as most effective way in the future to identify sheep that are both resistant and resilient to worm infection[16]. The advantages of MAS techniques include:

1. No need to subject animals to prolonged worm challenge, so production cost associated with the test procedure will be avoided as well as costly and messy sampling procedures.
2. Higher possibility to determine an animal's genetic status at an early age.
3. Overcome the problem associated with recessive or over-dominant genes under traditional selection pressure.
4. Valuable in overcoming the antagonistic association between low FEC and some production traits.

Immunological control

The induction of immunity through vaccination is a very cost-effective control measure to protect ruminants from several important bacterial, viral and protozoan diseases. Earlier, it was thought that ruminants could not develop natural immunity to GIN because they were remain effectively outside the body and could not affected by the immune system. This was found to be not so and the possibility of developing vaccine against internal parasites suggest an additional possible alternate mean of control. Attempts to control helminthic diseases using vaccine have met with more limited success. Lung worm (*Dictyocaulus filarial*) can be controlled effectively in sheep using an attenuated larval vaccine[93]. Successful vaccines have also been developed to control larval stages of dog tape worms *Taenia ovis* and *Echinococcus granulosus*[43]. To make a vaccine against a gut parasite, it is necessary to identify the components of the worm (antigens) that are critical to stimulate a protective immune response in the host. Because the worms are composed of hundreds of different proteins and carbohydrates, it is not a trivial task to identify the necessary molecules. This is one of the reasons why vaccine development has been slow. Over the past decade, several research groups, have tried both native antigens (harvested from worms) and recombinant antigens (synthesized e.g. by bacteria) from *Haemonchus, Trichostrongylus* and *Ostertagia* spp. as vaccines [40, 49, 71, 103, 119].

These immunogenic antigens are of two types: concealed antigens (hidden), which are not normally "seen"; and conventional or excretory / secretory (ES) antigens that are "seen" by the host immune system during natural infection.

Conventional antigen vaccines

To get adequate all round protection of our livestock, it will be necessary to have vaccines that are effective against both abomasal and intestinal worms. Concealed antigen vaccines found to have no role against parasites which browse on the gut mucosae like *T. colubriformis, O. circumcincta* etc [84, 103]. These vaccines have not been able to generate high level of protective responses that are acquired following natural infection. The high levels of protective immunity following natural infection in adult sheep are associated with allergic type response. The research carried out at CSIRO Livestock Industries with neonatal lambs indicated that trickle and truncated infections more effectively immunize suckling lambs than weaned lambs against *Haemonchus* and *Trichostrongylus*. Striking protective effects have been found against *H. contortus* using 2 proteins of 15 and 24 kD from adult ES products. In contrast to the situation with hidden antigens, immunity to natural antigens will be boosted by field exposure to the parasites and this may enhance the practical usefulness of natural antigens as vaccine candidates.

Concealed antigen vaccine

These are based on proteins that have been isolated from the surface of the worm intestine. In response to antigen, the produced antibodies circulate in blood and ingested by the parasites and adhere to the gut of worm leading to interruption in normal digestive process and death of worms. Novel antigen from gut of *H. contortus* (H 11), a prominent 110 kD integral membrane glycoprotein[73,119] is able to induce high level of protective immunity in a variety of sheep breeds, become the vaccine candidate. The major problem that has so far prevented commercialization of a vaccine based on H11 is the inability to obtain adequate level of protective immunity after vaccination with recombinant version of H11 protein. Further, as the antigens are hidden the immunity acquired through vaccination will not be boosted by exposure to natural infection which implies the need for regular revaccination. Unfortunately it is not cost - effective to harvest the candidate vaccine proteins from *H. contortus* for use in a commercial product.

Research in the field of worm vaccine has been disappointing because control of internal parasites with vaccines is not as high as the level of control afforded by anthelmintics. Unlike broad spectrum anthelmintics, vaccines are likely to be effective against only one species of parasite, so that the ultimate goal, a multivalent worm vaccine, is most likely to be achieved by combining several individual vaccines - a prolonged process. There are presently no commercial vaccines for any gastrointestinal worm parasite species of any host[49]. In the last few years, however, a number of potentially useful antigens have been discovered for each of the main worm species of sheep. In the early 1990s there were a number of nematode vaccines under study and some showed some degree of success (Table 4).

Table 4. Experimental nematode vaccines tested in the early 1990s

Nematode	Animal Species	Antigen Component	Source of Antigen	Recombinant Antigen	Reduction (%)	
					Worms	FEC
T. colubriformis	G. pigs	Tropomyosin	L_3 larvae	+	43-51	-
	G. pigs	Exsheathing fluid	L_3 larvae	-	<50	-
	Sheep	Excretory / secretory fluid	Adult	+	<50	-
	Sheep	Secretory fluid	Adult	-	31	0
H. contortus	Sheep	Exsheathing fluid	L_3 larvae	-	60	46-66
	Sheep	Tropomyosin	L_3 larvae	-	54	46
	Sheep	Gut H11	Adult	+	78-95	78-99
	Sheep	Intestine H-gal-GP	Adult	+	54-93	58-97
	Sheep	Surface	L_3 +L_4 larvae	-	-	-
	Sheep	Somatic	Adult	-	69	21
O. circumcincta	Sheep	Excretory/secretory	L_3 larvae	-	58	<50
O. radiatum	Calves	Excretory/secretory	L_3 larvae	-	99	75

Modulation of host immune system

It has been found that among animals within population, the pathogenic effect of worms is variable indicating that there is a genetically determined propensity in some animals to mount an exaggerated allergic type response to nematode antigens. It is true to say that "almost without exception, worms do not harm stock, it is the response of stock to the worms that cause harm". An alternative approach for hypersensitive animals would be to ameliorate the damaging effects of worms by reducing or "down - regulating" the immune response which occur. In this approach, potentially dangerous allergic responses are either modified / dampened or may disappear altogether[104]. It is accomplished by targeting the important intermediate molecules like cytokines and achieved by giving graded doses of an allergen so the immune response to natural challenge is altered. Using this approach in lamb at Lincoln University, it was found that worm burden of 300. 000 have not led to loss of appetite or other harmful effects.

Bioactive forages and role of condensed tannins (CTs)

Beneficial effect on the ability of ruminants to withstand internal parasitic infection has been found with tannin - containing plants[44]. CTs are polyphenolic secondary plant metabolites, well recognized for their ability to form complex with soluble protein at normal ruminal pH (when fed at concentration less than 10% of diet). This restrict the normal break down of protein to ammonia in the rumen and allow more dietary protein to pass unchanged to the small intestine, acting as by-pass protein. The increased protein supply to animals leads to increased resistance to infection[69, 81] as evident by suppression of growth and fecundity of worms, reduced worm burden and FEC and improved status of protein and phosphorus[4, 20].

On *in-vitro* trials, it has been observed that CTs significantly inhibited the development of eggs to infective larvae. The study conducted at CSWRI, Avikanagar revealed significant reduction in larval development rate in faeces from CT fed animals[100]. CTs could bind to the protein available in the nutrient media for larvae leading to larval starvation and death as well as developing larvae can ingest CT, which bind to the intestinal mucosa and cause autolysis. CTs may also inactivate hatching enzymes in worm eggs (causing reduction in egg hatching) and can paralyze the body musculature (pharyngeal muscles) of larvae (leading to reduced viability of larvae). Hence, sufficient evidences are available that suggests that CTs can disrupt the life cycle of nematodes and help in worm management. On *in-vivo* study[75, 76, 77] up to 50% reduction in worm burden was observed in lambs fed with sulla (*Hedysarum coronarium*) or great trefoil (*Lotus pedunculatus*) leaves containing CTs. Around 40-60% reduction in FEC with marginal rise in haemoglobulin and packed cell volume were observed in sheep (infected with GIN) fed CTs (@ 5 % of DM) either as leaves of Khejri (*Prosopis. cineraria*) or as extract (acetone:water) of leaves compared to those offered diet having no CT[100]. The production losses caused by GIN were found to be alleviated by feeding the animals on White clover[78].

The consumption of CTs may not result in economic gain because of the negative effect of excessive CT consumption on the performance of host in term of reduced feed intake and digestibility, impairment of rumen metabolism and mucosal toxicity[34]. The economic benefits of CTs were more pronounced in grazing studies with parasitized animals, however it is not certain whether the improvement in performance was the direct consequence of a reduced larval intake, an anthelmintic like effect or grazing of high quality fodder. The letter might have been the case as the majority of tanniferous forages are leguminous and usually are of higher nutritive value than conventional grazing pastures. It is therefore possible that improved performance observed in animals grazing such forages is a consequence of improved resilience rather than the consequence of improved resistance.

Phyto-anthelmintics

These have been used for centuries by farmers and traditional healers to treat parasitism and improve performance of livestock and many modern commercial medicines are derived from plants. Through improved understanding of nematode parasites and diseases they cause, new opportunities are being created to develop anthelmintic therapies using natural molecules (small peptides, enzymes etc) to interfere with natural processes in worm (metabolism, egg production, larval development etc). In India, under NATP a large number of crude extracts from different plants were evaluated for antiparasitic activities and it has been found that *Malia azedarach* (Bakain), *Butea frondosa* (Palas), *Artemesia martima* (Kirmala), *Mallotus philippinensis* (Kamala), *Azadirachta indica* (Neem), *Embelia ribes* (Baberang), *Tagetes patula* (Gaatakadi), *Andrographis peniculata* (Kalmegh), *Vernonia anthelmintica* (Kaljiri), *Vitex negundo* (Bana), *Xanthium stramarium* (Chotta Gokhru) etc posses low to high level of antiparasitic activity against different stages of GI nematodes[3, 94, 113, 123, 124]. Till date, it has been felt that scientific evidences on the antiparasitic effect of most plant products are limited, regardless of their wide ethno-veterinary usages. Scientific validation of the antiparasitic activities and possible side effects of phytoproducts in ruminants is necessary prior to their adoption as an alternate mean of worm control[82]. In most cases the active material has to be extracted from plants and *in-vitro* conditions and concentrations used are not always comparable to those *in-vivo*, thus often result can differ in the two assays. One may question the relationship of *in-vitro* studies with *in-vivo* efficacy. Considerable physiological differences are expected to be present between the *in-*

vitro conditions and in the predilection site of worm within their animal hosts including factors that affect bioavailability of the active compounds.

This approach involves high-throughput screening of natural products. While based on traditional methods, automation of screening has greatly increased the number of biological extracts which can be screened, potentially leading to more rapid identification of promising compounds. A variety of other technical breakthroughs have enhanced the ability of researchers to develop biological anthelmintics[43]. These include powerful "phage display" strategies to screen a huge variety of peptides or recombinant antibodies for those capable of binding to specific targets[32]. Mixtures of bacteriophage displaying different peptides on their surface are exposed to a biological target and phage that binds are selectively enriched by affinity. Since the phage contains the DNA that encodes the binding peptide, this peptide can be characterized simply by sequencing the DNA. If desired, the peptide or antibody can be re-expressed in a different recombinant protein expression system. Such peptides could be used to identify exposed proteins, or available binding sites, present on complex biological targets such as a worm surface. Alternatively, the peptides could become biological anthelmintics as targeting sequences for bioactive molecules such as protein toxins or as receptor binding agonists or antagonists[43].

Grazing/ pasture management

As a mean of limiting host-parasite contact has been utilized for several years. It is difficult to apply in many extensive production system and those with common grazing. On the other hand in many intensive systems there may not be sufficient land for conservation or adequate number of non - susceptible stock to provide a sufficient reduction in the number of parasites on pasture. The role of grazing management in worm control programme based on epidemiological knowledge is simply to provide "low risk" pastures (those with the fewest worm larvae) for the most susceptible sheep weaner and lactating ewes). The aim of grazing management is to have vulnerable sheep exposed to fewer larvae on pasture which not only supposed to reduce drench frequency but also provide better nutrition at the same time, thus allow sheep to better deal with parasites. The various aspects of grazing management had been reviewed thoroughly[9, 10, 126] the various forms of grazing management. Good grazing management is making the most efficient use herbage grown on a farm. The success of the strategy depends upon the relation between the annual cycle of the number of worm larvae on the pasture and when the grazing management practice is applied. It is most successful if applied when larval numbers are at low levels and this is determined by the season. If pasture larvae burden is high during winter than better to target the summer for reducing larvae on the pasture (in trichostrongylosis) while, in haemonchosis prone areas where high pasture larvae burdens occurred during spring and monsoon, efforts must be made to limit contamination of pasture during autumn and spring. The practices of grazing / pasture management are as follows:

Ratio of Stock Classes

Young or susceptible animals are generally responsible for the vast majority of pasture contamination on a farm[125]. Therefore contamination rates and parasitic disease may be reduced simply by reducing the proportion of young or susceptible stock on a farm[15].

Level of Feeding

Optimal levels of nutrition are essential in combating parasitism and achieving good levels of production in its presence for all classes of stock. It is stated that "Drenching is not a substitute for good feeding" and "There is no better anthelmintic than good quality green grass". To optimize feeding levels, knowledge of feed requirements and optimum pasture covers for susceptible classes of stock is essential. Grazing management decisions should aim at providing these, or if unachievable, high quality supplements should be fed. Good levels of feeding of pregnant and early lactating ewes, in particular multiple bearing ewes and poor conditioned ewes will help prevent the temporary breakdown in their immunity and the periparturient rise in faecal egg counts[116]. This will result in lower levels of pasture contamination than otherwise would have been the case.

Reduction in infectivity of pasture

On *in-vitro* trials,) it has been demonstrated that ovicidal and larvicidal property of urea against pre-parasitic stages of *H. contortus*. The addition of urea (@ 5.0 g N%) to faeces resulted in complete inhibition of development of larvae from eggs on coproculture while spray of urea (@ 1.25 g N%) on pasture contaminated with faeces from *H. contortus* infected sheep lead to around 80% reduction in larval translation on pasture[112].

Pasture rotation system

Rotational grazing by itself is not an effective means of internal parasite control[15], but it does have some inherent advantages. The main one is that it also allows for "safer' pasture to be prepared, especially if the spelling interval is adequate. This is an excellent tool from the productivity point of view as animals will consume a higher proportion of the available forage which stimulated pasture regrowth[10]. It is a grazing management technique involving subdivision of pasture in which each paddock is grazed for a short time and then rested for as relatively much longer time. The resting time should be framed in such a way that most of the infective larvae originating from the previous grazing to have died off. It is achievable in tropical climate where larval survival times are short compared to temperate environment where 3-9 months resting period is required to reduce the pasture infectivity. The advantages of this practice are comprised of continuous reduction in pasture larval burden, reduced use of anthelmintics and risk of anthelmintic resistance and better pasture utilization. Initial higher capital investment on fencing, availability of larger pasture area and increased labour are major limiting factors involved in this approach. In Rajasthan, the grazing system of farmers is automatically of rotational in nature depending upon the availability of fodder (although not perfectly systematized). It has been observed that farmers put their animals on community grazing land / forest with the onset of monsoon where animals are drenched during mid-monsoon and continued on the same up to October. After harvesting of kharif crop the flocks are moved on such field to utilize crop stubbles for a period of one month. From November onward the flocks were again moved to community grazing land, forest or maintained on top feed with intermittent grazing on crop stubbles after rabi harvesting[100].

Safe Pasture system

The safe pasture concept is based on the fact that the numbers of larvae in the pasture are reduced over time by resting the pasture during the period when they are normally being recontaminated or through the growing and harvesting of a crop of hay or silage followed by a period of regrowth. The advantages of this practice are comprised of reduced pasture larval burden, reduced use of anthelmintics and better pasture utilization. This approach is only suitable in combined crop and livestock production system but if combined with the use of anthelmintics (dose and move system) it may increase the selection pressure for anthelmintic resistance.

Pasture resting/ Grazing interval

It consist of preventing the animals from grazing in the same field or paddock. Resting of pasture allow grasses in a more vegetative stage and tall enough to provide better nutrition to keep the healthier, strengthening the immune system to prevent adult worms from producing eggs. The infective larvae on pasture are non-feeding and actually survive on stored metabolites. In cooler temperatures, some of these larvae can live for up to 6-8 months (even longer under certain conditions), while in warmer temperatures survival may be only 2-3 months. Rotational grazing utilises the concept of pasture spelling. The study conducted in semi-arid Rajasthan (using tracer lambs) revealed that sheep grazed during monsoon on spring contaminated-summer ungrazed pasture had very low FEC and worm count compared to those placed on continuously grazed contaminated pasture. The pasture larval burden was higher in general pasture than summer ungrazed pasture. Thus, resting the contaminated pasture during summer found to effectively control the pasture larval burden[97]. Though pasture rotation is recommended strategy for controlling internal parasites because it allows the use of safe or safer pastures but intensive rotational grazing generally does not help to reduce parasitism unless rest periods are long enough. In rotational grazing system, sheep and goat should not be returned to the same pasture for 2-3 months.

Grazing behaviour/Alternate grazing system

Goats, due to their particular grazing behaviour, may well get more advantage than sheep or cattle from rotational grazing[15]. It is well known that goats tend to graze the pasture "from the top down" more than sheep and cattle. Because the highest concentrations of nematode larvae are found in the lowest strata of the pasture, it has been suggested that a well planned grazing rotation which makes it possible to keep a bank of fresh pasture ahead of goats may reduce their rate of intake of worm larvae. To the extent that two or more host species in any given environment do not share common parasite species, alteration between the species can be a successful tool of improving worm control. Using alternate grazing for parasite control is based on different age groups of the same species (taking advantage of higher resistance in older animals) or different species (little cross infectivity between species and reduced susceptibility of different host species) grazing the pasture in sequence. It can not be recommended as a stand alone strategy[11]. In area where *T. axei* is not a major concern, alteration between small ruminants and cattle would be successful. In *H. contortus* predominant area, though the parasite infects cattle but fail to cause any disease, thus small ruminants and cattle alteration works. However in environment where *H. placei* infect cattle great caution should be exercised if their pastures are used for grazing by small ruminants, as this species is pathogenic in all hosts[10]. This approach is less efficient in controlling parasitism in temperate climate compared to tropical and subtropical regions because cool and moist weather prolongs the larval survival. In this approach older animals or a different species of livestock will act as a vacuum cleaner thereby reducing the number of infective larvae on pasture. It has been observed that intensity of GIN in farmer's flock was low compared to farm flocks. One of the reasons behind this finding is that in rural area farmers graze all types of animals on same community grazing land in contrast to mono-specific grazing in organized farms[96].

Pasture mass and height

One indisputable advantage of rotational grazing over a set stocked situation is that animals will be offered a "bank" of fed of relatively long pasture and high pasture mass. This has the double advantage of diluting the infective larvae by the greater amount of feed on offer and the taller pasture will offer a grazing horizon above the greatest concentration of larvae near the base of the pasture and the bottom 5 cm of the sward. Parasite infection and multiplication are prevented by allowing animals graze only 10 cm from the ground in a field where there are parasites. However, on new pasture animal should be grazed very close to the ground so that the sun can dry the pat quickly and thus diminish the chances of survival of parasite brought in with the animals[48].

Pasture Species and Specialised Forages

Pastures modify the microclimate which may directly affect larval development and survival, affect egg and larval predators and pathogens, and/or alter the rate of faecal decomposition. Rate and extent of larval migration may be affected by pasture plants differing in morphology and composition. Some species appear to retard the migration of L3 larvae on plants and thereby minimize the subsequent infection by grazing ruminants[70] Legumes such as lucerne, clover and herbs such as chicory, inhibit the vertical migration of larvae in comparison to grasses but with lucerne the larvae are carried higher in the sward than with grasses[74]. If larvae can be prevented from reaching the regions of the sward that are the most consistently grazed (26-125 mm above soil level), then ingestion of L_3 larvae by livestock is substantially reduced[70]. Another major advantage to legumes is that larval survival is lower.

The success of the different approaches of grazing/pasture management in worm control programme depends on a great deal of ability of the farmers to understand and implement them. In particular, he must be able to provide clean / safe pasture whenever they are required. More ever these options work best in situation where adequate epidemiological information exists. They are therefore inapplicable under the predominantly traditional and extensive system of management in India but can be applied at least on the organized farms provided such farms have the necessary skilled personnel and pasture.

Biological control

It is defined as the action of natural enemies which maintain a host population at levels lower than would occur in the absence of enemies[129]. Biological control agents rarely eliminate the target organism, but reduce the numbers to acceptable levels and maintain a balance between the pathogen and the antagonist[127]. In contrast to chemical control of nematode parasites which is directed entirely at the parasitic stage within the host, biological control focuses on the free-living stages on pasture.

Candidates for Biological Control of Nematode Parasites

Dung beetles are found throughout the world and these are often capable of rapid and often complete dung removal and thus are indirectly responsible for the significant reductions in the number of free-living stages of parasites[128]. However, such dung dispersal activity is very dependent on ideal weather conditions, therefore little opportunity exists to exploit these organisms in attempts to achieve cost-effective and reliable biological control of nematode parasites.

Earthworms take over the role of dung beetles in the cool, moist regions of the world[42]. Both of these invertebrates break up dung pats and carry some of the faecal material below ground The individual effects of these activities are, first, to expose some nematode eggs and larvae to dehydration and, secondly, to remove others of them from the immediate vicinity of pasture. The potential of these invertebrates as serious options for controlling parasite larvae numbers appears slim and have not been persevered with as effective options.

Fungi that exhibit anti-nematode properties have been known for a long time. They consist of a great variety of species which include nematode-trapping (predacious) fungi, endoparasitic fungi, fungi that invade nematode eggs and fungi that produce metabolites that are toxic to nematodes[14]. The most important groups of nematophagous fungi are:

- *Predacious fungi:* These fungi produce specialized hyphal trapping devices, such as adhesive networks, knobs, and constricting or non-constricting rings. Fungi in this class may also produce nematode chemo attractant and/or chemotoxic substances[128]. Within a short period of time following capture of the nematode, the fungus penetrates the worm and destroys it.
- *Endoparasitic fungi:* These fungi invade the nematode from adhesive spores on the cuticle, from spores that are ingested by the nematode, or from motile spores in water[127].

Fungi from these two classes are found in all environments throughout the world. Research has shown that many species of nematophagous fungi can move from the soil into faeces soon after it is deposited on the pasture. Furthermore, many of these fungi are able to reduce the number of worms which develop to the infective stage when added to fresh faeces containing worm eggs. As r result of renewed interest and intensified research in biological control during last two decades, a convincing amount of evidences on potential for this approach has been gathered[61]. Most of the work has been carried out in Europe and Australia and recently initiated in USA, Latin America, Africa, South-East Asia and Far East. While there are many (>100) species of nematode destroying fungi, those whose spores can survive passage through the gastrointestinal tract would be most useful in livestock parasite control programmes[49]. Research in Denmark and Australia has identified the fungus *Duddingtonia flagrans* as the most promising candidate for biological control[43, 49, 62]. European field trials have demonstrated that pastures grazed by animals given fungal spores may have up to 80% less infective larvae on pasture and this can reduce by 80% the worm burdens in sheep that subsequently graze these pastures[49]. CSIRO field trials have also established that fungal treatment of young weaned lambs in autumn resulted in lower faecal egg counts and greater live weight gains over the following winter and spring (Kahn & Watson, 2001). It is possible that fungal spores may be delivered to animals through on farm addition to grain supplements, incorporation into lick blocks or in a controlled-release device[59, 130, 131]. These delivery systems are most effective when spores are kept dry prior to ingestion, since spores that have already germinated have diminished survival through the gastrointestinal tract. In India, the straw based complete feed blocks containing *D. flagrans* chlamydospores has been developed and tested at CSWRI, Avikanagar with encouraging results in controlling the pasture infectivity[100].

In India, nematophagous fungi were surveyed in Gujarat, Rajasthan and Chattishgarh[52, 89] and *Duddingtonia flagrans* is most suitable candidate followed by *Arthobotrys oligospora, A. oviformis* and *A. mesiformis*[22]. *D. flagrans* has a chlamydospore which is sufficiently robust that it can survive passage through the ruminant gastrointestinal tract and pass safely though the gut to the faeces[22, 53, 54, 55, 89]. In controlled experiments with small ruminants infected with *H. contortus* and fed on *D. flagrans* chlamydospores, the inoculated fungus would be recovered from faeces of fungus fed animals and could significantly prevented larval translation on herbage[55, 90, 92].

D. flagrans on 2% corn meal agar (CMA) showed excellent net-formation and trapping of larvae. All *H. contortus* larvae were effectively trapped within 2-3 days (Khan *et al.*, 2001a). Addition of chlamydospores of *D. flagrans* (@ 20 / g faeces) in to faecal culture caused only 3.9% reduction in larval number compared to control. However, 56.3% and 92.0% reduction in larval recovery was obtained with concentration of fungal spores to the tune of 200 and 2000 / g faeces[53]. Similarly addition of conidia of *A. oligospora* to faecal culture caused reduction in larval number to the tune of 40.0, 73.7, 78.7 and 84.2% at the concentration of 10, 100, 1000 and 2000 / g faeces, respectively[55].

Environmental studies conducted to determine the effects of deployment of these fungi in faecal material on improved pastures recorded no detectable detrimental effects on soil nematode populations or other microfauna[49]. Researchers are hopeful that a commercial product incorporating the fungus *D. flagrans* can be developed within the next five years which will give farmers an additional tool for parasite control to further reduce reliance upon anthelmintic treatments.

Use of Models for Internal Parasite Control

The role of modeling rose to prominence during the decade 1990-2000 especially in examining strategies such as sequential, rotational or simultaneous use of anthelmintics[68] and in examining anthelmintic resistance[64, 65]. Models are merely a mathematical representation of a real situation or hypothetical and a series of equations are put together to mimic the scenario[63]. The population dynamics of the internal parasites of grazing sheep and their interactions with weather, management, immunity and drenching is a good example of a system that was simply too complex to model in its entirety before the advent of the computer[8]. Using epidemiological information, life cycle of parasite, agroclimatic conditions of Rajasthan a zone wise computer based simulation / forecasting programme - **FROGIN** was developed to predict the intensity of *H. contortus* in sheep which can be used to design and improve worm management schemes. It is a computer based mathematical modelling of *H. contortus* population tailored for different zones of Rajasthan. The programme is based on Climatic data, development, dissemination and survival of larvae on pasture, Mortality rate / establishment rate of larvae in sheep, adult worms in sheep, fecundity of parasite, stocking density on pasture and farm management. This programme is capable of precisely forecasting the magnitude of parasitism for 60 days in advance and is extremely useful in decision making to organize deworming camps for sheep. Further, the programme is capable of shifting the scheduled drench in face of changing weather pattern and managemental modifications made by farmer[100].

One of the great advantages of models is that it is possible to systematically change each variable in turn, leaving all else unchanged, and hence observes the effect each variable has on the output of interest. Simulation model of the population dynamics can be used to explore the potential effectiveness of control strategies[12, 13, 35, 36]. Models were originally designed as research tools[63] and are still mainly used in the realms of research[68], but in future they could become increasingly important as one of the main extension or educational tools in internal parasite control programmes available to wider range of practitioners.

Conclusions

Worm control for the last 3-4 decades has depended on anthelmintics. Now, threats of multiple anthelmintic resistance, demand for residue free animal products and regulations of General Agreement on Tariffs and Trades (GATT) have forced us to reconsider our worm control options. There are many options available but some of them are practically or socially not feasible. In India limited efforts have been made so far to use environmental friendly alternatives to anthelmintics including genetic resistance of host, vaccine, biological control, role of nutrition and mathematical model based decision support system. Now there is growing awareness for importance of parasites and its sustainable control. Although chemotherapy will remain as part of integrated worm management but TST may play very crucial role for application of chemotherapeutics. Similarly genetic selection of host based on MAS may rapidly yield results to create resistance stock to reduce the dependence on chemotherapy. Biological control through nematophagous fungi has given very encouraging results. Therefore in years to come the technology may be put to practical use. The success of the different approaches of pasture management in worm control programme depends on a great deal of ability of the farmers to understand and implement them. More ever these options work best in situation where adequate epidemiological information exists. They are therefore inapplicable under the predominantly traditional and extensive system of management in India but can be applied at least on the organized farms provided such farms have the necessary skilled personnel and pasture. In Indian perspective the work on the role of nutrition in the management of parasitic diseases is yet to be initiated. This is partly because the animal sector in India is such that the role of nutrition for augmenting the productivity is not well defined. Most of the animals are under nourished and under low state of productivity. Under these situations the effect of pathophysiology of parasitic diseases increases many folds. The sporadic Indian studies in this aspect need to be systematize and perused vigorously to translate them in fruitful technology.

References

1. Albers G A A, Gray G D, Piper L R, Barker J S F, Le Jambre L F and Barger I A. 1987. The genetics of resistance and resilience to *Haemonchus contortus* infection in young Merino sheep. *International Journal for Parasitology,* 17: 1355–1363.

2. Amynalem H, Tembely S, Anindo D O, mukasa-Mugerwa E, Rege J E O, Alemu Yami and Baker R L. 2002. Effects of breed and dietary protein supplementation

on the response to gastrointestinal infections in Ethiopian sheep. *Small Ruminant Research,* 44: 247-261.

3. Annon 2003. Final Report on NATP "*Identification and Evaluation of Medicinal Plants for Control of Parasitic Diseaes on Livestock*", Central Sheep and Wool Research Institute, Avikanagar (Rajasthan) pp. 1-136.

4. Athanasiadou S, Kyriazakis F, Jackson F and Coop R L. 2001. Direct anthelmintic effects of condensed tannin towards different gastrointestinal nematodes of sheep: *in-vitro* and *in-vivo* studies. *Veterinary Parasitology,* 99: 205-219.

5. Baker R L, Watson T G, Bisset S A, Vlassof A and Douch P G C. 1991. Breeding sheep in New Zealand for resistance to internal parasites: research results and commercial application. In: G D Gray and R R Woolaston (eds.) *Breeding for Disease Resistance in Sheep.* Australian Wool Corporation, Melbourne, pp.19–32.

6. Ballic A. Bowles V M and Meeusen E N. 2000. The immunobiology of gastrointestinal infection in ruminants. *Advances in Parasitol,* 45: 181-241.

7. Bang K S, Familton A S and Sykes A R. 1990. Effect of copper oxide wire particle treatment on establishment of major gastrointestinal nematodes in lambs. *Research in Veterinary Science,* 49: 132-137.

8. Barger I A. 1997. Models as a guide to sustainable worm control. In*: Sustainable control of internal parasites in ruminants - an Animal Industry Workshop.* (Ed. G K Barrell), Animal and Veterinary Science Group. Lincoln University, Lincoln, pp 203-13.

9. Barger I A. 1998. Control by management. *Veterinary Parsitology,* 72 : 493-500.

10. Barger I A. 1999. The role of epizootiological knowledge and grazing management for helminth control in small ruminants. *International Journal for Parsitology,* 29 : 41-47.

11. Barger I A, Siale K, Banks D J D and Le Jambre L F. 1994. Rotational grazing for control of gastrointestinal nematodes of goats in a wet tropical environment. *Veterinary Parasitology,* 53*:* 109-116.

12. Barnes E H, Dobson R J and Barger I A. 1995. Worm control and anthelmintic resistance. Adventures with a model. *Parasitology Today,* 11*:* 56-63.

13. Barnes E H, Dobson R J, Stein P A, Le Jambre L F and Lenane I J. 2001. Selection of different genotype larvae and adult worms for anthelmintic resistance by persistent and short-acting avermectin/milbemycins. *International Journal for Parasitology,* 31*:* 720-727.

14. Barron G L. 1977. The nematode destroying fungi. *Topics in Mycology* No.1, Canadian Biological Publications Ltd, Guelph, Ontario.

15. Bisset S A, Vlassoff A and Pulford H. 1991. Rotational grazing/grazing interchange systems. *Proceedings Sheep and Beef Cattle Society of the New Zealand Veterinary Association,* 21*:* 47-54.

16. Bisset S A, Morris C A, McEwan J C and Vlassoff A. 2001. Breeding sheep in NZ that are less reliant on anthelmintics to maintain health and productivity. *New Zealand Veterinary Journal,* 49*:* 236-246.

17. Bouix J, Krupinski J, Rzepecki R, Nowosad B, Skrzyzala I, Robarzynski M, Fuldalewicz-Niemczyk W, Shalzka M, Malezewski A and Gruner L. 1998. Genetic resistance to gastrointestinal nematode parasites in Polish long-wool sheep. *International Journal of Parasitology,* 28: 1797–1804.

18. Bown M D, Poppi D P and Sykes A R. 1991. Nitogen transactions along the digestive tract of lambs concurrently infected with *Trichostrongylus colubriformis* and *Ostertagia circumcincta. British Journal of Animal Nutrition,* 66: 237-249.

19. Bradley R E, Radhakrishnan C V, Patil-Kulkarni V G and Loggins P E. 1973. Responses in Florida Native and Rambouillet lambs exposed to one and two doses of *Haemonchus contortus. American Journal of Veterinary Research,* 34: 729-735.

20. Butter N L, Dawson J M, Wakelin D and Buttery P J. 2000. Effect of dietary tannin and protein concentration on nematode infection (*Trichostrongylus colubriformis*) in lambs. *Journal of Agricultural Science,* 134: 89-99.

21. Chartier C, Etter E, Hoste H, Pors I, Koch C and Dellac B. 2000. Efficacy of copper oxide needles for the control of nematode parasites in dairy goats. *Veterinary Research Communications,* 24: 389-399.

22. Chauhan J B, Sanyal P K, Mukhopadhyaya P N and Subramaniam R B. 2002. Evaluation of an Indian isolate of *Arthrobotrys musiformis,* a possible biocontrol candidate against round worms. *Journal of Veterinary Parasitology,* 16: 17-21.

23. Coltman D W, Wilson k, Pilkington J G, Stear M J and Pemberton I M. 2001. A micro satellite polymorphism in the gamma interferon gene associated with resistance to gastrointestinal nematodes in a naturally parasitized population of Soay sheep. *Parasitology,* 122: 571-582.

24. Coop R L and Field A C. 1983. Effect of phosphorus intake on growth rate, food intake and quality of the skeleton of growing lambs infected with the intestinal nematode *Trichostrongylus vitrinus. Research in Veterinary Science,* 38: 175-181.

25. Coop R L, Huntley J F and Smith W D. 1995. Effect of dietary protein supplementation on the development of immunity to *Ostertagia circumcincta* in growing lambs. *Research in Veterinary Science,* 59: 24-29.

26. Courtney C H, Parker C F, Maclure K E and Herd R P. 1984. A comparison of the peri-parturient rise in faecal egg counts of exotic and domestic ewes. *International Journal of Parasitology,* 14: 377–381.

27. Courtney C H, Parker C F, Maclure K E and Herd R P. 1985a. Resistance of exotic and domestic lambs to experimental infection with *Haemonchus contortus. International Journal of Parasitology,* 15: 101–109.

28. Courtney C H, Parker C F, Maclure K E and Herd R P. 1985b. Resistance of non lambing exotic and domestic ewes to naturally acquired gastrointestinal nematodes. *International Journal for Parasitology,* 15: 239–243.

29. Crawford A M. 2001. A review of QTL experiment in sheep. In: *Proc. Association for the Advancement of Animal Breeding and Genetics,* 14: 33-38.

30. Crawford A M and McEwan J C. 1999. Identification of animals resistant to nematode parasite infection. Patent application 330201, Assignee:NZ, New Zealand Pastoral Agriculture Research Institute Limited, pp 45.

31. Cummings L J, Thompson R L, Yong W K, Riffkin G G, Goddard M E, Callinan A P L and Saunders M J. 1991. Genetics of *Ostertagia* selection lines. In: G D Gray and R R Woolaston (Eds.), *Breeding for Disease Resistance in Sheep.* Australian Wool Corporation, Melbourne, 11–18.

32. Cwirla S E, Peters E A, Barrett R W and Dower W J. 1990. Peptides on phage: a vast library of peptides for identifying ligands. *Proc. The National Academy of Sciences,* 87*:* 6378-6382.

33. Datta F U, Nolan J V, Rowe J D, Gray G D and Crook B J. 1999. Long-term effects of short-term provision of protein-enriched diets on resistance to nematode infection and live weight gain and wool growth in sheep. *International Journal for Parasitology,* 29: 479-488.

34. Dawson, J M, Buttery P J, Jenkins D Wood C D and Gill M. 1999. Effects of dietary Quebracho tannin on nutrient utilization and tissue metabolism in sheep and rats. *Journal of Science for Food and Agriculture,* 79: 1423-1430.

35. Dobson R J and Barnes E H. 1999. Mathematical modeling of parasite populations and the use of drenches that are not fully effective. *Proc. Annual Conference of the Australian Society of Sheep Veterinarians,* Hobart. Pp 6-9.

36. Dobson R J, Le Jambre L F and Gill J H. 1996. Management of anthelmintic resistance: Inheritance of resistance and selection with persistent drugs. *International Journal for Parasitology,* 26*:* 993-1000.

37. Dominik S. 2005. Quantitatiove trait loci for internal nematode resistance in sheep: a review. *Genetic Selection Evolution,* 37 (suppl. 1): 83-96.

38. Donaldson J, van Houtert M F J and Sykes A R. 2001. Effect of dietary fish meal supplementation on parasite burdens of periparturient sheep. *Animal Science,* 72: 149-158.

39. Eady J J, Woolaston R R and Barger I A. 2003. Comparison of genetic and non-genetic strategies for control of gastrointestinal nematodes of sheep. *Livestock Production Science,* 81*:* 11-23.

40. Emery D L. 1996. Vaccination against worm parasites of animals. *Veterinary Parasitology,* 64: 31-45.

41. Gamble H R and Zajac A M. 1992. Resistance of St. Croix lambs to *Haemonchus contortus in* experimentally and naturally acquired infections. *Veterinary Parasitology,* 41: 211– 225.

42. Gronvold J. 1987. Field experiment on the ability of earthworms *(Lumbricidae)* to reduce the transmission of infective larvae of *Cooperia oncophora* (Trichostrongylidae) from cow pats to grass. *The Journal of Parasitology,* 73*:* 1133-1137.

43. Hein W R, Shoemaker C B and Heath A C G. 2001. Future technologies for control of nematodes of sheep. *New Zealand Veterinary Journal,* 49*:* 247-251.

44. Hoskin S A, Barry T N, Wilson P R, Charleston W A G and Kemp P P. 1999. Growth and carcass production of young farmed deer grazing sulla, chicory or perennial ryegrass/white clover pasture in New Zealand. *New Zealand Journal of Agricultural Research,* 42*:* 83- 92.

45. Hoste H, Chartier C and Le Frileux Y. 2002. Control of gastrointestinal parasitism with nematodes in dairy goats by treating the host category at risk. *Veterinary Research,* 33: 531-545.

46. Jackson F and Miller J M. 2006. Alternative approaches tocntrol – Quo Vadit? *Veterinary Parasitology,* 139: 371-384.

47. Jilek A F and Bradley R E. 1969. Haemoglobin types and resistance to *Haemonchus contortus* in sheep. *American Journal of Veterinary Research*, 30: 1773 – 1778.

48. Jones J. 1993. Grazing to control worms. *The New Farm*, 15 : 52 - 53.

49. Kahn L P and Watson D L. 2001. A summary of recent and current research on control of internal parasites of sheep. *In: Sustainable control of internal parasites of sheep,* pp 29.

50. Kahn L P, Knox M R and Gray G D. 1999. Enhancing immunity to nematode parasites in pregnant and lactating sheep through nutrition and genetic selection. In : *Recent Advances in Animal Nutrition in Australia* (Ed. J L Corbett), 12 : 15-22.

51. Ketzis J K, Vercruysse J, Stromberg B E, Larsen M, Athanasiadou S and Houdijk J G M. 2006. Evaluation of efficacy expectations for novel and non – chemical helminth control strategies in ruminants. *Veterinary Parasitology,* 139: 321 – 335.

52. Khan F A, Swarnkar C P, Singh D, Bhagwan P S K and Dubey S C. 2001. Nematophagus fungi isolated from sheep faeces: a promising biological control of *Haemonchus contortus. Indian Journal of Animal Science,* 71: 834-836.

53. Khan F A, Singh D, Swarnkar C P, Bhagwan P S K and Dubey S C. 2001a. Biological control of gastrointestinal nematodes of ruminants using nematophagus fungi. *Journal of Veterinary Parasitology,* 15: 107-112.

54. Khan F A, Swarnkar C P, Singh D, Bhagwan P S K and Dubey S C. 2002. Predacious activity of *Duddingtonia flagrans* against *Haemonchus contortus* larvae in faeces after passage through the gastrointestinal tract of sheep. *Indian Journal of Animal Science,* 72 : 559-560.

55. Khan F A, Singh D, Swarnkar C P, Bhagwan P S K and Dubey S C. 2004. Nematode trapping fungus, *Arthrobotrys oligospora* for its use in bio- control of *Haemonchus contortus. Journal of Veterinary Parasitology,* 18: 13-16.

56. Knight R A, Vegors H H and Glimp H A. 1973. Effect of breed and date of birth of lambs on gastrointestinal nematode infections. *American Journal of Veterinary Research,* 34: 323 – 327.

57. Knox D P. 2000. Development of vaccines against gastrointestinal nematodes. *Parasitology,* 120*:* 43-61.

58. Knox M R. 2002. Effectiveness copper oxide particles for *Haemonchus contortus* control in sheep. *Australian Veterinary Journal,* 80: 224-227.

59. Knox M R and Faedo M. 2001. Biological control of field infections of nematode parasites of young sheep with *Duddingtonia flagrans. Veterinary Parasitology,* 101: 155-160.

60. Knox M R and Steel J W. 1999. The effects of urea supplementation on production and parasitological responses of sheep infected with *Haemonchus contortus* and *Trichostrongylus colubriformis. Veterinary Parasitology,* 83: 123-135.

61. Larsen M. 2002. Biological control in global perspective – A review with emphasis on *Duddingtonia flagrans* in biological control of nematode parasites of small ruminants in Asia. *FAO Animal Production and Health Paper,* pp 19-37.

62. Larsen M, Faedo m, Waller P J and Hennessy D R 1998. The potential of nematophagous fungi to control the free-living stages of nematode parasites of sheep: Studies with *Duddingtonia flagrans. Veterinary Parasitology,* 77: 121 – 128.

63. Leathwick D M. 1991. Modelling parasite epidemiology and control. *Proc. 21st Seminar of the Sheep and Beef Society of the New Zealand Veterinary Association:* 73-78.

64. Leathwick D M and Sutherland I A. 2002. Heads or tails - which drench do I choose? *Proc. 32nd Seminar, Sheep and Beef Cattle Society of the New Zealand Veterinarian Association :* pp 32.

65. Leathwick D M, Pomroy W and Heath A C G. 2001. Anthelmintic resistance in New Zealand. *New Zealand Veterinary journal, 49:* 227-235.

66. Loggins P E, Swanson L E and Kager M. 1965. Parasite levels in sheep as affected by heredity. *Journal of Animal Science,* 24: 286 – 287.

67. McEwan J C, Mason P, Baker R L, Clarke J N, Hickey S M and Turner K. 1992. Effect of selection for productive traits on internal parasite resistance in sheep. *Proceedings of New Zealand Society for Animal Production,* 52: 53-56.

68. McPherson W B. 2002. Influencing sheep and cattle parasite control - options, opinions, opportunities. *Proc. 32nd Seminar of the Sheep and Beef Society of the New Zealand Veterinary Association:* 139-145.

69. Molan A I, Waghorn G C, Min Br and Mc Nabb W C. 2000. The effect of condensed tannins from seven herbage on *Trichostrongylus colubriformis* larval migration *in-vitro. Folio Parasitology,* 47 : 39-44.

70. Moss R A and Vlassoff A. 1993. Effect of herbage species on gastrointestinal roundworm populations and their distribution. *New Zealand Journal of Agricultural Research,* 36*:* 371- 375.

71. Munn E A. 1997. Rational design of nematode vaccines - hidden antigens. *International Journal for Parasitology,* 27*:* 359-366.

72. Nari A. 2005. The FAMACHA© assay as a component of integrated pest management. In: *Proc. FAO Symposium on "Integrated Animal Parasite Management: From Acedemic Interest to Reality"* pp. 95 – 102.

73. Newton S E and Munn E A. 1999. The development of vaccines against gastrointestinal nematode parasites, particularly *Haemonchus contortus. Parasitology Today,* 15*:* 116-122.

74. Nicol A M and Everest P G. 1997. Integrated control systems for the management of internal parasites in ruminants. *In: Sustainable control of internal parasites in ruminants.* (Ed. G K. Barrell), Animal Industries Workshop, pp 263-281.

75. Niezen J H, Wahgorn T S, Charleston W A G and Waghorn G C. 1995. Growth and gastrointestinal nematode parasitism in lambs grazing lucern (*Medicago sativa)* or sulla (*Hedysarum coronarium)* which contasins condensed tannins. *Journal of Agricultural Science,* 125: 281-289.

76. Niezen J H, Charleston W A G, Hodgson J, Miller c M, Wahgorn T S and Robertson H A. 1998 a. Effect of plant species on the larvae of gastrointestinal nematodes which parasitise sheep. *International Journal for Parasitology,* 28: 791-803.

77. Niezen J H, Robertson H A, Wahgorn G C and Charleston W A G. 1998 b. Production, faecal egg counts and worm burdens of ewe lambs which grazed six contrasting forages. *Veterinary Parasitology,* 80: 15-27.

78. Niezen J H, Robertson H A, Sidex A and Wilson S R. 2002. The effect of pasture species on parasitism and performance of lambs grazing one of three grass-white clover pasture swards. *Veterinary Parasitology,* 105: 303-315.

79. Nimbkar C. 2005. Importance of genetic resistance to gastrointestinal nematodes in sheep in developing countries with special reference to the Garole breed. In: *Proc. FAO Symposium on "Integrated Animal Parasite Management: From Acedemic Interest to Reality"* pp. 34 – 40.

80. Pachlag S V and Kumar P N. 1974. Studies on the gastrointestinal helminthiasis in sheep. In : *Annual Report,* CSWRI, Avikanagar.

81. Paolini V, Bergaud J P, Grisez C, Prevot F, Drochies P and Hoste H. 2003. Effects of contensed tannins on goats experimentally infected with *Haemonchus contortus. Veterinary Parasitology,* 113: 253-261.

82. Pessoa, L M, Morais S M, Bevilaqua C M L and Luciano J H S. 2002. Anthelmintic activity of essential oil of *Ocimum gratissimum* Linn. And euganol against *Haemonchus contortus. Veterinary Parasitology,* 109: 59 – 63.

83. Piper L R. 1987.Genetic variation in resistance to internal parasites. In: Mc Guirk B J (eds) *Merino Improvement Programme in Australia.* Australian Wool Corporation, Melbourne, 351–363.

84. Pomroy W E. 2000. A perspective on nematode control in NZ. *Proc. 20th Seminar of the Sheep and Beef Cattle Society of the New Zealand Veterinarian Association:* 89-98.

85. Poppi D P, Sykes A R and Dynes R A. 1990. The effects of parasitism on host nutrition – the implications for nutrient manipulation. *Proc. New Zealand Society for Animal Production,* 50: 237-243.

86. Preston J M and Allonby E W. 1978. The influence of breed on the susceptibility of sheep and goats to a single experimental infection with *Haemonchus contortus. Veterinary Record,* 103: 509-512.

87. Radhakrishnan C V, Bradley R E and Loggins P E. 1972. Host response of worm-free Florida Native and Rambouillet lambs experimentally infected with *Haemonchus contortus. American Journal of Veterinary Research,* 33: 817–823.

88. Sanyal P K. 1988. The epizootology and control of gastrointestinal nematodiasis of sheep. *Proceedings of 2nd National Seminar on Sheep and Goat Diseaes*, CSWRI, Avikangar.

89. Sanyal P K. 2000. Screening for Indian isolated of predacious fungi for use in biological; control against nematode parasites of ruminants. *Veterinary Research Communications*, 24: 55 – 62.

90. Sanyal P K. 2001. Biological control of *Haemonchus contortus* in sheep by nematophagous fungus, *Duddingtonia flagrans. Indian Veterinary Journal*, 78: 1091-1094.

91. Sanyal P K. 2005 Biological; control of parasitic gastroenteritis in ruminants by nematode trapping fungi. In: *Proc. FAO Symposium on "Integrated Animal Parasite Management: From Acedemic Interest to Reality"* pp. 55 – 63.

92. Sanyal P K and Mukhopadhayaya P N. 2003. Fungal spread on pasture and faecal decomposition as parameter to evaluate short-term environmental impact of *Duddingtonia flagrans. Journal of Veterinary Parasitology,* 17: 32-36.

93. Sharma R L, Bhat T K and Dhar D N. 1988. Control of sheep lungworm in India. *Parasitology Today,* 4: 193-198.

94. Sharma S R, Singh D, Khan F A, Swarnkar C P and Bhagwan P S K. 2003. *In* vitro anthelmintic potential of *Xanthium strumarium* leaves against *Haemonchus contortus* of sheep. *Indian Journal of Animal Science,* 73 : 342-344.

95. Shaw K L, Nolan J V, Lyneh J J, Coverdole O R and Gill H S. 1995. Effect of weaning supplementation and gender on acquired immunity to *Haemonchus contortus* in lambs. *International Journal for Parasitology,* 25: 381-387.

96. Singh D and Swarnkar C P. 2005. Epidemiology and pasture management for the control of ovine gastrointestinal nematodosis. In: *Proc. FAO Symposium on "Integrated Animal Parasite Management: From Acedemic Interest to Reality"* pp. 23 – 33.

97. Singh D, Swarnkar C P Khan F A and Bhagwan P S K. 1997. Infectivity of spring contaminated – summer ungrazed pasture for gastrointestinal nematodes in sheep. In: *Proc. IX National Congress of Veterinary Parasitology*, Punjab Agricultural University, Ludhiana (India), pp. 64.

98. Singh D, Swarnkar C P, Khan F A, Jayasankar J and Bhagwan P S K. 1999. Heritability of faecal egg counts in Avikalin sheep. *Indian Journal of Animal Sciences,* 69: 983-985.

99. Singh D. Swarnkar C P and Khan F A. 2002. Anthelmintic resistance in gastrointestinal nematodes of livestock in India. *Journal of Veterinary Parasitology,* 16 : 115-130.

100. Singh D, Swarnkar C P and Khan F A. 2005. Annual Progress Report on *All India Network Programme on Gastrointestinal Parasitism.* Central Sheep and Wool Research Institute, Avikanagar (Rajasthan) India.

101. Singh D, Swarnkar C P and Khan F A. 2006. Annual Progress Report on *All India Network Programme on Gastrointestinal Parasitism.* Central Sheep and Wool Research Institute, Avikanagar (Rajasthan) India.

102. Singh D, Swarnkar C P, Kumar Sushil, Kumar S and Arora A L. 2006a. Variability of sheep resistance to natural gastrointestinal nematode infection. In : *Proc. XVII National Congress of Veterinary Parasitology*, Rajiv Gandhi College of Veterinary and Animal Sciences, Pondicherry, (India).

103. Smith W D. 1993. Protection in lambs immunized with *Haemonchus contortus* gut membrane proteins. *Research in Veterinary Science,* 54*:* 94-101.

104. Stirling R G and Chung K E. 2000. Future treatments of allergic diseases and asthma. *British Medical Bulletin,* 56*:* 1037-1053.

105. Strain S A J and Stear M J. 2001. The influence of protein supplementation on the immune response to *Haemonchus contortus. Parasite Immunology,* 23: 527-531.

106. Suttle N F, Knox D P, Angus K W, Jackson F and Coop R L 1992 a. Effect of dietary molybdenum on nematodes and host during *Haemonchus contortus* infection in lambs. *Research in Veterinary Science,* 52: 230-235.

107. Suttle N F, Knox D P, Jackson F, Coop R L and Angus K W. 1992 b. Effect of dietary molybdenum on nematodes and host during *Trichostrongylus vitrinus* infection in lambs. *Research in Veterinary Science,* 52: 224-229.

108. Swarnkar C P and Singh D. 2005. Nutritional intervention in management of gastrointestinal nematodes in small ruminants. In: *Proc. FAO Symposium on "Integrated Animal Parasite Management: From Acedemic Interest to Reality"* pp. 41 – 54.

109. Swarnkar C P, Khan F A, Singh D, Dixit S and Bhagwan P S K. 1997a. Genetic variation in resistance to nematode parasites of sheep. *Proceedings IXth National Congress Veterinary Parasitology*, PAU, Ludhiana, 6-8 October.

110. Swarnkar C P, Singh G, Singh D, and Khan F A. 1997b. A pilot study on genetic variation in resistance to nematode parasites in fine wool sheep. *Proc. IIIrd National Seminar on Small Ruminant Diseases,* CIRG, Makhdoom, 2-3 December.

111. Swarnkar C P, Sanyal P K, Singh D, Khan F A and Bhagwan P S K. 2001. Anthelmintic resistance in an organised sheep farm in India. *Tropical Animal Health and Production*, 33 : 305-312.

112. Swarnkar C P, Singh D, Khan F A, Bhagwan P S K and Dubey S C. 2002. Effect of urea application on pre-parasitic stages of *Haemonchus contortus* and its role in reducing pasture infectivity. *Journal of Veterinary Parasitology,* 16: 23-26.

113. Swarnkar C P, Singh D, Khan F A, Bhagwan P S K and Dubey S C. 2004. *In-vitro* evaluation of *Melia azedarach* (Bakain) against pre-parasitic stages of *Haemonchus contortus. Indian Journal of Animal Science,* 74 : 940 – 943.

114. Sykes A R. 1982. Nutritional and physiological aspects of helminthiasis in sheep. In: *Biology and Control of Endoparasites* (Eds. L E A Ymons, A D Donald and J K Dineen), Academic Press, New York, pp. 217-230.

115. Sykes A R. 1994. Parasitism and production in farm animals. *Animal Production,* 59: 155-172.

116. Sykes A R. 1997. Effects of nematode parasitism on ruminant animal performance. *In: Sustainable control of internal parasites in ruminants* .(Ed. G K. Barrell), Animal Industries Workshop, pp 81-94.

117. Todd K S, Mansfield M E and Levine N D. 1978. *Haemonchus contortus* infection in Targhee and Targhee-Barbados Blackbelly cross lambs. *American Journal of Veterinary Research,* 39: 865–866.

118. Torres-Acosta J F, Aguilar-Caballero A J, Le Bigat C, Hoste H, Canal-Ku H L, Santos-Ricalde R and Gutierrez-Segura I. 2005. Comparing different formulae to test for gastrointestinal nematode resistance to different benzimidazoles in small holders goat farms in Mexico. *Veterinary Parasitology*, 134: 163 – 173.

119. Tavernor A S, Smith T S, Langford C F, Graham M and Munn E A. 1992. Immune response of Clun Forest sheep to vaccination with membrane glycoproteins from *Haemonchus contortus. Parasite Immunology,* 14*:* 671-675.

120. Van Houtert M F J and Sykes A R. 1996. Implication of nutrition for the ability of ruminants to withstand gastrointestinal infection. *International Journal for Parasitology,* 26: 1151-1167.

121. Van Houtert M F J, Barger I A, Steel J W, Windon R G and Emery D L. 1995. Effect of dietary protein intake on response of young sheep to infection with *Trichostrongylus colubriformis. Veterinary Parasitology,* 56: 163-180.

122. Van Wyk J A andBath G F 2002. The FAMACHA© system for managing haemonchosis in sheep and goats by clinically identifying individual animal for treatment. *Veterinary Research,* 33: 509-529.

123. Vihan V S, Kumar A and Arora N. 2002a. Evaluation of anthelmintic activity of *Melia azedarach* (Bakain) against *Haemonchus contortus* infection in goats. In: *X Internatinal Congress of Indian Association of Animal Production*, New Delhi.

124. Vihan V S, Arora N, Kumar A and Sharma D K. 2002b. Evaluation of anthelmintic activity of *Butea frondosa* (Palas) against *Haemonchus contortus* in goats. In: *National Symposium on Recent Trends in Diagnostics and Therapeutics of Animal Diseases,* Rajasthan Agricultural University, Bkaner.

125. Vlassoff A, Leathwick D M and Heath A G C. 2001. The epidemiology of nematode infections of sheep. *New Zealand Veterinary Journal,* 49*:* 213-221.

126. Waller P J. 1997a. Sustainsble helminth control of ruminants in developing countries. *Veterinary Parasitology,* 71 : 195 – 207.

127. Waller P J. 1997b. Biological control of parasitism. *In: Sustainable control of internal parasites in ruminants.* (Ed. G K. Burrell), Animal Industries Workshop. pp 215-213.

128. Waller P J and Faedo M. 1993. The potential of nematophagous fungi to control the freeliving stages of nematode parasites of sheep: screening studies. *Veterinary Parasitology,* 49*:* 285-297.

129. Waller P J and Faedo M. 1996. The prospects for biological control of the free-living stages of nematode parasites of livestock. *International Journal for Parasitology,* 26*:* 915- 925.

130. Waller P J, Faedo M and Ellis K. 2001 a. The potential of nematophagous fungi to control the free-living stages of nematode parasites of sheep: towards the development of a fungal controlled release device. *Veterinary Parasitology,* 102*:* 299- 308.

131. Waller P J, Knox M R and Faedo M. 2001 b. The potential of nematophagous fungi to control the free-living stages of nematode parasites of sheep: feeding and block studies with *Duddingtonia flagrans. Veterinary Parasitology,* 102*:* 321- 330.

132. Watson T G, Baker R L and Harvey T G. 1986. Genetic variation in resistance or tolerance to internal nematode parasites in strains of sheep at Rotomahana. *Proceedings of New Zealand Society for Animal Proceeding,* 46: 23-26.

133. Windon R G and Dineen J K. 1984. Parasitological and immunological competence of lambs selected for high and low responsiveness to vaccination with irradiated *Trichostrongylus colubriformis* larvae. In: J K Dineen and P M Outteridge (eds) *Immunogenetic Approaches to the Control of Endoparasites.* Melbourne, 13–28.

134. Yadav C L, Grewal H S and Banerjee D P. 1993. Susceptibility of two cross breeds of sheep to *Haemonchus contortus. International Journal for Parasitology,* 23: 819-822.

135. Yazwinski T A, Goode L, Moncol D J, Morgan G W and Linnerud A C. 1979. Parasite resistance in straight bred and cross bred Barbados Blackbelly sheep. *Journal of Animal Science,* 49: 919-926.

136. Zajac A M, Herd R P and McClure K E. 1988. Trichostrongylid parasite population in pregnant or lactating and unmated Florida Native and Dorset / Rambouillet ewes. *International Journal for Parasitology,* 18: 981–985.

137. Zajac A M, Krakowka S, Herd R P and McClure K E. 1990. Experimental *Haemonchus contortus* infection in three breeds of sheep. *Veterinary Parasitology,* 36: 221–235.

-ﾑﾑﾑ-

19

Protozoan Diseases of Small Ruminants

D K SHARMA

Division of Animal Health
Central Institute for Research on Goats
Makhdoom, Farah, Mathura (U.P.) - 281 122

Protozoans are unicellular organisms. Their bodies are constituted by eukaryotic cells having a distinct nucleus, mitochondria, golgi apparatus and endoplasmic reticulum in the cytoplasm. The organisms differ from bacteria which are prokaryotic in nature having indistinct nucleus and lack membrane system in their cytoplasm. Some of the protozoan are responsible for serious diseases in domestic animals and are of considerable veterinary importance. Protozoan diseases at times appear with severity resulting in mortality, however, in chronic or acute from they lead to serious production losses through reduced milk, meat and wool production.

The protozoan diseases in small ruminants viz. sheep and goats are far more serious than they are considered so far. While diseases like coccidiosis has a direct impact on their production, trypanosomiasis leads to immune suppression, a condition which encourage other infections to set in. Moreover, most of the protozoan diseases are of public health importance as they are communicable to human beings. The new emerging diseases like cryptosporidiosis, sarcocystosis and toxoplasmosis are zoonotic in nature in view of their public health importance. Here we will discuss some important protozoan disease of sheep and goats. Major protozoan diseases in small ruminants are:

Trypanosomiasis

The disease is caused by a haemoprotozoan flagellate, *Trypanosoma,* occurring in blood and tissue fluids as intercellular parasite. The disease is important and prevalent in different parts of the world affecting various species of animals like horses, camels, cattle, buffaloes, dogs, goats, sheep and pig and is transmitted by blood sucking insects or through mechanical transmission. The importance of trypanosomiasis can be gauged from the fact as it has been considered as a unique disease which by itself has denied a vast area of the land to all domestic animals other than poultry in Eastern Africa[31].

The prevalence of *Trypanosoma* infection in goats and sheep has been reported by various researchers[36,42,48,57]. Asonganyi, 1992, reported a sero-prevalence of 72.1% in goats Cameroon. In India, an outbreak of *Trypanosoma evansi* was reported from Maharashtra 39. Though reports of natural trypanosomiasis in sheep and goats are not so common, the two animal species are considered to be important as they might be acting as reservoir host of trypanosome Spp for other animals and man[58].

In sheep and goats 4 species of *Trypanosoma* have been reported. They are *T. congolense. T. vivex, T. brucei,* and *T. evansi.*

Reports on pathogenicity of Trypanosoma spp in sheep and goats are contradictory. Many authors have reported trypanosomal infection in these animals as sub acute and of mild type leading to anaemia with a low or undetectable parasitaemia of cryptic nature. In experimental infections the disease syndrome was characterized by manifestation of fever, dullness, weight loss, weakness, enlarged lymph nodes, oedema of head, scrotum, limbs, tail, ventral surface of thorax and abdomen followed by progressive recumbancy32. However, the disease is important as it leads to immunosuppression in the host and can aggravate the damage caused by the concurrent infection if any[62].

The disease diagnosis in sheep and goats is difficult through conventional methods. As the infection remains sub clinical and show very low level of parasitaemia. Animal inoculation or immunodiagnostic tests like IFTA, IHA, ELISA are more reliable and advanced tests 67. PCR being the most advanced tool for diagnosis of *Trypanosoma* infection.

Treatment is still based on trypanocidal drugs like quinopyramine, suramines and diamidines etc. Prophylaxis through vaccine still remained obscure and unattainable.

Babesiosis

Babesiosis is caused by a haemoprotozoan parasite, *Babesia* an organism affecting erythrocytic cells of the host. *Babesia* organisms are usually round in shape but piriform, amoeboid, rod, like or pleomorphic shapes are also common. It is a non pigment (haemozoin) forming haemoparasite without extra-erythrocytic multiplication. The parasite multiplies by binary fission in schizogony in red blood corpuscles. Babesiosis is a tick borne infection occurring in domestic animals like cattle, sheep, goats and other species of animals.

There are 4 species of *Babesia* commonly infecting sheep and goats. There is one large form and 3 small forms. They are *Babesia* foliata, *B. motasi, B. ovis* and *B. taylori*. Of these, *Babesia foliata* and *B. ovis* has been recorded from India.

Prevalence of babesiosis is determined by prevalence of tick vector in an area. In serological surveys, the prevalence in sheep ranged from 6.1 to 88.9 percent[15,14,70]. Through blood smear examination, the prevalence in sheep and goats has been reported to be 26.1 and 14.8%, respectively. Goats seem to be more resistant to *Babesia* infection and show lower level of parasitaemia[55]. *Babesia ovis* infection rate was more common than *B. motasi*, though, double infection was not uncommon[54]. In India, *Babesia* infection in sheep and goats has been reported by several workers[1,9,13,33,52,63]. *Babesia motasi* and *B. foliata* are the commonly occurring species.

Dermacentor silverum, Haemophysalis punctatum Rhipicephulus bursa, R. sanguineus, Hyalomma anatolicum anatolicum, H. marginatum are common tick vectors to transmit the disease.

In vertebrate host, parasite multiplie in erythrocyte by binary fission or endodygony or endopolygony, budding process or schizogony to form 2, 4 or more trophozoites. These trophozoites are liberated from red blood cells and invade new developing erythrocytes. The process of multiplication is repeated in the blood. Red cells some time may carry multiple infection and show large number of trophozoites. These trophozoites, the erythrocytic forms, are transmissible by mechanical means to other animals, however, these are transmitted through ticks in natural case.

In ticks, the development and further transmission of erythrocytic forms are of two types ie. trans-ovarian and stage to stage depending on the nature of ticks involved. The trans-ovarian transmission is dependent on atmospheric temperature[56].

Babesia ovis is more pathogenic than *B. motasi*. The disease seems to occur in acute or chronic form. The acute form of disease is characterized by high fever, haemoglobinuria and marked anaemia. The affected animals may show weakness and prostration. The clinical signs in sheep and goats are not very apparent. About 8 % sheep and 6.8% goats were reported to show clinical symptoms[53].

In India, *Babesia foliata* has been reported in sheep and *B. taylori*, from goats. Their pathogenesis is not clear but *B. taylori* is known to cause benign type of disease with haemoglobiuria.

Clinical disease is not very apparent in sheep and goats. Conventionally Giemsa stained blood smear examination is effective method to detect erythrocytic forms of parasite. However, the new techniques like Indirect Immunofluorscent Antibody Assay (IFA) and nucleic acid based diagnostic tests like PCR and FISH (Fluorescent In-situ Hybridization Assay) are popular and effective tools for detection of *Babesia* infection. Nucleic acid based tests are more efficient as all serological tests detect antibodies which usually develop after clinical symptoms are over.

Antibiotics like Clindamycin, Zithromax are effective and quinine being the good alternative. Drugs like imidacarb and parvoquonone are also efficiently used to treat Babesiosis.

Theileriosis

Theileriosis is caused by a non pigment forming pleomorphic haemoparasite known as *Theileria*. It occurs as minute round, ovoid, rod, shaped or in irregular forms in lymphocytes, histiocytes or erythrocytes of vertebrate hosts. It differs from other genus of haemoparasite, Babesia in having extra-erythrocytic multiplication (Schizogony) and only trans-stedial transmission.

The disease is more prevalent and pathogenic in dairy animals, specially the cross bred. In sheep and goats, two species *Theileria hirci* and *T. ovis* have been reported world wide[4,8, 22]. The sero-prevalence of 59.9% has been reported in sheep 6 and up to 34.3% in goats 4. In India, the disease has been reported in sheep and goats by several workers[10,13,49,51].

The parasitic forms are transmitted by various species of hard tick viz. *Rhipicephalus bursa, Hyalomma anatolicum anatolicum, Haemophysalis impeltatum* are common ticks involved in transmission.

Life cycle is similar to *Theileria annulata* in cattle. Koch blue bodies develop in lymphocytes and reticulo-endothelial tissues like spleen and have 2-4 bodies. The erythrocytic forms are round and oval and develop to micro and macro gamonts. These gamonts are picked up by ticks while sucking and transmitted to new host.

Theleria hirci is pathogenic and causes malignant theileriosis in sheep and goats in different parts of the world. The disease occurs in acute and chronic forms. The acute form which is more common is characterized by presence of high fever, nasal discharge, jaundice, anaemia, atony of rumen and weakness leading to listlessness. Haemoglobinuria is often transitory. Ulceration of abomasum, marked enlargement of spleen with infarction in kidneys are common post-mortem lesions observed. *Theileria ovis* is almost non pathogenic and may cause slight fever and swelling of lymph nodes.

The diagnosis is through blood smear examination in common practice. However serological technique like IFA and recently used nested PCR and PCR are effective tools for diagnosis[3,5,46].

There is no cross immunity among two species. Recovered animals become premune. The vaccination against *Theileria* is based on parasites derived from cell culture. The cell culture vaccine is, however, considered not protective. The infection with sporozoites and followed by treatment provide good protection. A recombinant subunit vaccine for East Coast Fever (ECF) based on the major surface protein of the sporozoites is now on trial. However, a vaccine for small ruminants in the market is still awaited.

Coccidiosis

Coccidiosis in sheep and goats is an economically important disease. It presents serious threat to lambs and kids up to 6 months of age through out the world. Lambs and kids maintained with adults are likely to get infected by oocysts (sporulated) during suckling, licking and through contaminated food and water. The disease in adults is symptom less and infected animals work as reservoir of infection. The disease has been observed as seasonally occurring before rain and during winter[11], however, author recorded the disease as all season disease in young ones.

Coccidia or Eimeria spp. are typically intracellular parasites of intestine of vertebrates. They infect epithelial cell lining in general though some other body organs may also be involved exceptionally.

Important *Eimeria* species infecting goats and sheep in India.

Goat	**Sheep**
Eimeria alijevi	*Eimeria ahsata*
E. apsheronica	*E. crandalis*
E. arloingi	*E. faurie*
E. caprina	*E. granulosa*
E. christenseni	*E. ovina*
E.hirci	*E. parva*
E. jolchijevi	*E. ovinoidalis*
E. ninalkohlyakimovae	*E. gilruthi*
E. gilruthi	*E. intricata*
E. kocharii	

It is a world widely distributed disease and no management system available is perfect to provide coccidian free condition in herd.

Coccidia are intracellular parasites. They multiply within cells lining of gastrointestinal tract of their hosts. The oocysts passed out in the faeces undergo sporulation in presence of oxygen and moisture. The sporulated oocysts (infective stage) contain sporozoites in sporocysts. Hosts are infected by taking the sporulated oocysts with contaminated food or water. Sporozoites are released from sporocysts in the intestine and invade cell-lining of the intestinal tract. They develop intracellularly to form schizonts which are specific to different species. The schizonts are asexual stages which produce merozoites through multiple fissions. The merozoites released from schizonts further invade the epithelial cells of intestine and produce second generation of schizonts. The cycle may be repeated several times depending on species. After a pre-determined number of asexual generations, the merozoites differentiate into male and female gametes and fertilization of gametes fusion results in the formation of zygotes. The zygotes form oocysts which are passed out with faeces. In some species the cycle may take place in other organs also.

In coccidiosis, the damage done to the host is essentially the result of intestinal cell destruction occurring when coccidian cysts are released from host cells. In young lambs and kids, coccidiosis is clinically manifested by the presence of most conspicuous sign of diarrhea. Faeces is loose, foul smelling, discoloured with or without blood. Animals may show abdominal pain, anaemia, inappetance, and weight loss. The look of animal is unthrifty with rough coat. Author recorded hypothermia in most of the animals suffering from coccidiosis but slight rise of body temperature may be possible11. Affected lambs or kids may show indifferent behaviour and prefer to lie down due to weakness. Clinical disease is common after the stress of weaning, feed change or transport. In stall fed herd or in a large group of sheep and goats, the disease usually occurs in the form of outbreak, affecting large number of the animals.

The clinical signs may be correlated with oocystic faecal counts. Though presence of oocysts does not indicate disease, yet the clinical disease is always associated with large number of oocysts in faeces. The count of over 5000 oocysts/g of faeces is considered significant and generally results in clinical disease condition in lambs and kids. In severe outbreaks counts in excess of 100,000/g are common. However, If a herd of lambs and kids with apparent clinical coccidiosis, does not responds to treatment, should be investigated for other gastrointestinal diseases like salmonellosis and *E. coli* enteritis etc.

Coccidiosis, if not severe or acute, is self limiting and clinical signs usually subside spontaneously. The immunity developed will reduce, but not eliminate, the number of coccidian living in the gut. Proper farm management practices reduce the rate of infection and development of new cases. Isolation of sick animals, hygienically maintained feed and water trough, removal of soiled bedding are important and effective management practices.

One or more coccidiostats with supporting therapy, are effective in reducing or inhibiting coccidias in intestine. Combination of sulfonamides (e.g. sulfamezathine, sulfaquinoxaline) with ionophores (monensin, lasalocid) may be helpful. Amprolium and monensin, presently appear to be most popular and effective drugs against coccidiosis. While, most of the coccidiostats are approved for use in cattle, sheep, monensin has recently been approved for use in goats in United States. Similar drugs like lasalocid and decoquinate have been shown to be effective in coccidiosis in sheep and goats.

All the attempts for development of vaccine against coccidiosis are limited to poultry coccidiosis and the problem in small ruminants could not get due attention. Live attenuated vaccine 'Supracox' performed well in terms of mortality, cost and overall profit[64]. Synthetic peptide vaccine from poultry Eimeria spp. Induced high level of antibody and cellular responses[65]. DNA vaccine delivered in attenuated *Salmonella typhinurium* has recently been new development in bacterial - vectored oral delivery system 17. Use of live attenuated multivalent ionophore - resistant Eimeria vaccine in poultry coccidiosis is quite promising[37].

Cryptosporidiosis

It is an important enteric disease of sheep and goats resulting in diarrhea. It is caused by *Cryptospordium parvum*. Primarily the disease affects new born lambs and kids[27] less than 30 days of age and is usually milder than coccidiosis The affected young ones are often active, alert and nursing. The disease carry importance as it is communicable to human beings. Almost 80 % of the patients with AIDS die from infections other than human immunodeficiency virus (HIV) and *Cryptosporidium parvum* is one the most important of those infections.

In India cryptosporidial oocysts have been reported from buffaloes and cattle calf in and around Bareilly, Uttar Pradesh[20]. Human cryptosporidiosis has been reported from, Vellore, Chandigarh and many other places[47]. There is no report known on cryptosporidiosis in sheep and goats in India so far.

Transmission of *Cryptosporidum* is through oral route by ingestion of oocysts with food and drinking water. Air borne transmission is also possible. Transmission by arthropods associated with faeces may be another source of infection. Close association with affected animal is important to transmission.

Cryptosporidium spp. develop in gastrointestinal tract of vertebrates through out or partial part of their life. Fayer reported cryptosporidial infection in various species of animals in detail. The life cycle resembles other coccidia life cycle. Sporulated oocysts containing 4 sporzoites are passed out in faeces and work as infective stage. Sporozoites when ingested, parasitize epithelial surface of gastro-intestinal or respiratory tract. The organism is intracellular but lies close to luminar part of cell membrane and looks superficially attached. The schizogony results in two types of schizonts. First multiply to form merozoites which in turn attack new host cells while second type, perhaps start gametogony and results in zygotes. The zygotes undergo sporulation and form oocysts with 4 sporozoites which are infective when passed out in faeces. Cycle in respiratory tract is same and sporulated oocysts released here are carried out with nasal secretion. Some sporulated oocyst may again start schizogony in intestine with sporozoites contained it lead to auto infection.

Severity of infection is variable and depends on immune status of the host. *Cryptosporidium parvum* particularly develop in immunocompromised hosts. It is a significant cause of diarrhoeal disease. The immunodeficient host can become chronically and terminally ill. Enteric infection affects absorption of nutrient after digestion.

Experimental cryptosporidiosis in kids was observed with decreased appetite, watery faeces with mucus and colour change. The most severe lesions were found in jejunum and ileum characterized by villus atrophy, villus blunting, crypt hyperplasia and metaplasia of mucosal epithelium[35].

Effectiveness of DNA vaccine to confer protection against cryptosporidiosis has been evaluated 59. A 15kDa surface protein from *Cryptosporidium parvum* (CP-15 DNA) vaccine through nasal immunization in pregnant dam resulted in immunized kids. Colostrum and milk from immunized ewes were successful to significantly lower the oocysts shedding and severity of diarrhea[40].

To date, there is no universal treatment for cryptosporidiosis but recent study with probiotic like *Lactobacillus acidophilus* and *L. reuter* have shown that they detrimentally affect *C. parvum* and *C. hominis*[29]. Field trial of paromomycin in natural infection in lambs and kids successfully reduced cryptosporidial oocysts output and severity of signs[68]. As disease is self limiting, intravenous hydration therapy with parenteral nutrition improve the situation. Drugs like furazolidon, diloxanide, quinine with Clindamycin, Azithromycin are efficiently used against *Cryptosporidium*.

Sarcocystosis

It is caused by cyst forming isosporian coccidia, Sarcocystis Spp. The parasite differs from common eimerian species as it shows heteroxezenous life cycle involving two hosts. Out of these two, one is prey and other is predator. The life cycle consists of asexual and sexual phase. Asexual phases of life cycle only occurs in intermediate host, which is usually a prey animal. The sexual phase of life cycle occurs in definitive host, always a carnivore predator. The intermediate hosts include a wide range of animals including cattle, buffaloes, sheep, goats, birds and other animals, while definitive hosts are canidae, felidae and primates.

In sheep, there are 4 species of *Sarcocystis* viz. *Sarcocystis tenella, S. arieticanis, S. gigantea* and *S. medusiformis*. The first two are transmitted by dogs while the later two are transmitted by cats[18].

Several reports have been published on prevalence of sarcocysts sheep and goats in different countries viz. France, 95%[16], Ethiopia, 93%[69], Turkey, 90%[45], Slovakia, 87.6%[38], Iran 97.5%[53]. In India, *S. tenella* and *S. arieticanis* have been reported from muscles of sheep[60]. There are 3 speices of *Sarcocystis* infecting goats. These are *S. capracanis, S. hircicanis* and *S. moulei,* the first being in India.

Species	Intermediate host	Definitive host	Reference
Sarcocystis tenella	Sheep	Dog	18
S. areiticanis	Sheep	Dog	60
S. gigantea	Sheep	Cat	
S. medusiformis	Sheep	Cat	
S. capracanis	Goat	Dog	18
S. hircicanis	Goat	Dog	61
S. moulei	Goat	Cats	18

In typical life cycle, the definitive hosts pick up infection by ingesting infected muscles or neural tissue containing sarcocysts filled with the bradyzoites. The bradyzoites get liberated in the GI tract after digestion and then actively penetrate mucosa of the small intestine. Here, they transform into male and female gamonts popularly known as micro and macro gamonts respectively. The micro gamonts form microgametes which out number the female gametes, the macro gametes. The gametogony takes place, in goblet cells, or near the intestine surface, in enterocytes and rarely in mesenteric lymph nodes. The gametes form zygotes after fertilization which in turn form oocysts in lamina propria.

Sporulated oocysts being thin walled rupture during its passage from intestine and sporcysts containing 4 elongated sporozoites are passed out in faeces depending on species.

The intermediate host picks up infection by ingesting sporocysts in feed and water. The sporozoites so released in intestine start schizogony in endothelial cells forming schizonts. Further generation of schizonts develop in endothelial cells and arteries, through out the body and in kidney. The schizonts give rise to merozoites through multiple fission. The merozoites are found in blood extracellularly or in mononuclear cells. Later on these merozoites, from terminal generation, develop to form sarcocysts having bradyzoites in muscular or neuronal tissue, which act as infective stages for the definitive hosts.

Sarcocystosis in sheep and goats is pathogenic. It is manifested clinically by weight loss, fever, anaemia, loss of wool, abortion and premature birth, nervous symptoms, myositis and some time death. *S. tenella* in sheep and *S. capracanis* in goats are comparatively more pathogenic than other species. Pregnant sheep and goats can abort due to *Sarcocystis* infection. The mechanism of *Sarcocystis* induced abortion is not known[19].

Sarcocystosis is an important disease from public health point of view. In India, man serves as definitive host to *S. hominis* (intermediate host cattle) and *S. suihominis* (intermediate host pig). Sarcocysts are found in muscles of oesophagus, diaphragm, tongue and heart of intermediate host. Uncooked meat eating is common source of infection. The disease is mildly pathogenic in human cases and causes nausea, stomachache and diarrhea.

Antigen preparations developed from *Sarcocystis* species are highly non specific and show cross reactivity with antibodies against non pathogenic species and can not differentiate between the non pathogenic and pathogenic ones[66]. The new techniques in immunology, protein chemistry and molecular biology have facilitated further diagnostic studies in this regard.

Toxoplasmosis

The protozoan parasite, *Toxoplasma gondii,* is a matter of serious concern in sheep and goats that leads to fetal mortality in them. The parasite has cat as primary and sheep, goats along with many other mammals, as intermediate host. *Toxoplasma gondii* has very wide range of intermediate hosts including man.

The prevalence of toxoplasmosis in sheep and goats has been reported by several workers who used different diagnostic techniques[25,26,43]. The sero-prevalence up to 56.2 in sheep and 24 percent in goats has been reported.

The parasite undergoes sexual reproduction through gametogony and then sporogony follows in definitive host. The infection in primary host is initiated by ingestion of cysts having (i) tachyzoites, (ii) bradyzoites and (iii) sporulated oocysts. The released zoites, in the stomach and intestine, penetrate the epithelial cells of intestine and multiply forming different stages (A to E). The process of schizogony is followed by gametes formation. The male and female gametes resulted in gametogony fuse to form zygotes which further turn into oocysts. The oocysts undergo soporulation after they are passed out in the faeces. Some zoites in cat may spread to extra - intestinal tissues via lymph and blood and develop as in non feline species.

The primary infection in sheep and goats is initiated by oocysts produced in cats. The development in intermediate is asexual and through multiple fission. The oocysts released sporozoites enter intestinal epithelial cells and lymph nodes and multiply intra-cellularly and form tachyzoites. The tachyzoites spreads to different parts of the body via blood and lymph and develop as bradyzoites cysts in brain, skeletal, cardiac and psoas muscles and diaphragm etc. On ingestion of bradyzoites and tachyzoites, the cat becomes infected.

The primary infection of *Toxoplasma* can establish in placenta and fetus in pregnant sheep and goats and results in fetal death followed by resorption, abortion or still birth 12. The infection is characterized by the development of small white necrotic foci in placental cotyledons. There may be fever, anorexia, diarrhea and respiratory distress and death may follow in goats.

It is aided by clinical observations, presence of characteristic small while necrotic foci in cotyledons, presence of mummified foetus if any, fetal serology and histology. PCR test specific for *T. gondii* is the most modern tool for diagnosis.

Vaccination of sheep with live vaccine is a effective measure. Decoquinate in feed may be a useful treatment.

Neosporosis

Neospora caninum is a protozoan parasite which creates *Toxoplasma like* infection in sheep and goats. Though it does not appear to cause frequent losses in them, experimental infection can readily be induced.

The parasite has been reported from several geographical areas of the world including USA, Canada, Europe, Australia. In sheep and goats, a sero-prevalence of 9 and 6.4 percent respectively has been reported[25,26].

The life cycle of *Neospora* is not known, however, its close similarity with *Toxoplasma* suggests that it is coccidian and a carnivore acts as definitive host. Tachyzoites and bradyzoites containing tissue cysts are only known stages reported in intermediate host like sheep and goats. Neospora caninum cysts have been found in brain and spinal cord of one week old lamb[21].

Neospora caninum infection has been associated with abortion in sheep and goats[20,21,41]. In an experimentally induced neosporosis in pregnant ewes resulted abortion in all ewes inoculated on gestation day 65 but inclusion lately result in weak or normal lambs[41]. However, its role has been controversial as an abortion causing organism. A survey conducted to determine whether *Neospora caninum* is a significant cause of ovine abortion in England and Wales resulted to the conclusion that parasite was not associated with ovine abortion. Clinical and pathological features of neosporosis in sheep and goats closely resemble those of ovine toxoplasmosis.

A killed *Neosopora caninum* tachyzoites vaccine provided nearly 2 fold protection against fetal loss associated with experimental neosprosis in pregnant ewes[34].

References

1. Acharya, R. S. and Bhaskar Rao, P. 1981. A note on Babesiosis in sheep. Livestock advisor, 6: 51.

2. Akbar, A.P., Elamin, E.A. , Osman A.Y., Oman, H. M. 2004. Coccidian infection of sheep in Khartoum, Sudan Journal of Animal and Veterinary Advances 3 (10): 649-652.

3. Aktas, M., Altay, K. and Dumanli, N. 2005. Survey of *Theileria* parasites of sheep in eastern Turkey using polymerase chain reaction. Small Ruminant Research 60 (3): 289-293.

4. Al-Amery, M.A.Y., and Hasso, S.A. 2002. Laboratory diagnosis of novel species of *Theileria hirci, Eimeria caprovina* and *E. pallida* in goats in Iraq. Small Ruminant Research 44 (2): 163-166.

5. Altay, K., Dumanli, N., Holman, P.J., and Aktas, M. 2005. Detection of *Theileria ovis* in naturally infected sheep by nested PCR. Veterinary Parasitology 127 (2): 99-104.

6. Alyasino, Y and Griener, M. 1999. Serosurvey of theileriosis in Awassi sheep in Syria. Veterinary Parasitology 81 (4): 275-280.

7. Asonganyi, T. 1992. Prevalence of Trypanosomes in goats in sleeping sickness endemic areas in Cameroon. Recent Advances in goat production. V International conference on Goats, March1992, pp1665-1668.

8. Bell-Sakyi, L., Koney, E.B.M., Dogbey, O. and Walker, A.R. 2004. Incidence and prevalence of tick born haemoparasites in domestic ruminants in Ghana. Veterinary Parasitology 124 (1-2): 25-42.

9. Bhashar Rao, P., Surendran, N.S. and Ramchandra Rao, P.V. 1989. Study of outbreak of Babesiosis in Andhra Pradesh. I.V.J. 66 : 348-51.

10. Bhaskar Rao, P., Acharya, R.S. and Ramachandra Rao, P.V. 1982. Ovine Theleriaosis in Andhra Pradesh. Livestock Advisor 7 (1) 33-34.

11. Bhatia, B. B. 2000. Text book of Veterinary Protozoology. Published by Directorate of Information and Publication of Agriculture, ICAR, New Delhi p 368.

12. Buxton, D. 1998. Protozoan infection (*Toxoplasma gondii, Neospora caninum* and *Sarcocystis* spp. In sheep and goats: recent advances, Vet. Res. 29 (3-4): 289-310.

13. Chandra, D., Singh, K.P., Srivastava, V.K. and Bansal, G.C. 2001. Haemoprotozoan infections in sheep of an organized farm in Uttar Pradesh. J. Vet. Parasitol. 15(2): 145-46.

14. Cicek, H. Duzgun, A. Emre Z., Karaer, Z. 2004. Seroprevalence of Babesia ovis in sheep around Afyon. Turk Veterinarlik ve hayvancilik Dergisi 28 (4): 683-686.

15. David F., Castella J. and Gutierrez, J.F. 1998. Sero-prevalence of Babesia ovis in sheep in Catalonia, northeastern Spain. Veterinary Parasitology 79 (4) 275-281.

16. Diez, P and Banos, P.D. 1978. The prevalence of Sarcosporidiosis in sheep in province of Leon, with a comparative study of various diagnosis methods Anales de la faculted de veterinaria de Leon, 24: 195-199.

17. Du., A and Wang, S. 2005. Efficacy of a DNA vaccine delivered in attenuated *Salmonella typhinurium* against *Eimeria tenella* infection in chickens. International Journal of Parasitology 35 (7): 777-785.

18. Dubey, J.P, Spear, C.A. and Fayer, R. 1989. Sarcocystosis of animals and man. CRC Press, Florida, USA pp 215.

19. Dubey, J.P. and Fayer, R. 1987. Sarcocystosis present and future. Journal of Veterinary Parasitology 1: 1-6.

20. Dubey, J.P., Acland, H.M., and Hamir, A.N. 1992. *Neospora caninum* (Apicomplaxa) in still born goat. J. Parasitol, 78 (3): 532-04.

21. Dubey, J.P., Hartley WJ., Lindsay, D.S. and Topper, M. 1990. Fatal congenital *Neospora caninum* infection in lambs. J. Parasitol. 76: 127-30.

22. El-Azazy, O.M.E., El-Metenaw, T.M. and Wassef, H.Y. 2001. *Hyalomma impetatum* (Acari Ixodidae) as a potential vector of malignant theileriosis in sheep in Saudi Arabia. Veterinary Parasitology 99 (4): 305-309.

23. Fayer, R. Speer, C.A. and Dubey, J.P. 1991 General biology of *Cryptosporidium*: Cryptosporidiosis of Man and Animals. (Eds) J. P. Dubey, C.A. Speer and R. Fayer, CRC Press, Boca Raton, Florida, pp- 1-29.

24. Ferrer, D., Castella, J. and Gutierrez, J.F. 1998. Seroprevalence of *Babesia ovis* in sheep in Catalonia, north eastern Spain. Veterinary Parasitology 79 (4): 275-281.

25. Figliualo, L.P.C. Kasai, N, Ragozo, A.M.A., Paula, V. S.D. de Dias, R.A. Sluza, S.L. Gennari S.M. 2004. Prevalence of anti *Toxoplasma gondii* and *Neospora caninum* anti bodies in ovine from Sao Paulo state of Brazil. Veterinary Parasitology 123 (3/4) 161-166.

26. Figliuolo, L.P.C.; Rodrigues, A.A.R.; Viana, R.B; Aguiar, D.M; Kasai, N; Gennari, S.M. 2004. Prevalence of anti *Toxoplama gondii* and anti *Neospora caninum* antibodies in goats from Sao, Saulo state, Brazil, Small Ruminant Research 55 (1/3) 29-32.

27. Foreyt, W.J. 1990. Coccidiosis and cryptosporidiosis in sheep and goats. Vet. Clin. North Am. Food Anim. Pract. 6(3): 655-70.

28. Gauly, M, Reeg, J, Bauar, C., Erhardt, G. 2004. Influence of production system in lambs on the *Eimeria* oocysts output and weight gain. Small Ruminant Research 55(1/3) : 159-167.

29. Glass, M. D., Courtney, P. D., LeJeune J. T. and Ward L. A. 2004. Effects of *Lactobacillus acidophilus* and *Lactobacillus reuteri* cell free supernatants on *Cryptosporidium* viability and infectivity in vitro. Food Microbiology 21(4) 423-429.

30. Gutierrez, C., Corbera, J.A., Morales, M, Buscher, P 2004. Performance of serological tests for *Trypanosoma evansi* in experimentally inoculated goats. Annals of the New York Academy of Sciences 1026: 152-153.

31. Hornby, H.F. 1952. Animal trypanosomiasis in eastern 1949. H.N. Stat. Off. London.

32. Ikede, B.O. and Losos, G.J. 1975. Pathogenesis of *Trypanosoma brucei* infections in sheep. I. Clinical signs. J. Comp. Path., 85: 23-44.

33. Jagannath, M.S., Hedge, K.S. Shivram, K. and Nagraja, K.V. 1974. An outbreak of babesiosis in sheep and goats and its control. Maysore J. Agri. Sci., 8: 441.

34. Jekins, M., Tuo, and Dubey, J. 2004. Vaccination of sheep with *N. caninum* protein produces against fetal loss associated with experimental Neosporosis. Am. J. Vet. Res. 65: 1404-1408.

35. Koudela, B and Jiri, V 1997. Experimental cryptosporidiosis in kids. Veterinary Parasitology, 71(4) : 273-281.

36. Kramer, J. W. 1966. Incidence of Trypanosomes in the West African Dwarf sheep and goat in Nsukka, Eastern Nigeria. Bull. Epizoot. Dis. Afr., 14: 423-28.

37. Li, G.Q., Kanu, S., Xiang, F.Y., Xiao, S.M., Zhang, L. Chen, H.W. and Ye, H.J. 2004. Isolation and selection of ionophore tolerant *Eimeria* precocious lines: *E. tenella, E. maxima* and *E. acervulina*. Veterinary Parasitology 119(4): 261-276.

38. Mala, P. and Baranova M. 1995. Diagnosis of *Sarcocystis* infection in slaughter animals at veterinary meat inspection. Vet. Med. 40: 97-100.

39. Manjrekar, S.L. 1950. Trypanosomiasis in goats- A preliminary note. Indian Vet. J., 27: 103-104.

40. Martin-Gomez, S. Alvarez-Sanchez M.A. and Rojo-Vazquez F.A. 2005. Immunization protocols against *Cryptosporidium parvum* in ovines: protection in suckling lambs. Veterinary Parasitology 129(1-2) 11-20.

41. McAllister, M.M., McGuire, A. M., Jolley, W.R., Lindsay, D.S., Trees, A.J. and Stobart, R.H. 1996. Experimental neosporosis in pregnant ewes and their offspring. Vet. Pathology, 33 (6): 647-655.

42. Mornet, P. 1954. The pathogenic trypanosomes of French West Africa. Their distribution and frequency and the incidence of domestic animal infections. ISCTR 5th meeting, Pretoria, pp- 37-51.

43. Negash, T., Tilahun, G. Patton S. Prevot, F., Dorchies, P 2004. Serological survey on Toxoplasmosis in sheep and goats in Nazareth, Ethiopia. Reveu de Demdcine Veterinaire 155 (10): 486-487.

44. Otter, A., Wilson, B.W., Scholes, S.F., Jeffrey, M., Helmick, B. and Trees, A.J. 1997. Results of a survey to determine whether *Neospora* is a significant cause of ovine abortion in England and wales. The Veterinary Record, 140 (7): 175-177.

45. Ozturk, G. 1994. Incidence of ovine Sarcosporidiosis in the myocardium of sheep. Saglik Bilimleri Dergisi, 8:66-69.

46. Papadopolos, B., Brossard, M. and Marie Perie, N. 1996. Piroplasms of domestic animals in Macedonia region of Greece, 3 Piroplasma of small ruminant. Veterinary Parasitology 63(1-2): 67-74.

47. Parija, S.C. 1990. Review of Parasitic Zoonosis, AITBS Publishers Distributors, Delhi, pp 463.

48. Picaud, G. 1909. Note surles trypanosomiases des petes animaux domestiques Bas- Daho-mey. Bull. Soc. Path. Exot. 2: 127-28.

49. Prabhakar, K.S. and Heregoudan, L.S. 1977. A note on the prevalence of theileriosis among sheep in Karanataka. Ind. J. Anim. Sc. 47: 848-850.

50. Rabo, J.S., Oyejide, A., Karumi, Y. 2003. The cerebral pathology and clinical course of *Trypanosoma congolense* (Gbokostrain) infection in West African dwarf sheep. Studies and Researcher in Veterinary Medicines, Supplement I, 47-55.

51. Raghavachari, K and Reddy M.K. 1959. Acute theleriosis in sheep in Hyderabad (India) Ind. Vet. J. 29:1.

52. Ray, H.N. and Raghavachari, K. 1941. Observations on Babesia foliate new species sp from a sheep. Indian Journal of Veterinary Science 11: 239-42.

53. Razavi, S.M., Shekarforoush, S. S., Farahani, M,, Sarihi, K. 2003. Prevalence of sarcocyst in slaughtered sheep in Shiraz, Iran. Journal of Veterinary Parasitology, 17 (2) 139-141.

54. Razmi, G.R., Naghibi, A., Aslani, M.R., Dastjerdi, K and Hossieni, H. 2003. A epidemiological study on *Babesia* infection in small ruminant in Mashhad suburb, Khorasan province, Iran, Small Ruminant Research 50 (1-2): 39-44.

55. Razmi, G.R., Naghibi, A., Aslani, M.R., Fathivand M. and Dastjerdi K. 2002. An epidemiological study on ovine Babesiosis in the Mashhad suburb area, province of Khorasan, Iran Veterinary Parasitology, 108 (2) 109-115.

56. Riek, R.F. 1964. The life cycle of *Babesia bigemina* (Smith and Kilborne, 1893) in tick vector *Boophilus microplus* (Canestrini). Australian Journal of Agriculture Research,15:802-21.

57. Robson, J. and Ashkar, T. S. 1972. Trypanosomiasis in domestic livestock in Lamb we Valley area and a field evaluation of various diagnostic techniques. Bull. Weld. Health Org., 47: 727-734.

58. Robson, J. and Rickman, L.R. 1973. Blood incubation with infectivity test results for *Trypanosoma brucei* sub-group. Trop. Anim. Hlth. Prod., 5: 187-191.

59. Sagodira S., Buzoni-Gatel D., Iochmann S., Naciri M. and Bout D. 1999. Protection of kids against *Cryptosporidium parvum* infection after immunization of dams with CP15-DNA. Vaccine 17 (19) 2346-2355.

60. Saleque, A; Bhatia, B.B. and Daya Shankar 1992. Prevalence of two species of Sarcocystis in sheep in Uttar Pradesh. Ind. Vet. J. 69: 841-42.

61. Shankar, D. and Bhatia, B.B. 1993. *Sarcocystis* infection in goats in Uttar Pradesh. Ind. J. Anim. Sci. 63: 284-87.

62. Sharma, D.K., Chauhan, P.P.S., and Agrawal, R.D. 2000. Interaction between *Trypanosoma evansi* and *Haemonchus contortus* infection in goats. Veterinary Parasitology 92: 261-267.

63. Singh, C. P., Srivastava, S.M. and Verma, M.C. 1985. Incidence of haemoprotozoan infection amongst sheep. Ind. Vet. Med. Journal, 9: 42-47.

64. Suo, X, Zhang, J.X., Li, Z.G., Yang, C.T., Min., Q.R., Xu, L.T., Liu. X. and Zhu X.Q. 2006. The efficacy and economic benefits of supercox, a live anticoccidial vaccine in a commercial trail in broiler chickens in China Veterinary Parasitology, 142: (1-2): 63-70.

65. Talebi A. and Mulcahy G. 2005. Partial protection against *Eimeria acervulina* and *Eimeria tenella* induced by synthetic peptide vaccine. Experimental Parasitology 110(4): 342-348.

66. Tenter, A.M. 1995. Current reseach on *Sarcocystis* species of domestic animals. International J. Parasitology 25 (11): 131-1330.

67. Verma and Gautam, O.P. 1977. Serological diagnosis of experimental bovine surra (*Trypanosoma evansi*) infection. A comparison of passive haemagglutination, Gel diffusion and indirect florescent antibody tests. Indian Veterinary Journal 54:809.

68. Viu, M., Quilez, J., Sanchez-Acedo, C., Cacho E. del and Lopez-Bernad F. 2000. Field trial on the therapeutic efficacy of paromomycin on natural *Cryptosporidium parvum* infections in lambs Veterinary Parasitology 90 (3) 163-170.

69. Woldemeskel, M. and Gebreab, F. 1996. Prevalence of sarcocysts in livestock of north west Ethiopia. J. Vet. Med. Series B. 43: 55-58.

70. Yeruham, I, Hadani, A, Galkar, F. and Rosan, S. 1995. A study on an enzootic focus of sheep Babesiosis (*Babesia ovis*, babes, 1892). Veterinary Parasitology 60 (3-4): 349-354.

–ΔΔΔ–

20

Molecular Approaches for the Diagnosis of Infectious Diseases

V K GUPTA

Animal Health Division
Central Institute for Research On Goats
Makhdoom, PO - Farah, Mathura (U.P.) - 281 122

Molecular diagnosis is most appropriate for infectious agents that are difficult to detect, identify, or test for susceptibility in a timely fashion with conventional methods. Livestock diseases, particularly those of a transboundary nature and impacting on trade, pose a major challenge to the production and distribution of food of animal origin. Progressive control and eradication of these diseases requires action at national and international level to develop and apply improved techniques and strategies for diagnosis and surveillance, and to set national guidelines and standards for disease control. As molecular techniques for identifying and detecting microorganisms in the clinical microbiology laboratory have become routine, questions about the cost of these techniques and their contribution to animal care need to be addressed.

The tools of molecular biology have proven readily adaptable for use in the clinical diagnostic laboratory and promise to be extremely useful in diagnosis, therapy, and epidemiologic investigations and infection control. Although technical issues such as ease of performance, reproducibility, sensitivity, and specificity of molecular tests are important, cost and potential contribution to patient care are also of concern. Molecular methods may be an improvement over conventional microbiologic testing in many ways. Currently, their most practical and useful application is in detecting and identifying infectious agents for which routine growth-based culture and microscopy methods may not be adequate. Nucleic acid-based tests used in diagnosing infectious diseases use standard methods for isolating nucleic acids from organisms and clinical material and restriction endonuclease enzymes, gel electrophoresis, and nucleic acid hybridization techniques to analyze DNA or RNA. Because the target DNA or RNA may be present in very small amounts in clinical specimens, various signal amplification and target amplification techniques have been used to detect infectious agents in clinical diagnostic laboratories. Although mainly a research tool, nucleic acid sequence analysis coupled with target amplification is clinically useful and helps detect and identify previously uncultivatable organisms and characterize antimicrobial resistance gene mutations, thus aiding both diagnosis and treatment of infectious diseases. Automation and high-density oligonucleotide probe arrays (DNA chips) also hold great promise for characterizing microbial pathogens.

Molecular Epidemiology

Laboratory characterization of microbial pathogens as biologically or genetically related is frequently useful in investigations. Several different epidemiologic typing methods have been applied in studies of microbial pathogens. The phenotypic methods have occasionally been useful in describing the epidemiology of infectious diseases; however, they are too variable, slow, and labor-intensive to be of much use in most epidemiologic investigations. Newer DNA-based typing methods have eliminated most of these limitations and are now the preferred techniques for epidemiologic typing. The most widely used molecular typing methods include plasmid profiling, restriction endonuclease analysis of plasmid and genomic DNA, Southern hybridization analysis using specific DNA probes, and chromosomal DNA profiling using either pulsed-field gel electrophoresis (PFGE) or PCR-based methods. All these methods use electric fields to separate DNA fragments, whole chromosomes, or plasmids into unique patterns or fingerprints that are visualized by staining with ethidium bromide or by nucleic acid probe hybridization. Molecular typing is performed to determine whether different isolates give the same or different results for one or more tests. Epidemiologically related isolates share the same DNA profile or fingerprint, whereas sporadic or epidemiologically unrelated isolates have distinctly different patterns. If isolates from different patients share the same fingerprint, they probably originated from the same clone and were transmitted from patient to patient by a common source or mechanism.

Molecular typing methods have allowed investigators to study the relationship between colonizing and infecting isolates in individual patients, distinguish contaminating from infecting strains, document nosocomial transmission in hospitalized patients, evaluate reinfection versus relapse in patients being treated for an infection, and follow the spread of antimicrobial-drug resistant strains within and between hospitals over time. Most available DNA-based typing methods may be used in studying nosocomial infections when applied in the context of a careful epidemiologic investigation. In contrast, even the most powerful and sophisticated typing method, if used indiscriminately in the absence of sound epidemiologic data, may provide conflicting and confusing information.

Recombinant DNA Technology

Recombinant DNA technology is one of the many genetic modification techniques we have developed over the centuries. In nature and in the lab, recombinant DNA is made by combining genetic material from different sources.Humans began to preferentially combine the genetic material of domesticated plants and animals thousands of years ago by selecting which individuals would reproduce. By breeding individuals with valuable genetic traits while excluding others from reproduction, we changed the genetic makeup of the plants and animals we domesticated. Techniques for making selective breeding more predictable and precise have been evolving, especially since we discovered the genetic basis of heredity in the early 1900s. Now, in addition to using selective breeding to combine valuable genetic material from different organisms, we combine genes at the molecular level using the more precise techniques of recombinant DNA technology. Genetic modification through selective breeding and recombinant DNA techniques fundamentally resemble each other, but there are important differences:

- Genetic modification using recombinant DNA techniques allows us to move single genes whose functions we know from one organism to any other.
- In selective breeding, large sets of genes of unknown function are transferred between related organisms.

By making our manipulations more precise and our outcomes more certain, we decrease the risk of producing organisms with

Table 1. Genotypic methods for epidemiologic typing of microorganisms

Method	Examples	Comments
Plasmid analysis	Staphylococci Enterobacteriaceae	Plasmids may be digested with restriction endonucleases Only useful when organisms carry plasmids
Restriction endonuclease analysis of chromosomal DNA with conventional electrophoresis	Enterococci *Staphylococcus aureus* *Clostridium difficile* *Candida* spp.	Large number of bands Difficult to interpret Not amenable to computer analysis
PFGE	Enterobacteriaceae Staphylococci Enterococci *Candida* spp.	Fewer bands Amenable to computer analysis Very broad application.
Genome restriction fragment length polymorphism analysis: ribotyping, insertion sequence probe fingerprinting	Enterobacteriaceae Staphylococci *Pseudomonas aeruginosa* *Mycobacterium tuberculosis* *Candida* spp.	Fewer bands Computer analysis Sequence-based profiles Automated
PCR-based methods: repetitive elements PCR spacer typingselective amplification of genome restriction fragments, multilocus allelic sequence based typing	Enterobacteriaceae *Acinetobacter* spp. Staphylococci *M. tuberculosis* HCV	Crude extracts and small amounts of DNA may suffice
Library probe genotypic hybridization schemes: multilocus probe dot-blot patterns, high-density	*Burkholderia cepacia* *S. aureus* *M. tuberculosis*	Unambiguous yes-no result Less discrimination than other methods Couple with DNA chip

unexpected traits and avoid the time-consuming, trial-and-error approach of selective breeding. By increasing the breadth of species from which we can obtain useful genes, we can access all of nature's genetic diversity.

Cloning

Cloning technology allows us to generate a population of genetically identical molecules, cells, plants or animals. Because cloning technology can be used to produce molecules, cells, plants and some animals, its applications are extraordinarily broad. Any legislative or regulatory action directed at "cloning" must take great care in defining the term precisely so that the intended activities and products are covered while others are not inadvertently captured.

Molecular Cloning

Molecular, or gene, cloning, the process of creating genetically identical DNA molecules, provides the foundation of the molecular biology revolution and is a fundamental and essential tool of biotechnology research, development and commercialization. Virtually all applications of recombinant DNA technology, from the Human Genome Project to pharmaceutical manufacturing to the production of transgenic crops, depend on molecular cloning.

In molecular cloning, the word *clone* refers to a gene or DNA fragment and also to the collection of cells or organisms, such as bacteria, containing the cloned piece of DNA. Because molecular cloning is such an essential tool of molecular biologists, in scientific circles "to clone" has become synonymous with inserting a new piece of DNA into an existing DNA molecule.

Cellular Cloning

Cellular cloning produces cell lines of identical cells and is also a fundamental tool of biotechnology research, development and product manufacturing. All these applications depend on producing genetically identical copies of cells: the therapeutics and diagnostic uses of monoclonal antibody technology; the regeneration of transgenic plants from single cells; pharmaceutical manufacturing based on mammalian cell culture; and generations of therapeutic cells and tissues, which is known as *therapeutic cloning.*

Animal Cloning

Animal cloning has helped us rapidly incorporate improvements into livestock herds for more than two decades and has been an important tool for scientific researchers since the 1950s. Although the 1997 debut of Dolly, the cloned sheep, brought animal cloning into the public consciousness, the production of an animal clone was not a new development. Dolly was considered a scientific breakthrough not because she was a clone, but because the source of the genetic material that was used to produce Dolly was an adult cell, not an embryonic one.

Recombinant DNA technologies, in conjunction with animal cloning, is providing us with excellent animal models for studying genetic diseases, aging and cancer and, in the future, will help us discover drugs and evaluate other forms of therapy, such as gene and cell therapy. Animal cloning provides zoo researchers with a tool for helping to save endangered species.

Protein Engineering

Protein engineering technology is used, often in conjunction with recombinant DNA techniques, to improve existing proteins, such as enzymes, antibodies and cell receptors, and to create proteins not found in nature. These proteins may be used in drug development, food processing and industrial manufacturing. The most pervasive uses of protein engineering to date are applications that alter the catalytic properties of enzymes to develop ecologically sustainable industrial processes. Enzymes are environmentally superior to most other catalysts used in industrial manufacturing, because, as biocatalysts, they dissolve in water and work best at neutral pH and comparatively low temperatures. In addition, because biocatalysts are more specific than chemical catalysts, they also produce fewer unwanted byproducts. The chemical, textile, pharmaceutical, pulp and paper, food and feed, and energy industries are all benefiting from cleaner, more energy-efficient production made possible by incorporating biocatalysts into their production processes. The characteristics that make biocatalysts environmentally advantageous may, however, limit their usefulness in certain industrial processes. For example, most enzymes fall apart at temperatures above 100° F. Scientists are circumventing these limitations by using protein engineering to increase enzyme stability under harsh manufacturing conditions. In addition to industrial applications, medical researchers have used protein engineering to design novel proteins that can bind to and deactivate viruses and tumor-causing genes; create especially effective vaccines; and study the membrane receptor proteins that are so often the targets of pharmaceutical compounds. Food scientists are using protein engineering to improve the functionality of plant storage proteins and develop new proteins as gelling agents.

Biosensors

Biosensor technology couples our knowledge of biology with advances in microelectronics. A biosensor is composed of a biological component, such as a cell, enzyme or antibody, linked to a tiny transducer—a device powered by one system that then supplies power (usually in another form) to a second system. Biosensors are detecting devices that rely on the specificity of cells and molecules to identify and measure substances at extremely low concentrations.

When the substance of interest binds with the biological component, the transducer produces an electrical or optical signal proportional to the concentration of the substance. Biosensors can, for example,

- measure the nutritional value, freshness and safety of food.
- provide emergency room physicians with bedside measures of vital blood components.
- locate and measure environmental pollutants.
- detect and quantify explosives, toxins and biowarfare agents.

Nanobiotechnology

Nanotechnology, which came into its own in 2000 with the birth of the National Nanotechnology Initiative, is the next stop in the miniaturization path that gave us microelectronics, microchips and microcircuits. The word *nanotechnology* derives from *nanometer*, which is one-thousandth of a micrometer (micron), or the approximate size of a single molecule. Nanotechnology—the study, manipulation and manufacture of ultra-small structures and machines made of as few as one molecule—was made possible by the development of microscopic tools for imaging and manipulating single molecules and measuring the electromagnetic forces between them. Nanobiotechnology joins the breakthroughs in nanotechnology to those in molecular biology. Molecular biologists help nanotechnologists understand and access the nanostructures and nanomachines designed by 4 billion years of engineering—cell machinery and biological molecules. Exploiting the extraordinary properties of biological molecules and cell processes, nanotechnologists can accomplish many goals that are difficult or impossible to achieve by other means. For example, rather than build silicon scaffolding for nanostructures, DNA's ladder structure provides nanotechnologists with a natural framework for assembling nanostructures; and its highly specific bonding properties bring atoms together in a predictable pattern to create a nanostructure. Nanotechnologists also rely on the self-assembling properties of biological molecules to create nanostructures, such as lipids that spontaneously form liquid crystals. DNA has been used not only to build nanostructures but also as an essential component of nanomachines. Most appropriately, DNA, the information storage molecule, may serve as the basis of the next generation of computers. As microprocessors and microcircuits shrink to nanoprocessors and nanocircuits, DNA molecules mounted onto silicon chips may replace microchips with electron flow-channels etched in silicon. Such biochips are DNA-based processors that use DNA's extraordinary information storage capacity. Conceptually, they are very different from the DNA chips discussed below. Biochips exploit the properties of DNA to solve computational problems; in essence, they use DNA to do math. Scientists have shown that 1,000 DNA molecules can solve in four months computational problems that require a century for a computer to solve.

Other biological molecules are assisting in our continual quest to store and transmit more information in smaller places. For example, some researchers are using light-absorbing molecules, such as those found in our retinas, to increase the storage capacity of CDs a thousandfold.

Some more imminent applications of bio-nanotechnology include

- increasing the speed and power of disease diagnostics.
- creating bio-nanostructures for getting functional molecules into cells.
- improving the specificity and timing of drug delivery.
- miniaturizing biosensors by integrating the biological and electronic components into a single, minute component.
- encouraging the development of green manufacturing practices.

Microarrays

Microarray technology is transforming laboratory research because it allows us to analyze tens of thousands of samples simultaneously. Researchers currently use microarray technology mostly to study gene structure and function. Thousands of DNA or protein molecules are arrayed on glass slides to create DNA chips and protein chips, respectively. Recent developments in microarray technology use customized beads in place of glass slides.

DNA Microarrays

DNA microarrays are being used to

- detect mutations in disease-related genes.
- monitor gene activity.
- diagnose infectious diseases and identify the best antibiotic treatment.
- identify genes important to crop productivity.
- improve screening for microbes used in bioremediation.

DNA-based arrays will be essential for converting the raw genetic data provided by the Human Genome Project and other genome projects into useful products. Gene sequence and mapping data mean little until we determine what those genes do—which is where protein arrays come in.

Protein Microarrays

While going from DNA arrays to protein arrays is a logical step, it is by no means simple to accomplish. The structures and functions of proteins are much more complicated than that of DNA, and proteins are less stable than DNA. Each cell type contains thousands of different proteins, some of which are unique to that cell's job. In addition, a cell's protein profile varies with its health, age, and current and past environmental conditions.

Protein microarrays will be used to

- discover protein biomarkers that indicate disease stages.
- assess potential efficacy and toxicity of drugs before clinical trials.
- measure differential protein production across cell types and developmental stages, and in both healthy and diseased states.

- study the relationship between protein structure and function.
- assess differential protein expression in order to identify new drug leads.
- evaluate binding interactions between proteins and other molecules.

The fundamental principle underlying microarray technology has inspired researchers to create many types of microarrays to answer scientific questions and discover new products.

Tissue Microarrays

Tissue microarrays, which allow the analysis of thousands of tissue samples on a single glass slide, are being used to detect protein profiles in healthy and diseased tissues and validate potential drug targets. Brain tissue samples arrayed on slides with electrodes allow researchers to measure the electrical activity of nerve cells exposed to certain drugs.

Whole-Cell Microarrays

Whole-cell microarrays circumvent the problem of protein stability in protein microarrays and permit a more accurate analysis of protein interactions within a cell.

Small-Molecule Microarrays

Small-molecule microarrays allow pharmaceutical companies to screen ten of thousands of potential drug candidates simultaneously.

Although most clinicians and microbiologists enthusiastically welcome the new molecular tests for diagnosing infectious disease, the high cost of these tests is of concern. Despite the probability that improved patient outcome and reduced cost of antimicrobial agents and length of hospital stay will outweigh the increased laboratory costs incurred through the use of molecular testing, such savings are difficult to document. Much of the justification for expenditures on molecular testing is speculative; however, the cost of equipment, reagents, and trained personnel is real and substantial, and reimbursement issues are problematic. Given these concerns, a facility's need for molecular diagnostic testing for infectious diseases should be examined critically by the affected clinical and laboratory services. In many instances, careful overseeing of test ordering and prudent use of a reference laboratory may be the most viable options.

References

1. Abd-Elsalam KA, Aly IN, Abdel-Satar MA, Khalil MS, Verreet JA (2003). PCR identification of *Fusarium* genus based on nuclear ribosomal-DNA sequence data. Afr. J. Biotechnol. 2: 82-85.

2. Agindotan BO, Thottappilly G, Uwaifo A, Winter S (2003). Production of monoclonal and polyclonal antibodies against a nigerian isolate of banana streak virus. Afr. J. Biotechnol. 2: 171-178.

3. Cho SN (1990). Contemporary approaches to the serology of tuberculosis. Bull. Int. Union Tuberc. Lung Dis. 66: 59-62.

4. Crowther JR (1998). Detection of viruses in livestock. Parasitology 117: S29-40.

5. Khan JA, Aminuddin, Raj SK, Singh BP (1998). Detection of plant viruses—biotechnological and molecular advances. Indian J. Exp. Biol. 36:546-552.

6. Nantulya VM (1991). Molecular diagnosis of parasites. Experientia 47:142-145.

7. Picozzi K, Tilley A, Fèvre EM, Coleman PG, Magona JW, OdiitM, Eisler MC, Welburn SC (2002). The diagnosis of trypanosome infections: applications of novel technology for reducing disease risk. Afr. J. Biotechnol. 1: 39-45.

8. Rocha-Pena MA, Lee RF (1991). Serological techniques for detection of citrus tristeza virus. J. Virol. Methods 34: 311-331

9. Siegel DL (2002). Recombinant monoclonal antibody technology. Transfus. Clin. Biol. 9:15-22.

10. Van Regenmortel MH (1984). Monoclonal antibodies in plant virology. Microbiol. Sci. 1: 73-78.

–ΔΔΔ–

21

Biotechnology in the Diagnosis of Small Ruminant Diseases

G BUTCHAIAH AND P X ANTONY

Rajiv Gandhi College of Veterinary and Animal Sciences
Kurumbapet, Puducherry - 605 009

INTRODUCTION

Laboratory diagnosis of animal diseases has traditionally relied on methods of detecting the pathogens or antibodies, using a variety of tests such as neutralisation, agar gel immunodiffusion, complement fixation, enzyme-linked immunosorbent assay etc. In the past fifteen years, veterinary diagnosticians have improved older techniques through the use of recombinant antigens, monoclonal antibodies, synthetic peptides and incorporated new molecular techniques such as polymerase chain reaction (PCR) and Nucleic Acid (NA) hybridisation. Although conventional diagnostic assays are still used routinely, new molecular techniques have broadened the scope of disease diagnosis and provided powerful new tools that enabled rapid and specific diagnosis of animal diseases.

Conventional immunoassays for the diagnosis of animal diseases are based on the detection of antibody to the pathogen of interest, using techniques such as virus neutralisation, agar gel immunodiffusion, complement fixation and enzyme-linked immunosorbent assay (ELISA). These assays generally rely on the interaction of serum polyclonal antibodies against the agent of interest, followed by the use of a detection system such as the presence of cytopathic effect in cell culture, the haemolysis of red blood cells due to fixation of complement as a result of antibody-antigen interaction, or the colour change of a reaction medium due to the chemical reaction of an indicator enzyme. New biotechnological methods such as the cloning of genes, use of expression vectors, over expression of immunogens and peptide synthesis have made possible the production of specific proteins or peptides that serve as target antigens or positive control reagents in existing and newly-developed immunoassays. The use of these improved antigens can increase the specificity or sensitivity of immunoassays by providing a more defined target for binding antibody and can reduce lot-to-lot variation of test kit performance due to the use of a more homogenous and well-defined antigen to capture antibodies.

The use of synthetic peptides or recombinant antigens in the detection of animals infected with pathogens reduces the risk involved in the production of native antigens from pathogens and production of kits with antigen that has not been completely inactivated and, therefore, remains potentially infectious.

The cloning and expression of specific proteins produced by a pathogen have enabled the development of assays that can differentiate vaccinated from non-vaccinated (infected) animals. This can be accomplished by expressing a single immunogen in a vectored vaccine, or by deleting the expression of a single immunogen in a vaccine, followed by the development of a complementary serological assay.

Sandwich ELISA using PPR specific monoclonal antibody to an epitope of nucleocapsid protein was develope[19]. The test uses polyclonal sera to capture the antigen from clinical samples (swabs and tissues). Captured antigens from clinical samples are detected using PPR specific monoclonal antibody. The test is specific to PPR as it failed to detect rinderpest vaccine virus (RBOK strain). PPR outbreaks from various parts of India have been confirmed using the test. Findings suggested that the newly developed ELISA is suitable for PPR diagnosis under field conditions

A stable Vero cell line expressing peste des petits ruminants virus (PPRV) haemagglutinin (H) protein was assessed for potential use as diagnostic antigen in ELISA. PPRV H gene of the vaccine strain (Sungri-96) was amplified by RT-PCR, cloned into a eukaryotic expression vector (pTarget(TM)) and subsequently transfected and expressed in Vero cells[3,4]. The cell line was developed after 20 repeated passages by using G418 antibiotic selection pressure. The integration of PPRV H gene in Vero cell genome and its genomic transcription was confirmed by PCR and RT-PCR assays, respectively and the 70-kDa PPRV H protein was characterized by SDS-PAGE and Western blot. The recombinant protein reacted specifically with PPRV anti-H neutralizing monoclonal as well as polyclonal antibody in competitive, sandwich and indirect ELISA, indicating that the protein expressed was in native form.

Their study indicated that the PPRV H protein could be a sustainable source of safe antigen in non-endemic countries without the need to handle infectious virus for sero-diagnosis.

NUCLEIC ACID BASED DIAGNOSTICS

In recent years, molecular identification and characterization have favoured analysis that reflects one of the most fundamental properties of an organism, its genetic information.

Polymerase Chain Reaction

The use of nucleic acid-based diagnostics in veterinary medicine has increased exponentially in recent years. These techniques have redefined the level of information available for animal disease control programmes. In addition, modifications of nucleic acid detection techniques such as polymerase chain reaction (PCR) have led to the development of rapid, sensitive and specific assays for diagnosis as well as differential diagnosis.

The molecular technique having the widest and varied application in disease diagnosis is PCR. The strength of this technique is its ability to make millions of copies of target DNA. This amplification enables the desired target to be readily detected by other techniques such as electrophoresis and sequencing. Initial use of PCR in disease diagnosis was for specific genomic detection; such as that of Foot and Mouth disease virus. The role of PCR has expanded, and it was used for the rapid and large-scale diagnosis in disease outbreaks.

A simple, rapid and specific diagnostic PCR for sheep poxvirus identification was developed[13,14]. The primers used were from the sequenced genomes of the capripox viruses KS-1 and InS-1. Six different sheep pox isolates were tested against two orf (parapox) and three animal herpesviruses as controls. Material from uninfected cell cultures was also used as control. The sensitivity of the PCR was approximately equivalent with each of the two primers and for the six sheep pox virus isolates.

The detection of viral RNA by PCR is accomplished by the use of reverse transcription PCR (RT–PCR). This assay initially makes a complementary DNA copy of the original viral RNA before final amplification and is more sensitive than the traditional Northern blot method of RNA detection. RT-PCR has been used extensively in the detection of Blue tongue[8,14,20] and peste des petits ruminants virus (PPRV)[5,10].

Nested PCR

Nested PCR refers to the application of a second set of primers targeting a shorter area on the first-stage amplified product (DNA). This approach increases the sensitivity of the PCR and generates amplified products for confirmation purposes. This technique has been used to detect a number of pathogens of veterinary interest. This has been used in the diagnosis of PPRV[5].

A nested polymerase chain reaction based assay, for detection of Bluetongue virus (BTV) RNA in cell culture and tissue samples, was developed by Aradiab *et al.* 1991. Two pairs of oligonucleotide primers (BTV1 and BTV4 and BTV2 and BTV3), selected from non-structural protein 1 (NS1) gene of BTV-17, were used for the nested PCR in two amplification steps. First a 826-bp product was amplified using an outer primer pair BTV1 and BTV4. The second amplification, using nested or internal primer pair BTV2 and BTV3, produced a 517-bp PCR product.

PCR- ELISA

PCR-ELISAs are capture assays for nucleic acids that mimic enzyme linked immunosorbant assays. In this assay, PCR products are hybridized to an immobilized capture probe. The assay thus measures sequences internal to the PCR product and is a less expensive assay and could be an alternative to real time PCR.

A highly sensitive N gene-based PCR-ELISA for the detection of PPRV was developed[17]. The RT-PCR yielded a digoxigenin (DIG)-labeled product of 336 bp comprising a sequence from PPRV N gene, which was then detected by ELISA. The assay could detect the viral RNA in PPRV-infected tissue culture fluids with a titer as low as 0.1 $TCID^{50}$/ml. The assay is 10,000 times more sensitive than a classical RT-PCR combined with agarose gel electrophoresis.

Multiplex PCR

PCR reactions using multiple primers directed at different targets in a single assay are referred to as multiplex PCR assays. In multiplex PCR, various infectious agents can be detected and differentiated in a single reaction vessel at the same time. The different PCR targets amplified in a standard PCR assay are identified based on PCR product size.

Multiplex RT-PCR-based assay for simultaneous detection and differentiation of North American serotypes of BTV and epizootic hemorrhagic disease virus (EHDV) in cell culture and clinical samples was developed[2]. Two pairs of primers (B1 and B4) and (E1 and E4) were designed to hybridize to non-structural protein 1 (NS1) genomes of (BTV-11) and (EHDV-1), respectively. The multiplex PCR-based assay utilized a single tube-PCR amplification in which EHDV and BTV primers were used simultaneously in a multiplex format.

A fast and simple method for capripoxvirus species identification was introduced[15]. The method is based on multiplex polymerase chain reaction (MPCR) with species-specific primers and does not require nucleotide sequencing or restriction analysis of PCR products. To differentiate vaccine stains used in Russia and countries of the former Soviet Union from epizootic isolates of sheep pox virus, a method based on restriction analysis of the ankyrin-repeat protein gene fragment amplified by PCR has been developed. Being highly specific, both methods may be used for routine diagnosis of capripoxvirus-associated diseases.

A single-tube one-step multiplex RT-PCR was standardized[3,4] to amplify both 337 bp and 191 bp fragments of N and M genes of PPRV, respectively, and only a 337 bp fragment of N gene of Rinderpest virus (RPV).

PCR-RFLP

The PCR-Restriction fragment length polymorphism (PCR-RFLP) is a technique wherein DNA sequence variation is identified by amplification of the region using PCR, followed by digestion of amplified product with a restriction endonuclease. The restriction fragments vary in size and can be revealed as different sized bands on agarose/acrylamide gels.

Sheep pox and Goat pox are highly contagious viral diseases of small ruminants. These diseases were earlier thought to be caused by a single species of virus, and are difficult to distinguish serologically. P32, one of the major immunogenic genes of Capripoxvirus, was isolated and sequenced from two Indian isolates of goat poxvirus (GPV) and a vaccine strain of sheep poxvirus (SPV). The sequences were compared with other P32 sequences of capripoxviruses available in the database. Sequence analysis revealed that sheep pox and goat poxviruses share 97.5 and 94.7% homology at nucleotide and amino acid level, respectively. A major difference between them is the presence of an additional aspartic acid at 55th position of P32 of sheep poxvirus that is absent in both goat poxvirus and lumpy skin disease virus (LSDV). Further, six unique nucleotide substitutions were observed at positions 77, 275, 403, 552, 867 and 964 in the sequence of goat poxvirus, which can be taken as GPV signature residues. Similar unique nucleotide signatures could be identified in SPV and LSDV sequences also. Phylogenetic analysis showed that members of the Capripoxvirus could be delineated into three distinct clusters of GPV, SPV and LSDV based on the P32 genomic sequence. Using this information, Hosamani *et al.* 2004 designed a PCR-RFLP method for unequivocal genomic differentiation of SPV and GPV.

Real time PCR

Real-time PCR is the latest improvement in the standard PCR technique to be implemented in veterinary laboratories. This technique is a single-tube, closed assay that greatly decreases the problem of cross-contamination between samples. Detection and quantification of pathogens in positive samples is dependent on the amount of fluorescence released during amplification. These fluorescence readings are plotted by computer software and results can be transmitted electronically, eliminating the need for post-PCR reaction analysis by electrophoresis. The development of extraction methods such as the magnetic bead technique has made it possible to use real-time PCR to test large numbers of samples in a matter of hours during disease outbreaks.

A fast (about 3.5 h) and versatile diagnostic procedure able to detect and quantify BTV-RNA, has been developed using a Molecular Beacon (MB) fluorescent probe; PCR primers were designed to target 91 bp within the NS3 conserved region of the viral RNA segment 10[16].

A real-time RT-PCR assay for the detection of BTV in blood samples was reported. A combination of primers specific for a highly conserved region in RNA segment 5 and a DNA probe bound to 5'-Taq nuclease-3' minor groove binder (TaqMan MGB) was used to detect a range of isolates. This real-time RT-PCR assay could detect 5.4×10^{-3} tissue culture infectious doses ($TCID_{50}$) of virus per milliliter of sample, which was comparable to our current BTV diagnostic nested RT-PCR assay.

Real-time PCR has also been adapted for use in the field through the use of portable thermocyclers and lyophilised reagents. This approach may allow for more rapid decision-making during potential disease outbreaks.

Nucleic acid Probes

The detection of specific viral nucleic acid by hybridization using labeled viral DNA and RNA probes to detect nucleic acid has been used for rapid diagnosis of animal diseases. These molecular approaches have allowed pathogen detection directly from clinical samples, dramatically reducing the time required for identification.

BTV group-specific probe that hybridized to all BTV samples but not to epizootic hemorrhagic disease virus (EHDV) samples was developed. The double-stranded (ds) RNA genome segment 4 of BTV serotype 2 was cloned and used as a serogroup-specific complementary cDNA probe for BTV diagnosis. A cDNA probe representing a 60% copy of genome segment 4 BTV-2 prototype was produced. The specificity of the cDNA probe was determined by hybridizing this probe to a Northern blot of dsRNA (separated by polyacrylamide gel electrophoresis) of plaque-purified BTV-2 prototype. This cDNA probe was then used to hybridize to the RNA samples.

The morbilliviruses which infect ruminants (RPV and PPRV) are difficult to distinguish clinically and serologically. They can be distinguished by differential neutralisation tests and by the migration of the major virus structural protein, the nucleocapsid protein on polyacrylamide gels. Both these methods are time consuming and require the isolation of live virus for identification; and are not suitable for analysis of material directly from post-mortem specimens. A rapid method was developed for differential diagnosis of infections caused by RPV or PPRV, which uses specific cDNA probes, derived from the mRNAs for the nucleocapsid protein of each virus, which can be used to distinguish unequivocally the two virus types rapidly[9].

Fluorescent *in situ* hybridisation (FISH) is a technique that can localize nucleic acid sequences within cellular material. Peptide nucleic acids are molecules in which the sugar backbone has been replaced by a peptide backbone. These molecules are perfect mimics of DNA with high affinity for hybridisation that can be used to improve FISH techniques.

A new proprietary, conformationally restricted oligonucleotide known as 'locked nucleic acid' shows promise for hybridisation assays. New techniques utilising locked nucleic acids can better discriminate between correct and incorrect DNA/RNA target sequences and their use appears to be increasing in many applications where oligonucleotide probes are used.

BIOSENSORS

A new approach to the detection of either the antigens or antibodies is the development of biosensors. This type of assay involves the use of a receptor (usually an antibody) for the target pathogen or a disease-specific antibody and a transducer which converts a biological interaction into a measurable signal. Some of the transducer technologies under development include electrochemistry, reflectometry, interferometry, resonance and fluorimetry. Biosensors are frequently coupled to sophisticated instrumentation to produce highly-specific analytical tools, most of which are still in use only for research and development due to high cost of the instrumentation, high cost of individual sample analysis, and the need for highly trained personnel to oversee the testing.

PROTEOMICS

In addition to the use of proteomics to identify and characterise the protein produced by pathogenic agents, proteomic technologies have great potential in veterinary diagnostic applications because they target the patterns of protein expression of the target analyte whether it is viral, bacterial or parasitic. The standard proteomic approach involves the separation of proteins by two-dimensional gels with the staining of the proteins and molecular weight control. This protein 'pattern' or fingerprint is then analysed by performing image analysis.

Proteome maps can be compared in order to find proteins that may be up or down-regulated due to disease. A protein of interest can be cut from the gel and fully characterized using peptide-mass fingerprinting and/or mass spectrometry methods. In the future, veterinary diagnostics may make use of proteomics to identify or look for known disease markers or patterns with biochip technology and instrumentation that combines mass spectrometry with other separation chromatography or molecular techniques. These instrumentations are designed to specifically select, separate by molecular mass, and identify the complex mixture of proteins in a sample, which can then be compared to known samples for diagnostic purposes.

Microarray technology, originally developed for the mapping of genes, is being used to detect a wide variety of veterinary pathogens. Specific oligonucleotides are bound to small solid supports such as glass slides, silicon chips or nylon membranes. Extracted DNA or complementary DNA is labelled with a fluorescent dye and then hybridised with the microarray. Specific patterns of fluorescence are detected by a microarray reader which allows the identification of specific gene sequences found only in the animal pathogen of interest. This technology has the potential to identify the presence of agents of interest at the serotype or subspecies level, or to differentiate agents that cause similar clinical signs, for example, vesicular lesions.

NANOTECHNOLOGY

The term 'nanotechnology' is broadly defined as systems or devices related to the features of nanometre scale (one billionth of a metre). This scale of technology as it applies to diagnostics would include the detection of molecular interactions. The small dimensions of this technology have led to the use of nanoarrays and nanochips as test platforms. One advantage of this technology is the potential to analyse a sample for an array of infectious agents on a single chip. Applications include the identification of specific strains or serotypes of disease agents, such as the identification of specific influenza strains, or the differentiation of diseases caused by different viruses but with similar clinical signs, such as vesicular viral diseases. Many research groups are considering the use of chip assays that detect a number of agents in each sample. Small, portable platforms are being designed to allow pen-side testing of animals for diseases of concern.

Another facet of nanotechnology is the use of nanoparticles to label antibodies. These labelled antibodies can then be used in various assays to identify specific pathogens, molecules or structures. Examples of nanoparticle technology include the use of gold nanoparticles, nanobarcodes, quantum dots (cadmium selenide) and nanoparticle probes.

Nanotechnology is still primarily in the research stage, with the focus on the detection of bioterrorism agents. It is anticipated that many of the specific nanotechnologies will eventually be applied to the diagnosis of endemic animal diseases in the future.

Biotechnology and its applications hold great promise for improving the speed and accuracy of detection of pathogens of small ruminants. Further work will be required to realize the potential with well-characterized, validated assay systems that provide improved diagnostic capabilities to safeguard small ruminant health.

References

1. Aradaib, I.E., Schore, C.E., Cullor, J.S and Osburn, B.I. 1998. A nested PCR for detection of North American isolates of bluetongue virus based on NS1 genome sequence analysis of BTV-17. *Vet Microbiol.* 59:99-108.
2. Aradaib I.E, Smith, W.L., Osburn, B.I and Cullor JS. 2003. A multiplex PCR for simultaneous detection and differentiation of North American serotypes of bluetongue and epizootic hemorrhagic disease viruses. *Comp Immunol Microbiol Infect Dis.* 26 :77-87.
3. Balamurugan, V., Sen, A., Saravanan, P., Rasool, T.J., Yadav, M.P., Bandyopadhyay. S.K and Singh RK. 2006. Development and characterization of a stable Vero cell line constitutively expressing peste des petits ruminants (PPR) virus haemagglutinin protein and its potential use as antigen in ELISA for sero-surveillance of PPR. *Clin Vaccine Immunol.* Oct 2006.
4. Balamurugan, V., Sen, A., Saravanan, P., Singh, R.P., Singh, R.K., Rasool, T.J and Bandyopadhyay, S.K. 2006.One-step multiplex RT-PCR assay for the detection of peste des petits ruminants virus in clinical samples. *Vet Res Commun.* 30 :655-66.
5. Brindha, K., Raj, G.D., Ganesan. P.I., Thiagarajan, V., Nainar, A.M., Nachimuthu, K. 2001. Comparison of virus isolation and polymerase chain reaction for diagnosis of peste des petits ruminants. *Acta Virol.* 45 :169-72.
6. Chinsangaram, J., Hammami, S and Osburn, B.I. 1992.Detection of bluetongue virus using a cDNA probe derived from genome segment 4 of bluetongue virus serotype 2. *J Vet Diagn Invest.* 4: 8-12.
7. Clavero, M.A., Aguero, M., San Miguel Mayoral, T., Lopez, M.C., Ruano, M.J., Romero, E., Monaco, F., Polci, A., Savini, G and Gomez-Tejedor C. 2006.
8. Dangler CA, de Mattos CA, de Mattos CC, Osburn BI. 1990 Identifying bluetongue virus ribonucleic acid sequences by the polymerase chain reaction. J *Virol Methods.* 28:281-92.

9. Diallo, A., Barrett, T., Barbron, M., Subbarao, S.M and Taylor, W.P. 1989 Differentiation of rinderpest and peste des petits ruminants viruses using specific cDNA clones. *J Virol Methods.* 23 :127-36.

10. Forsyth, M.A and Barrett, T. 1995. Evaluation of polymerase chain reaction for the detection and characterisation of rinderpest and peste des petits ruminants viruses for epidemiological studies. *Virus Res.* 39:151-63.

11. High throughput detection of bluetongue virus by a new real-time fluorogenic reverse transcription-polymerase chain reaction: application on clinical samples from current Mediterranean outbreaks. *J Vet Diagn Invest.* 18:7-17.

12. Hosamani, M., Mondal, B., Tembhurne, P.A., Bandyopadhyay, S.K., Singh, R.K and Rasool, T.J. 2004 . Differentiation of sheep pox and goat poxviruses by sequence analysis and PCR-RFLP of P32 gene. *Virus Genes.* 29:73-80.

13. Mangana, V.O., Markoulatos, P., Koptopoulos, G., Nomikou, K., Bakandritsos, N and Papadopoulos, O. 1999. Sheep poxvirus identification by PCR in cell cultures. *J Virol Methods* 77 :75-9.

14. McColl, K.A and Gould, A.R. 1991. Detection and characterisation of bluetongue virus using the polymerase chain reaction. *Virus Res.* 21:19-34.

15. Orlova, E.S., Shcherbakova, A.V., Diev, V.I and Zakharov, V.M. 2006 Differentiation of capripoxvirus species and strains by polymerase chain reaction. *Mol. Biol* 40:158-64.

16. Orru, G., Ferrando, M.L., Meloni, M., Liciardi, M., Savini, G and De Santis P. 2006.Rapid detection and quantitation of Bluetongue virus (BTV) using a Molecular Beacon fluorescent probe assay. *J Virol Methods.* 137 :34-42.

17. Saravanan, P., Singh, R.P., Balamurugan, V., Dhar, P., Sreenivasa, B.P., Muthuchelvan, D., Sen, A., Aleyas, A.G., Singh, R.K and Bandyopadhyay, S.K. 2004. Development of a N gene-based PCR-ELISA for detection of Peste-des-petits-ruminants virus in clinical samples. *Acta Virol.* 48:249-55.

18. Schmitt, B and Henderson, L. 2005. Diagnostic tools for animal diseases. *Rev.sci.tech.int.Epiz.* 24: 243-250.

19. Singh, R.P., Sreenivasa, B.P., Dhar, P and Bandyopadhyay SK. 2004 A sandwich-ELISA for the diagnosis of Peste des petits ruminants (PPR) infection in small ruminants using anti-nucleocapsid protein monoclonal antibody. *Arch Virol.* 149 : 2155-70.

20. Tiwari, A.K., Kataria, R.S., Desai, G., Butchaiah, G and Bandopadhyay, S.K. 2000. Characterization of an Indian bluetongue virus isolate by RT-PCR and restriction enzyme analysis of the VP-7 gene sequence. *Vet Res Commun.* 24: 401-9.

–ΔΔΔ–

22

Recent Advances in Bacterial and Viral Vaccines

G K SHARMA

National Dairy Development Board
Anand - 388 001

Recent outbreaks of Highly Pathogenic Avian Influenza (HPAI), Severe Acute Respiratory Syndrome (SARS), Bovine Spongiform Encephalopathy (BSE) and Foot and Mouth Disease (FMD) in UK had raised enormous public concern about health issues and also had tremendous impact on economy, animal welfare and environment. Similarly the threat of bioterrorism has thrown a challenge to protect vast population swiftly against a variety of bio-terror agents. Debates have initiated on the strategies to control animal diseases and whether stamping out - the preferred methodology of animal disease control in the developed countries, is really the most ideal one. Vaccination is now emerging as a practical option that many countries would like to keep open for disease control. The proceedings of the OIE International Conference on Control of Infectious Animal Diseases by Vaccination, held in Argentina in 2004 states among other things under Recommendations that 'For ethical, ecological and economical reasons, it is no longer acceptable to control and eradicate disease outbreaks mainly by applying mass slaughter of animals'[55]. It also recommends that 'Greater emphasis should be placed on the use of vaccination for the control of food-borne and other zoonotic disease in animals to safeguard public health. This may include wildlife reservoirs of pathogens.'[55] The global acceptability of the use of vaccination for control of animal diseases is also evident from the fact that OIE has recognized 'FMD Free Countries with vaccination' and 'FMD Free Zones with vaccination' and developed guidelines for the same[50]. Therefore vaccines are likely to play significant role in future animal and human health policies.

More than 2 centuries ago close observations of Edward Jenner - an English Physician - who noticed in 1796 that the milkmaids who had suffered from cowpox were rarely the victims of smallpox laid the foundation of modern vaccines. By the beginning of 20th century vaccines for rabies, diphtheria, typhoid and plague were already in use. Smallpox was eradicated from the globe in 1980 for which vaccination played the crucial role.

There have been numerous developments in the field of vaccinology and it would be almost impossible to mention all of them here. However, the attempt here is to briefly mention the type of available vaccines; the advances that have taken place in FMD vaccine research to serve as an example of developments taking place in vaccine research; and in summary the status of a few important vaccines of veterinary importance.

TYPES OF AVAILABLE VACCINE

Depending on the nature of the agent included in the vaccine, following types of vaccines are currently available in the market.

Weakened or Attenuated Microbes

These contain live microbes that are either naturally incapable or are weakened in the laboratory so that they do not cause disease. Although cowpox virus produces severe disease in cattle, it does not cause disease in humans but protects against smallpox virus and thus can be used for vaccination against smallpox. On the other hand Kabate 'O' strain of Rinderpest (RP) was developed by Plowright by successive passages in bovine kidney cells[51]. Many viruses and bacteria have been adapted to new hosts and the attenuated virus has in many cases been found to replicate and afford immunity. Likewise, African Horse Sickness virus was adapted in mice and Canine Distemper virus (CDV) in ferrets. The technique of successive passaging in embryonated chicken egg was used for adaptation of CDV, Rabies and Bluetongue virus. Yellow Fever virus was adapted in mouse brain for human vaccine.

The vaccines for Tuberculosis, Measles, Mumps and Rubella are other examples of live vaccines for human. Similarly, the vaccines for

Anthrax, Brucellosis and Peste des petits ruminants are the example of live vaccines for veterinary use.

Killed Microbes

Many vaccines have been developed that incorporate killed organisms. Since these contain only inactivated microbes these can not cause disease but provide protection. Formaldehyde, Aziridin compounds (Binary ethylenimine-BEI) or â propiolactone (BPL) are more frequently used inactivants. The immunity provided by these vaccines is some what lower than live vaccines, and hence booster or repeated vaccinations are generally required. These vaccines usually incorporate adjuvant to enhance the immune response. Aluminum compounds (aluminum hydroxide gel) and mineral / non-mineral oils are the more frequently used adjuvants.

For human use the vaccines for Polio, Rabies, Hepatitis A and Japanese Encephalitis contain whole killed organisms. Likewise, the vaccines for Foot and Mouth Disease (FMD), Haemorrhagic Septicemia (HS), Black Quarter (BQ) etc. are vaccines containing inactivated organisms.

Inactivated Toxins

A few bacterial diseases are caused by the toxins produced by toxin producing bacteria like Tetanus and Diphtheria. Inactivated toxins have been used successfully to prevent these diseases.

Conjugate Vaccine

A few microbes are less immunogenic and thus do not render themselves suitable as vaccine candidate. A conjugate vaccine is developed by covalently attaching a poor antigen to a carrier protein, thereby conferring the immunological attributes of the carrier on the attached antigen. This technique for the creation of an effective immunogen is most often applied to bacterial polysaccharides for the prevention of invasive bacterial disease. *Haemophilus influenzae* type b, Meningococcus and Pneumococcal vaccines are conjugate vaccines. Diphtheria, Tetanus and Meningococcal proteins are usually used as carrier proteins.

Acellular Vaccine

Acellular vaccine is a preparation that contains bacterial fragments or components, but not the whole bacterial cell. The fragments are immunogenic in nature and are responsible for providing the immunity. Acellular Pertussis vaccine is available in the market for preventing whooping cough in babies.

Recombinant Vaccine

Recombinant vaccine is a preparation in which genes coding for the desired antigens are inserted in a vector that usually is bacteria, yeast or virus. The recombinant vector is grown in bulk and the antigenic protein is purified and used as vaccine, or the recombinant vector itself is used as vaccine. Examples of recombinant vaccine include Hepatitis B vaccine (Saccharomyces-HBsAg) in humans and oral rabies baits (Vaccinia-Rabies Glycoprotein) for wild-life.

A Fowl pox recombinant with cDNA insertion of H5 haemagglutinin gene from Avian Influenza (A/turkey/Ireland/83) was developed[63].

DNA Vaccine

DNA vaccine is usually a circular plasmid DNA that has the gene of interest inserted in it. The plasmid is usually injected in muscles where it integrates with the muscular DNA and produces the protein (antigen) of interest. The only DNA vaccine approved so far is for West Nile Virus infection in equines.

MARKER VACCINES AND DIVA CONCEPT

For those animal diseases where no vaccines exist, probably stamping out may be the right choice for disease eradication, but for other diseases more efforts should be made to control diseases by vaccination, considering the economic, animal welfare and ecological issues involved. However, one major obstacle in using vaccines for disease control is the failure of conventional disease diagnostic tests (serological tests) to differentiate the vaccinated animal from the infected animal. Hence, efforts are being directed to develop vaccines and the companion diagnostic tests that would help differentiate the vaccinated and infected animal. This is being termed as Differentiating Infected from Vaccinated Animal (DIVA) approach.

First marker vaccine was reported by Bartha in Hungary in 1961 for Pseudorabies that was spontaneously deleted gE glycoprotein mutant. Likewise, in Netherlands the gE deleted vaccine was developed for Infectious Bovine Rhinotracheitis (IBR) virus and also had a gE serology based companion ELISA test.

Subunit vaccines for Classical Swine Fever (CSF) were developed. These contained immunodominant F2 protein either expressed in baculovirus or vaccinia system. They also had a companion diagnostic test on non-structural protein (NS2). However, the protection afforded by these vaccines was not comparable to conventional live attenuated vaccines.

FMD vaccines with non-structural-protein (NSP) purification and a companion NSP ELISA, *Brucella abortus* RB 51 vaccine with conventional serological diagnostic tests are the example of vaccines suitable for DIVA approach. Likewise, heterologous Avian Influenza vaccines that have same haemagglutinin but differing neuraminidase subtype are fit for DIVA approach.

DEVELOPMENTS IN FMD VACCINE

(An example of various approaches taken for research on vaccine development)

The earliest vaccine used for FMD prevention was before the Second World War. 'Vallée-Schmidt-Waldman' type vaccine was obtained from tongue epithelium of deliberately infected cattle, adjuvanting with aluminum hydroxide and inactivating with formaldehyde[61]. Then came the 'Frenkel method' where tongue epithelium from healthy cattle was collected and grown in the laboratory in slices, on to which the virus was grown[30]. This allowed large scale production of vaccine. With the development of modern virological methods like cultivation of virus in cell culture, FMD virus was demonstrated to grow in BHK cells grown as monolayers by Mowat and Chapman[45], roller bottle[60] and subsequently as suspension culture[16]. Most modern vaccine manufacturers use large fermentors to grow BHK cells that are grown in suspension.

Earlier, formaldehyde was used to inactivate the virus but it was not realized that certain components in the culture medium can interfere with the inactivation process[8]. Some outbreaks were thought to have been caused due to improper inactivation of vaccine virus[36]. However, it was soon observed that aziridin compounds (Binary ethylenimine-BEI) produced first order of inactivation[4] and almost all FMD vaccine producers switched to BEI for inactivation. Further, a synergistic effect was noticed when formaldehyde and BEI were used simultaneously, providing a more than 100 fold increase in inactivation rate[9].

One constraint faced by aqueous aluminum hydroxide gel adsorbed vaccine was that it did not offer satisfactory protection in pigs. Subsequently oil emulsion vaccines were developed that protected all domestic species - cattle, buffalo, sheep, goat and pig[31,17]. Almost all the vaccine producers switched to oil adjuvanted vaccine as it generally provided a longer protection compared to aqueous vaccines.[2]

Marker Vaccine for FMD

The concentrated cell culture supernatant obtained after infecting the cells with FMD virus, is the source of antigen for vaccine preparation that invariably contains varying amounts of undesirable non-structural proteins (NSP) also. These proteins are produced during virus multiplication. Animals inoculated with this preparation develop antibodies to both structural proteins as well as NSPs. Antibodies to NSPs are also formed in animals that have been naturally infected with FMD. Thus, it becomes difficult to differentiate between vaccinated, infected or convalescent animal, reliably. Vaccine manufacturers are developing methodologies to purify the virus by removing the NSPs as much as possible with a view to facilitate differentiation of vaccinated animal from infected or convalescent animal. These vaccines are being called marker vaccine. At the same time NSP based diagnostic tests are being developed[12, 25]. These tests are available as kit in market as PANAFTOSA-I ELISA 3 ABC, Brescia 3 ABC TRAPPING ELISA, Ceditest FMDV-NS, CHEKIT FMD-3 ABC ELISA, UBI FMD NSP ELISA and SVANOVIR FMDV 3ABC-Ab ELISA.

Typical results of NSP ELISA test that help differentiate infected or convalescent animal from the one given purified FMD vaccine

ELISA	**Result by conventional serology (Virus neutralization, VP1 ELISA)**	**Result of NSP ELISA**
Vaccinated, non-infected	Positive	Negative
Vaccinated, infected	Positive	Positive
Unvaccinated, non-infected	Positive	Negative
Unvaccinated, infected	Positive	Positive

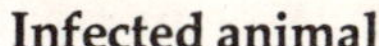
Infected animal

Purified vaccine

Conventional vaccine

SP = Antibodies against structural protein
NSP = Antibodies against non structural proteins

FMD vaccine Research

Considering that the VP1 is the most immunodominant capsid protein that has prominent surface exposure as well as the importance of VP1 G-H loop, scientists have focused their attention to make use of this knowledge for development of vaccines. Bacharach used isolated capsid proteins and measured the immune response[3]. Kleid cloned the viral protein and evaluated the same in cattle and pigs[38]. Strohmaier 1982 used VP1-derived peptides[57] and Bittle chemically synthesized the peptide that gave protection[15]. DiMarchi demonstrated protection of cattle with synthesized peptide[28]. Francis evaluated multiple antigen peptide[29]. Nargi showed protection of swine with one dose of all-D retro peptide[49].

Live vectors expressing VP1 fusion proteins (BHV1 expressing FMD epitopes) were developed by Kit[37]. Wong demonstrated that plasmids encoding FMD VP1 epitopes elicited immune response in mice and swine and were able to protect swine[67]. Wong in 2002 co-administered plasmid expressing VP1 epitope with IL-2 and found enhanced protection in swine[68]. Wigdorovitz demonstrated that VP1 can be expressed in alfalfa transgenic plants and oral or parenteral immunization induces a protective antibody response in mice[65]. He further demonstrated that foliar extract of plants infected with recombinant Tobacco Mosaic Virus (TMV) expressing the structural protein VP1 protected mice upon challenge[66].

However this approach represented only exposure of a limited subset of viral immunogen to the vaccinated animal and while these did generate good antibody response, the protection upon challenge was inferior to conventional vaccines[27,46,47]. The immunogenicity of these vaccines is due to the fact that they present a sequential epitopes from the immunologically important VP1 G-H surface loop. While they are immunodominant epitopes, they are not the only neutralizing epitopes[24]. With the use of limited subset of immunizing epitopes it is likely that the protection afforded will be partial and is likely to generate virus escape mutants[58].

Research for Live Vaccines

McKenna developed a FMD virus in which the RGD receptor binding site on VP1 was deleted[43]. This virus was unable to bind to cells and did not cause disease when given to mice or swine. Cattle given this virus as oil-emulsion vaccine did not develop any clinical disease but developed FMD specific neutralizing antibody response similar to those in animals inoculated with whole virus vaccine. These animals were totally protected. However there is a potential for development of virus variants that could enter nonintegrin receptors[5,6].

Chinsangaram developed a leaderless FMD virus for serotype A12 that was deficient in leader proteinase (L^{pro}) and used this virus as live as well as killed vaccine in cattle and swine with partial success[19].

Research for Empty Viral Capsids

Scientists have also taken the approach to produce intact virus particle lacking the infectious nucleic acid, as this would have all immunogenic epitopes in native conformation but would not be able to replicate and cause disease [13, 33, 39, 53]. This requires cloning of genes responsible for synthesis, processing and assembly of viral structural proteins into empty viral capsids. This approach has the advantage that animals vaccinated with this preparation could be easily distinguished from infected or convalescent animals as antibodies to NSPs would not be produced in the vaccinated animals. The empty viral capsids were expressed in *Escherichia coli* and in baculovirus-infected cell culture and inoculated to animals. However, the protection offered was less compared to current inactivated whole-virus vaccines[13, 33, 39].

Beard[10] and Chinsangaram[19] took the approach to directly express the empty viral capsids in the inoculated animal through a DNA vaccine. Ward[62] produced the cDNA that encoded the entire viral genome but contained a mutation in the cell binding site. While these preparations did produce a good antibody response in mice[10] the results were not very encouraging in swine[18]. Cedillo-Barron[18] further improved the preparation by co-administration of a plasmid encoding porcine granulocyte-macrophage colony stimulating factor along with FMDV empty capsid construct. While the DNA vaccine encoding empty viral capsids offer advantages, it did not offer the required protection.

Research for Adenovirus Vector Based Vaccines

Vaccinia virus vectors have been used in FMD vaccine research[1,13]. The mice inoculated with recombinant vaccinia virus encoding capsid proteins of Type C FMD virus had increased neutralizing antibody titre and also were protected after challenge[13]. Human Adenoviruses are also suitable for antigen delivery and are considered safe for immunization[59]. Replication defective Human Ad5 vectors are considered safer as these can multiply only in specific cell cultures that provide the missing functions. Mayer[42] and Moraes[44] found that swine inoculated with Ad5-A12 or Ad5-A24 that had wild type C^{pro} gene not only induced neutralizing antibody response but also provided protection either on contact or direct inoculation challenge. Similarly cattle inoculated with Ad5-A24 elicited a FMD virus specific neutralization response and were completely protected after direct challenge[32].

Emergency Vaccination Against FMD

Vaccine banks are maintained by many countries and organizations in an attempt to keep the option of using vaccination in an emergency situation. The first such bank was maintained by Denmark in 1976. Some 16 FMD vaccine banks are in operation of which four are international and the remaining are national. The inactivated, purified and concentrated FMD virus is stored in ultra-low temperatures in the quantities sufficient to produce millions of doses quickly. However, unlike routine prophylactic vaccination where the immune response takes weeks to develop, the urgency of emergency situation will demand a response that develops in days. One approach that has been found suitable is to increase the antigenic pay load and produce a vaccine of higher potency. Conventional vaccines have a minimum potency of 3 PD_{50} while the vaccines for emergency are targeted to have a minimum of 6 PD_{50}. New adjuvants like Montanide ISA 206 and ISA 25 have been found more suitable[22].

The other approach is to co-administer factors like interferon that are known to inhibit FMD virus replication[20]. Chinsangaram constructed a replication defective Ad5 containing porcine IFN-á (Ad5-pIFN-á)[21]. Swine inoculated with this preparation were completely protected upon challenge one day after immunization. These animals did not develop clinical disease, had no viremia and did not develop FMD specific immune response. This treatment can protect animal for three to five days. Moraes[44] demonstrated that inoculation of Ad5-pIFN-á and vaccination provided immediate and long-term protection in swine. However, the same approach did not work well with cattle and it is suspected that low production of interferon might be responsible for it[69].

Anti-sense technology has also been utilized by researchers who have demonstrated the inhibition of FMDV replication in cell culture[14, 34, 54]. However this approach has not yet been tried in animals.

STATUS OF OTHER IMPORTANT VETERINARY VACCINES

Viral diseases

Rinderpest

Initially the Kabate O strain was passaged in goats in India by Edwards which was found attenuated for native cattle. This work was done in Mukteswar and the attenuated virus was known as Mukteswar goat virus. However, this was not found suitable for cattle in Africa and European breeds.

Nakamura[48] adapted the virulent cattle RP strain to rabbits by IV injections. The live attenuated tissue culture Rinderpest vaccine (TCRV) was developed by Plowright by the serial passage of the virulent bovine RP strain Kabate 'O' (RBOK) in primary bovine calf kidney cells[51]. The OIE pathway for freedom from RP requires that vaccines shall not be used and serosurveillance should be used to identify if any virus is circulating in the population. Vaccination with TCRV induces a lifelong immunity that cannot be easily distinguished from the antibody response induced by the wild type virus infection. Hence, there is an urgent need to replace TCRV with a marker Rinderpest vaccine. New generation vaccines were developed by Yilma[72] and Yamanuchi [70] that were low cost, vaccinia based and were considered highly effective. However, none of these are commercialized.

One of the limitation of the Plowright vaccine was thermolability that required cold chain during transportation and storage till the vaccine is given to animal. Mariner[41] developed a production protocol that largely overcame this limitation.

Peste Des Petits Ruminants (PPR)

Homologous vaccine against PPR is a live vaccine cultured in Vero cells. The original strain of virus for the homologous vaccine against PPR is strain PPR 75/1, isolated in Nigeria in 1975. This is attenuated by passage in Vero cell cultures[26]. The strain provided for vaccine production is the 70th passage in Vero cells (PPRV 75/1 I.K6 BK2 Vero 70).

Rabies

Rabies vaccines for use in animals can be either live or inactivated or live recombinant. Live vaccine may incorporate live attenuated virus such as Flury low egg passage (LEP), Flury high egg passage (HEP) or Street-Alabama-Dufferin (Street).

The inactivated vaccines are inactivated by chemical or physical means. The virus is cultivated in the in embryonated eggs, or in cell cultures.

Rabies vaccines are usually lyophilized, but inactivated virus vaccines, preferably with an adjuvant, may be stored in liquid form.

For animals, live vaccines are also effective by the oral route and can be distributed in baits in order to immunize wild (or domestic) animals. Live recombinant vaccine (e.g. vaccinia rabies-glycoprotein recombinant) has also proved to be effective[35].

Avian Influenza (AI)

Influenza A viruses are classified into subtypes on the basis of their haemagglutinin (H) and neuraminidase (N) antigens. At present, 16 H subtypes (H1-H16) and 9 N subtypes (N1-N9) are recognized.

Conventional AI Vaccines

Conventionally, vaccines that have been used against Notifiable Avian Influenza (NAI) or Low Pathogenic Avian Influenza (LPAI) have been prepared from infective allantoic fluid inactivated by â-propiolactone or formalin and emulsified with mineral oil. Such vaccines have been used in Hong Kong, China and Indonesia.

Recombinant AI Vaccine

The recombinant fowlpox-AI-H5 vaccine is licensed in El Salvador, Guatemala, Mexico and the USA. Recombinant fowl poxvirus vaccine containing H5 HA have been prepared and used in the vaccination campaign against the H5N2 virus in Mexico, El Salvador and Guatemala.

Bluetongue

Live, attenuated vaccines that are serotype-specific are used in several countries, such as South Africa. Attenuated viruses are prepared by serial passage of field virus in embryonated chicken eggs or cultured cells. Attenuated BT virus vaccines are considered teratogenic and should not be administered to pregnant sheep during the first half of pregnancy as this may cause fetal death and abnormalities.

Japanese Encephalitis

The vaccine for Japanese encephalitis in horses is prepared by the inactivation of a virus suspension of Beijing-1 strain derived from infected mouse brains or primary culture of porcine kidney cell cultures. Formalin is used as inactivant and the preparation is usually adjuvanted with oil.

Sheep Pox & Goat Pox

All strains of capripox virus of ovine, caprine or bovine origin examined so far share a major neutralizing site, so that animals recovered from infection with one strain are resistant to infection with any other strain. Consequently, it is possible to use a single strain of Capri pox virus to protect both sheep and goats against all field strains of virus.

Live Capripox Vaccines

A number of strains of capripox virus had been used as live vaccines like 0240 Kenya sheep and goat pox strain used in sheep and goats, the Romanian and RM-65 strains used mainly in sheep, and the Mysore and Gorgan strains used in goats. Immunity in sheep and goats against capripox following vaccination with the 0240 strain lasts over a year, and will probably provide lifelong protection against lethal challenge.

Inactivated Capripox Vaccine

Inactivated capripox vaccines are made from tissue culture harvest that has undergone freeze-thaw cycles for release of virion from the cells. Inactivated vaccines produced from tissue culture are almost entirely naked virions, and when used as vaccines do not stimulate immunity to the membrane-bound virion. Inactivated vaccines are not as good as live vaccine in providing the immunity. Inactivated capripox vaccines provide, at best, only temporary protection.

A new generation of capripox vaccines is being developed that uses the capripox virus genome as a vector for the genes of other ruminant pathogens, for instance genes of RP and PPR viruses. The recombinant vaccine is likely to provide protection against capripox, RP and PPR in a single vaccination[52].

Bacterial Diseases

Anthrax

The most widely used vaccine for prevention of anthrax in animals was developed by Sterne in 1937. He derived a rough variant of virulent *B. anthracis* from culture on serum agar in an elevated CO2 atmosphere. This variant, named 34F2, was incapable of forming a capsule and was subsequently found to have lost the pX02 plasmid, which codes for capsule formation. It has become the most widely used strain world-wide for animal anthrax vaccine production.

Bovine Brucellosis *(Brucella abortus)*

Only live attenuated vaccines are available.

The most widely used vaccine for the prevention of brucellosis in cattle is the *Brucella abortus* S19 vaccine, which is normally given to female calves aged between 3 and 6 months as a single subcutaneous dose of 5-8 x 10^{10} viable organisms.

Since 1996, *B. abortus* strain RB51 has become the official vaccine for prevention of brucellosis in cattle in several countries. It affords good immunity in vaccinated animals. However, since the strain lacks the O chain in the lipopolyschharide (LPS) it does not have smooth colonial morphology. The immunized animals do not develop antibodies against O LPS and this is used as marker for differentiation between vaccinated and infected animal, as conventional diagnostic tests (serological) are based on detection of antibodies to sLPS.

Contagious Bovine Pleuropneumonia (CBPP)

Live attenuated vaccines are used for control of CBPP. After the 44th egg-passage of strain T1, the strain was found to have sufficiently attenuated and was able to protect cattle. The strain was named T1/ 44 and is one of the two strain widely used for CBPP vaccine production. The other vaccine strain was T1sr[71].

Haemorrhagic Septicemia (HS)

The three types of vaccines used against HS are bacterins, alum-precipitated vaccine (APV) and oil-adjuvanted vaccine (OAV). To provide sufficient immunity with bacterins, repeated vaccination is required. Administration of dense bacterins can give rise to shock reactions, which are less frequent with the APV and almost nonexistent with the OAV. OAV also enhance the duration of immunity.

Paratuberculosis

Use of vaccine is not popular, as vaccinated animals become reactor in delayed type hypersensitivity tests based on tuberculin testing. However, vaccines can be made against Paratuberculosis that are live, attenuated and are either incorporated with oil and pumice or lyophilized, that would need to be adjuvanted with oil after reconstitution. The strains used are *M. paratuberculosis* 316F or 2E (Weybridge) or *M. paratuberculosis* 3 and 5 or II (Canadian strains). Vaccination may cause a reaction at the site of injection and can interfere with the interpretation of DTH skin tests for bovine tuberculosis.

Caprine and Ovine Brucellosis *(Brucella Melitensis)*

The most widely used vaccine for the prevention of Brucellosis in sheep and goats is the *Brucella melitensis* Rev.1 vaccine. The Rev.1 vaccine is used as a freeze-dried suspension of live *B. melitensis* biovar 1 Rev.1 strain for the immunization of sheep and goats. It is normally given to lambs and kids aged between 3 and 6 months as a single subcutaneous injection. The standard dose is between 0.5 x 10^9 and

2.0×10^9 viable organisms. The vaccine provides lifelong immunity. However, it also develops antibodies against sLPS which gets detected in conventional diagnostic tests and could mislead the diagnostic interpretation. However, when this vaccine is administered conjunctivally to lambs and kids aged between 3 and 6 months it produces a similar protection without inducing a persistent antibody response, thus facilitating the application of eradication programmes combined with vaccination[40].

A single standard dose (10^9 colony-forming units) of the live *B. melitensis* Rev.1 vaccine, administered subcutaneously or conjunctivally, can be used safely and effectively in rams, for the prevention of *B. ovis* infection.

References

1 Abrams, C. C., A. M. King, and G. J. Belsham. 1995. Assembly of foot and-mouth disease virus empty capsids synthesized by a vaccinia virus expression system. J. Gen. Virol. 76:3089–3098.

2 Auge de Mello P. and I. Gomes. 1977. Anamnestic response in cattle after revaccination with oil adjuvanted foot-and-mouth disease vaccines. Bol. Centr. Panam. Fiebre Aftosa. 27-28:55-60.

3 Bachrach, H. L., D. M. Moore, P. D. McKercher, and J. Polatnick. 1975. Immune and antibody responses to an isolated capsid protein of foot-and mouth disease virus. J. Immunol. 115:1636–1641.

4 Bahnemann. H.G. 1975. Binary ethylenimine as an inactivant for foot-and-mouth disease virus and its application for vaccine production. Arch. Virol. 47:47-56.

5 Baranowski, E., C. M. Ruiz-Jarabo, F. Lim, and E. Domingo. 2001. Foot-and-mouth disease virus lacking the VP1 G-H loop: the mutant spectrum uncovers interactions among antigenic sites for fitness gain. Virology 288:192–202.

6 Baranowski, E., C. M. Ruiz-Jarabo, N. Sevilla, D. Andreu, E. Beck, and E. Domingo. 2000. Cell recognition by foot-and-mouth disease virus that lacks the RGD integrin-binding motif: flexibility in aphthovirus receptor usage. J. Virol. 74:1641–1647.

7 Barrett T., S. Parida, P. Walsh, J. Anderson, and M. Baron. 2003. New Rinderpest Vaccines: Marker Genes for the Plowright Vaccine. Proceedings of the Global Rinderpest Eradication Programme (GREP) Technical Consultation 2002, Food and Agriculture Organisation of the United Nations (FAO), Rome, Italy.

8 Barteling S.J., and R. Woortmeijer. 1984. Formaldehyde inactivation of foot-and-mouth disease virus. Conditions for the preparation of safe vaccine. Arch. Virol. 80:103.

9 Barteling, S.J. and N. Ismael Cassim. 2000. Formaldehyde increases the inactivation rate of foot-and-mouth disease virus at lease a hundred-fold. Pep. Res. Gr. Comm. Contr. FMD. Borovits, Bulgaria. FAO Rome, p.270

10 Beard, C., G. Ward, E. Rieder, J. Chinsangaram, M. J. Grubman, and P. W. Mason. 1999. Development of DNA vaccines for foot-and-mouth disease, evaluation of vaccines encoding replicating and non-replicating nucleic acids in swine. J. Biotechnol. 73:243–249.

11 Belsham, G. J., C. C. Abrams, A. M. King, J. Roosien, and J. M. Vlak. 1991. Myristoylation of foot-and-mouth disease virus capsid protein precursors is independent of other viral proteins and occurs in both mammalian and insect cells. J. Gen. Virol. 72:747–751.

12 Bergmann I.E., P. Auge de Mello, E. Neitzert, E. Beck and I. Gomes. 1993. Diagnosis of persistent aphthovirus infection and its differentiation from vaccination response in cattle by use of enzyme-linked immunoelectrotransfer blot analysis with bio-engineered nonstructural viral antigens. Am. J. vet. Res. 54:825-831

13 Berinstein, A.,, C. Tami, O. Taboga, E. Smitsaart, and E. Carrillo. 2000. Protective immunity against foot-and-mouth disease virus induced by a recombinant vaccinia virus. Vaccine 18:2231–2238.

14 Bigeriego, P., M. F. Rosas, E. Zamora, E. Martinez-Salas, and F. Sobrino. 1999. Heterotypic inhibition of foot-and-mouth disease virus infection by combinations of RNA transcripts corresponding to the 5• and 3• regions. Antiviral Res. 44:133–141.

15 Bittle, J. L., R. A. Houghten, H. Alexander, T. M. Shinnick, J. G. Sutcliffe, R. A. Lerner, D. J. Rowlands, and F. Brown. 1982. Protection against foot-and-mouth disease by immunization with a chemically synthesized peptide predicted from the viral nucleotide sequence. Nature 298:30–33.

16 Capstick P.B., R.C. Telling, W.G. Chapman and D.L. Stewart. 1965. Production of foot-and-mouth disease virus antigen from BHK 21 clone 13 cells grown and infected in deep suspension cultures. Nature (Lond.) 205:1135.

17 Casas Olascoaga R. 1978. summary of current research of Pan American foot-and-mouth disease centre on oil adjuvanted vaccines. Bull. Off. Int. Epiz. 89:1015-1054.

18 Cedillo-Barron, L., M. Foster-Cuevas, G. J. Belsham, F. Lefevre, and R. M. Parkhouse. 2001. Induction of a protective response in swine vaccinated with DNA encoding foot-and-mouth disease virus empty capsid proteins and the 3D RNA polymerase. J. Gen. Virol. 82:1713–1724.

19 Chinsangaram, J., C. Beard, P. W. Mason, M. K. Zellner, G. Ward, and M. J. Grubman. 1998. Antibody response in mice inoculated with DNA expressing foot-and-mouth disease virus capsid proteins. J. Virol. 72:4454–4457.

20 Chinsangaram, J., M. Koster, and M. J. Grubman. 2001. Inhibition of L-deleted foot-and-mouth disease virus replication by alpha/beta interferon involves double-stranded RNA-dependent protein kinase. J. Virol. 75:5498–5503.

21 Chinsangaram, J., M. P. Moraes, M. Koster, and M. J. Grubman. 2003. Novel viral disease control strategy: adenovirus expressing alpha interferon rapidly protects swine from foot-and-mouth disease. J. Virol. 77:1621–1625.

22 Cox SJ; P.V. Barnett, P. Dani and J.S. Salt. 1999. Emergency vaccination of sheep against foot and mouth disease: protection against disease and reduction in contact transmission. Vaccine, 17, 1855-1868.

24 Crowther, J. R., S. Farias, W. C. Carpenter, and A. R. Samuel. 1993. Identification of a fifth neutralizable site on type O foot-and-mouth disease virus following characterization of single and quintuple monoclonal antibody escape mutants. J. Gen. Virol. 74:1547–1553.

25 DeDiego M, E. Broochi, D. Mackay and F. DeSimone. 1997. The non-structural polyprotein 3ABC of foot-and-mouth disease virus as a diagnostic antigen in ELISA to differentiate infected from vaccinated cattle. Arch. Virol. 142:2021-2033.

26 Diallo A., W.P. Taylor, P.C. Lefevre, and A. Provost. 1989. Atténuation d'une souche de virus de la peste des petits ruminants: candidat pour un vaccin homologue vivant. Rev. Elev. Med. Vet. Pays Trop., 42: 311-319.

27 DiMarchi, R., G. Brooke, C. Gale, V. Cracknell, T. Doel, and N. Mowat. 1986. Protection of cattle against foot-and-mouth disease by a synthetic peptide. Science 232:639–641.

28 DiMarchi, R., G. Brooke, C. Gale, V. Cracknell, T. Doel, and N. Mowat. 1986. Protection of cattle against foot-and-mouth disease by a synthetic peptide. Science 232:639–641.

29 Francis, M. J., G. Z. Hastings, F. Brown, J. McDermed, Y. A. Lu, and J. P. Tam. 1991. Immunological evaluation of the multiple antigen peptide (MAP) system using the major immunogenic site of foot-and-mouth disease virus. Immunology 73:249–254.

30 Frenkel, H.S. 1951. Research on foot-and-mouth disease. II. The cultivation of the virus on a practical scale in explantations of bovine tongue epithelium. Am. J. Vet. Res. 12:187.

31 Graves, J.H., P.D. McKercher. H.E. Farris Jr. and K.M. Cowan. 1968. Early response of cattle and swine to inactivated foot-and-mouth disease vaccine. Res. Vet. Sci. 9. 35.

32 Grubman, M. J., and P. W. Mason. 2002. Prospects, including time-frames, for improved foot and mouth disease vaccines. Rev. Sci. Tech. Off. Int. Epizoot. 21:589–600.

33 Grubman, M. J., S. A. Lewis, and D. O. Morgan. 1993. Protection of swine against foot-and-mouth disease with viral capsid proteins expressed in heterologous systems. Vaccine 11:825–829.

34 Gutierrez, A., E. Martinez-Salas, B. Pintado, and F. Sobrino. 1994. Specific inhibition of aphthovirus infection by RNAs transcribed from both the 5' and the 3' noncoding regions. J. Virol. 68:7426–7432.

35 Kieny, M.P., R. Lathe, R. Drillien, D. Spehner, S. Skory, D. Schmitt, T. Wiktor, H. Koprowski, and J.P. Lecocq. 1984. Expression of rabies virus glycoprotein from a recombinant vaccinia virus. Nature, 312:163-166.

36 King, A.M.Q., B.O. Underwood, D. MaCahon, J.W.L. Newman, and Brown F. 1981 Biochemical identification of viruses causing the 1981 outbreaks of foot-and-mouth disease in UK. Nature. 293:479

37 Kit, M., S. Kit, S. P. Little, R. D. Di Marchi, and C. Gale. 1991. Bovine herpesvirus-1 (infectious bovine rhinotracheitis virus)-based viral vector which expresses foot-and-mouth disease epitopes. Vaccine 9:564–572.

38 Kleid, D. G., D. Yansura, B. Small, D. Dowbenko, D. M. Moore, M. J. Grubman, P. D. McKercher, D. O. Morgan, B. H. Robertson, and H. L. Bachrach. 1981. Cloned viral protein vaccine for foot-and-mouth disease: responses in cattle and swine. Science 214:1125–1129.

39 Lewis, S. A., D. O. Morgan, and M. J. Grubman. 1991. Expression, processing, and assembly of foot-and-mouth disease virus capsid structures in heterologous systems: induction of a neutralizing antibody response in guinea pigs. J. Virol. 65:6572–6580.

40 Marin, C.M., E. Moreno, I. Moriyon, R. Diaz, and J.M. Blasco. 1999. Performance of competitive and indirect enzyme-linked immunosorbent assays, gel immunoprecipitation with native hapten polysaccharide, and standard serological tests in diagnosis sheep brucellosis. Clin. Diagn. Lab. Immunol., 6:269-272.

41 Mariner, J.C., J.A. House, A.E. Sollod, C. Stem, M.C. van den Ende, and C.A. Mebus. 1990 Comparison of the effect of various chemical stabilizers and lyophilisation cycles on thermostability of a vero cell-adapted rinderpest vaccine. Vet Microbiol. 21:195-209.

42 Mayr, G. A., V. O'Donnell, J. Chinsangaram, P. W. Mason, and M. J. Grubman. 2001. Immune responses and protection against foot-and-mouth disease virus (FMDV) challenge in swine vaccinated with adenovirus-FMDV constructs. Vaccine 19:2152–2162.

43 McKenna, T. S., J. Lubroth, E. Rieder, B. Baxt, and P. W. Mason. 1995. Receptor binding site-deleted foot-and-mouth disease (FMD) virus protects cattle from FMD. J. Virol. 69:5787–5790.

44 Moraes, M. P., J. Chinsangaram, M. C. S. Brum, and M. J. Grubman. 2003. Immediate protection of swine from foot-and-mouth disease: a combination of adenoviruses expressing interferon alpha and a foot-and-mouth disease virus subunit vaccine. Vaccine 22:268–279.

45 Mowat, G.N. and W.G. Chapman. 1962. Growth of foot-and-mouth disease virus in a fibroblastic cell line derived from hamster kidneys. Nature, 194, 253.

46 Mulcahy, G., C. Gale, P. Robertson, S. Iyisan, R. D. DiMarchi, and T. R. Doel. 1990. Isotype responses of infected, virus-vaccinated and peptide vaccinated cattle to foot-and-mouth disease virus. Vaccine 8:249–256.

47 Mulcahy, G., L. A. Pullen, C. Gale, R. D. DiMarchi, and T. R. Doel. 1991. Mouse protection test as a predictor of the protective capacity of synthetic foot-and-mouth disease vaccines. Vaccine 9:19–24.

48 Nakamura, J., S. Wagatuma, K. Fukusho. 1938. On the experimental infection with rinderpest virus in rabbits. J Jap Soc Vet Sci. 17:185-204

49 Nargi, F., E. Kramer, J. Mezencio, J. Zamparo, C. Whetstone, M. H. Van Regenmortel, J. P. Briand, S. Muller, and F. Brown. 1999. Protection of swine from foot-and-mouth disease with one dose of an all-D retro peptide. Vaccine 17:2888–2893.

50 OIE : Terrestrial Animal Health Code. 2006. http://www.oie.int/eng/normes/mcode/en_chapitre_2.2.10.htm

51 Plowright, W. 1962. The application of monolayer tissue culture technique rinderpest research. II. The use of attenuated culture virus as a vaccine for cattle. Bull Off Int Epiz. 57:253-276.

52 Romero, C.H., T. Barrett, S.A. Evans, R.P. Kitching, P.D. Gershon, and D.N. Black. 1993. Single capripoxvirus recombinant vaccine for the protection of cattle against rinderpest and lumpy skin disease. Vaccine, 11:737-742.

53 Roosien, J., G. J. Belsham, M. D. Ryan, A. M. King, and J. M. Vlak. 1990. Synthesis of foot-and-mouth disease virus capsid proteins in insect cells using baculovirus expression vectors. J. Gen. Virol. 71:1703–1711.

54 Rosas, M. F., E. Martinez-Salas, and F. Sobrino. 2003. Stable expression of antisense RNAs targeted to the 5' non-coding region confers heterotypic inhibition to foot-and-mouth disease virus infection. J. Gen. Virol. 84:393–402.

55 Schudel, A., and M. Lombard. (eds). 2004. Control of Infectious Animal Diseases by Vaccination. Dev Biol. Karger. vol 119:443-446.

56 Sterne, M. 1937. The effect of different carbon dioxide concentrations on the growth of virulent anthrax strains. Onderstepoort J. Vet. Sci. Anim. Ind., 9:49-67.

57 Strohmaier, K., R. Franze, and K. H. Adam. 1982. Location and characterization of the antigenic portion of the FMDV immunizing protein. J. Gen. Virol. 59:295–306.

58 Taboga, O., C. Tami, E. Carrillo, J. I. Nunez, A. Rodriguez, J. C. Saiz, E. Blanco, M. L. Valero, X. Roig, J. A. Camarero, D. Andreu, M. G. Mateu, E. Giralt, E. Domingo, F. Sobrino, and E. L. Palma. 1997. A large-scale evaluation of peptide vaccines against foot-and-mouth disease: lack of solid protection in cattle and isolation of escape mutants. J. Virol. 71:2606–2614.

59 Top, F. H., E. L. Buescher, W. H. Bancroft, and P. K. Russell. 1971. Immunization with live types 7 and 4 adenovirus vaccines. II. Antibody response and protective effect against acute respiratory disease due to adenovirus type 7. J. Infect. Dis. 124:155–160.

60 Ubertini, B., L. Nardelli, S. Barei, G. Panina, and E. Lodetti. 1968. Controlli di efficacia su bovini di vaccine antiaftosi monovalenti e trivalenti preparati con virus da cellule BHK. Att Della Societa Italiana Della Scienze Veterinarie XXX:895-900.

61 Waldmann, D., K. Kobe, and G. Pyle. 1937. – Die aktiv Immunisierung des Rindes gegen Maul-und Klausenseuche. Zbl. Bakt. I. Orig. 138:401.

62 Ward, G., E. Rieder, and P. W. Mason. 1997. Plasmid DNA encoding replicating foot-and-mouth disease virus genomes induces antiviral immune responses in swine. J. Virol. 71:7442–7447.

63 Webster, R.G., Y. Kawaoka, J. Taylor, R. Weinberg, E. Paoletti. 1991. Efficacy of nucleoprotein and haemagglutinin antigens expressed in fowlpox virus as vaccine for influenza in chickens. Vaccine. 9:303-308.

64 Wesonga, H.O. and F. Thiaucourt. 2000. Experimental studies on the efficacy of T1sr and T1/44 vaccine strains of MmmSC against a field isolate causing contagious bovine pleuropneumonia in Kenya - Effect of a revaccination. Rev. Elev. Med. Vet. Pays Trop., 53:313-318.

65 Wigdorovitz, A., C. Carrillo, M. J. Dus Santos, K. Trono, A. Peralta, M. C. Gomez, R. D. Rios, P. M. Franzone, A. M. Sadir, J. M. Escribano, and M. V. Borca. 1999. Induction of a protective antibody response to foot and mouth disease virus in mice following

oral or parenteral immunization with alfalfa transgenic plants expressing the viral structural protein VP1. Virology 255: 347–353.

66 Wigdorovitz, A., D. M. Perez Filgueira, N. Robertson, C. Carrillo, A. M. Sadir, T. J. Morris, and M. V. Borca. 1999. Protection of mice against challenge with foot and mouth disease virus (FMDV) by immunization with foliar extracts from plants infected with recombinant tobacco mosaic virus expressing the FMDV structural protein VP1. Virology 264:85–91.

67 Wong, H. T., S. C. Cheng, E. W. Chan, Z. T. Sheng, W. Y. Yan, Z. X. Zheng, and Y. Xie. 2000. Plasmids encoding foot-and-mouth disease virus VP1epitopes elicited immune responses in mice and swine and protected swine against viral infection. Virology 278:27–35.

68 Wong, H. T., S. C. Cheng, F. W. Sin, E. W. Chan, Z. T. Sheng, and Y. Xie. 2002. A DNA vaccine against foot-and-mouth disease elicits an immune response in swine which is enhanced by co-administration with interleukin-2. Vaccine 20:2641–2647.

69 Wu, Q., M. C. S. Brum, L. Caron, M. Koster, and M. J. Grubman. 2003. Adenovirus-mediated type I interferon expression delays and reduces disease signs in cattle challenged with foot-and-mouth disease virus. J. Int. Cyt. Res. 23: 359–368.

70 Yamanouchi, K., K. Inui, M. Sugimoto, K. Asano, F. Nishimaki, R.P. Kitching, H. Takamatsu, and T. Barrett. 1993. Immunisation of cattle with a recombinant vaccinia vector expressing the haemagglutinin gene of rinderpest virus. Vet. Rec.132: 152-156.

71 Yaya A., R. Golsia, B. Hamadou, A. Amaro, and F. Thiaucourt. 1999. Essai comparatif d'efficacité des deux souches vaccinales T1/44 et T1sr contre la péripneumonie contagieuse bovine. Rev. Elev. Med. Vet. Pays Trop., 52:171-179.

72 Yilma, T., F. Aziz, S. Ahmad, L. Jones, R. Ngotho, H. Wamwayi, B. Beyene, M. Yesus, B. Egziabher, M. Diop. Jr., J. Sarr, and P. Veradi. 2003. Global Eradication of Rinderpest with Recombinant Vaccines and Rapid Diagnostic Kits Produced in Africa under the Auspices of AU/IBAR (African Union-Interafrican Bureau for Animal Resources). Proceedings of the Global Rinderpest Eradication Programme (GREP) Technical Consultation 2002, Food and Agriculture Organisation of the United Nations (FAO), Rome, Italy.

-ΔΔΔ-

23

Animal Health Information System through Monitoring and Surveillance and Forecasting of Small Ruminant Diseases

K PRABHUDAS

Project Directorate on Animal disease
Monitoring and Surveillance, Hebbal, Bangalore - 560 024

The Indian livestock industry is faced with a high magnitude of endemic diseases despite the country having a good network of veterinary clinics and diagnostic laboratories in the state animal husbandry and veterinary services. There are about 42,000 field veterinary clinics spread over an area of 3.29 million square kilometers to cover 6,36,000 villages involving nearly 573 million heads of livestock. On an average each veterinary clinic covers 15 villages, spread over an area of about 52 sq. km comprising of 14,000 heads of livestock (Annual Report 1996-97, Department of Animal Husbandry and Dairying, Government of India). The country has 593 districts (Administrative levels) in 29 states and 6 union territories; these should be effectively linked to the central disease monitoring, surveillance and policy planning and implementation units. The diagnostic laboratories under the private sectors and NGOs should also join network to have a comprehensive national coverage.

In India, at present the main method of collecting information on livestock diseases is through a passive disease reporting system. The occurrence of 10-20 percent of actual outbreaks is only recorded. A strong laboratory diagnostic service, subservient to the official veterinary service, and answerable to it, is absolutely essential for accurate diagnosis and also for active surveillance. Laboratory services are an essential backup to what is being done by field service staff. The Regional Disease Diagnostic Laboratories are set-up for this purpose.

However, the present reports provide the following information based on clinical diagnosis

- ❒ Which of the diseases are present in the country, and where located.
- ❒ Provides information to respond to the disease outbreaks.
- ❒ Meet the basic disease reporting requirements of OIE.

Cardinal Points for Outbreak Investigaton

The following cardinal points when observed during the period of outbreaks by the field veterinarian most of the aspects of data generation would be effectively covered.

1. Establish and verify diagnosis and define a "case".
2. Confirm that an outbreak is actually happening
3. Characterize the outbreak in terms of *time, animal and place*. This test involves measuring *disease frequency* and *documenting pattern*
4. Analyze the data. This step involves *factor-specific attack rate* and constructing attack rate table.
5. Undertake intensive follow-up investigations to identify *high risk groups* and possible future outbreaks.
6. Implement control and preventive measures
7. Formulate working hypothesis, in an attempt to identify *type of epidemic, the possible source and mode of spread.*
8. Report findings of the investigations with recommendation for dealing with future possible outbreak of the same disease.

Additional requirements

Strengthening of the national disease data reporting and recording through the following:

- ❒ Passive disease reports
- ❒ Meat inspection findings
- ❒ Diagnostic laboratory reports

- ❒ Livestock movement and migration profiles
- ❒ Vaccine production, distribution and delivery

To measure the disease prevalence through active surveillance

- ❒ Estimates of disease prevalence
- ❒ Estimates of disease incidence
- ❒ Develop computerized national relational databases for animal diseases
- ❒ Develop national library of GIS based livestock disease - information maps
- ❒ Develop specific livestock health systems for different species to promote production
- ❒ Develop national animal health information expert system

Organized Animal Health Management System requirements

The developed countries have well organized animal health information management systems through epidemiological surveillance, a priority area of international activity of the FAO. These systems essentially incorporate the following aspects.

1. Planning Survey, surveillance and monitoring Systems - Roles and Requirements.
2. Collection of active data by following stratified sampling techniques.
3. Statistical Considerations and Data Analysis of Survey, surveillance and monitoring Systems.
4. Sampling Survey to detect prevalence and incidence and determine status of the disease / freedom from infection.
5. Utilization of Simulation Models in surveillance and monitoring Systems.
6. Reports and Dissemination of Survey, surveillance, and monitoring findings.
7. Administrative and Human Resources Needed for Survey, surveillance, and monitoring Systems.

The above system helps to develop: Active monitoring through national random sample surveys; Develop computerized national relational data bases and networks; Analysis and GIS mapping of national epidemiological disease data and interactive national animal health information system and disease forecasting system for the benefit of end-users, contributors, planners, decision-makers and researchers.

Forecasting of Small Ruminant Diseases

PD_ADMAS has developed a system for forecasting of economically important animal diseases based on passive and active surveillance data and agro-ecological and meteorological parameters. The precipitating factors were identified for each disease and based on multiple regression analysis the diseases are being forecasted two months in advance. The important among them are PPR, sheep and goat pox and bluetongue. Now that the diagnostic and vaccines are available for PPR and Pox, necessary control measure by applying mass vaccination at the right place and at the right time during the year the forecasting helps controlling the diseases to a large extent.

How to achieve compilation of accurate data-base on animal health information

I. Village/Dist. level

1. Creating awareness to the field veterinarians, NGOs and farmers regarding benefits of active reporting and monitoring and surveillance of diseases and its utility in terms disease prevention and control. Provision of encouragement/incentives for reporting.
2. Networking of district, regional and state diagnostic laboratories.
3. Creation of linkages with state epidemiology and surveillance units.

II. State Level

1. Collection and collation of data from village, district levels.
2. All district level veterinary services to be linked to the state units.
3. Imparting periodical training on database, computer programs, networking and GIS programs.
4. Linkages with RDDLs, SAUs and NGOs engaged in animal health/welfare programs.
5. Identification of diseases (location-specific) which can be controlled /eradicated by effective measures and implementation.

III. Apex body /National level

1. Diseases are to be prioritized on a national level for control and eradication based on national level incidence/prevalence surveys based on use of approved diagnostics and structured surveys.

2. Identification of diseases of regional importance and act as advisory body for the states.

3. Good emergency management practices in case of an outbreak due to a new strain/or likelihood of reoccurrence of a new disease pandemic as the case may be.

-ΔΔΔ-

24

Genetic Resistance to Important Small Ruminant Diseases

P K ROUT AND R ROY

Genetics and Breeding Division
Central Institute for Research on Goats
Makhdoom, Farah, Mathura - 281 122

INTRODUCTION

Disease is economically important issue and has impact on human health, livelihood security of farmers and food safety. Disease is the major reason for the genotype and phenotype gap. Despite the traditional disease control measures, losses attributable due to diseases continue to impede the livestock industries. Economic losses from diseases of farm animals have been estimated to amount to 10-20% of the total production costs[1]. The costs of disease have been estimated to be 17% of turnover in the developed world and 35-50% of turnover in the developing world[2]. In India, the annual loss in livestock sector due to diseases is estimated as Rs. 13,830 crores. Genetic resistance to disease is a wide spread phenomenon and common in all species and has been recognized for a long time as an important factor in disease control. Genetic variations for disease resistance in domestic animals have been reported in cattle, sheep, Goat, pig and poultry[3,4,5]. It is a common observation that for most disease, some animals are more resistant than others. This signifies that there is phenotypic variation in the expression of resistance in a population. Where the nature of the variation is studied, it usually turns out that at least some of the observed variation is due to genetic factors (i.e resistance is heritable). Practically this suggests that it is usually possible to increase the average level of resistance by conventional selection programme. The knowledge of extent of genetic variation to a disease is an essential requirement for inclusion of disease resistance in breeding programme. The genetic variation in important diseases in small ruminants will give an idea about the nature of variation and inheritance of the trait in the population. This paper briefly reviews the genetic variation in important disease of small ruminants, association of markers, strategies for breeding for disease resistance and the national scenario.

WHY BREED FOR DISEASE RESISTANCE?

Disease is one of the most important biological constraints to small ruminant (sheep and goats) production in all the different agro-climatic zones. Loss of production performance, mortality and the cost of drugs are the major concerns for farmers. Presently the control strategies include vaccination, medication, isolation of animals from pathogens, improved sanitation and improved management systems. However in developing countries, it is not even possible for poor farmers to provide minimum veterinary services to keep the stock healthy. Moreover the development of resistance in pathogens to drugs and chemicals and lack of effectiveness of some vaccines are providing limited scope to improve management cost-effectively[6,7,8]. Again the cost of disease control i.e. preventive and therapeutic measures is high[9]. Recently. the consumer preference has been changed, as they do not want to consume the food with residues of drug and feed additives. In the present juncture, the alternative sustainable solution is to breed for disease resistance. Disease resistance has direct effect on production, product quality, food safety and environmental pollution. Disease resistance will enhance food security and significantly improve the intermediate trade and marketing. It has significant effect on efficient use of inputs and will reduce the disease risks to human and livestock.

Disease Resistance

The onset of disease is often the result of the complex interaction between an individual animal's genotype and the environment to which animal is exposed. Disease resistance is based on structural and immunological mechanisms. Host resistance to infection is a complex process governed by numerous genes, relatively few of which have been identified. Mouse models of host resistance provide reproducible system for studying mechanisms involved in important immunological functions. Defense against invading pathogens is mediated by both natural and acquired immunity and different protective mechanisms are stimulated depending on the nature of host's response to pathogens.

Generally two types of resistance have been described in all the bacterial, viral and parasitic diseases.

(1) Resistance to infection means that the mechanism that prevent the organisms to penetrate into living host cell i.e. first line of defense or cellular resistance. Examples of such resistance are (i) absence of receptor in the cell membrane for retroviruses of subgroups A, B, C, D and E[10,11]. (ii) Absence of cell receptor for K88 antigen in *pig*[12,13].

(2) Second type of resistance/ Second line of defense is the resistance to the disease process or effects of infestation in host in response to pathogen infection.

Similarly resistance in relation to parasitic disease described as resistance and resilience.

Resistance: It is defined as the initiation and maintenance of responses provoked in the host to suppress the establishment of parasite and /or eliminate parasite load. Resistant animals therefore harbour fewer parasites than susceptible animals.

Resilience: It is defined as the ability of host to survive and be productive in the phase of parasitic attack. Tolerant animals do not necessarily harbor fewer parasites and this can have important effects on the epidemiology of the disease and important negative effects on the more susceptible animals.

Measurement of Genetic Variation and Identification of Host Resistance Phenotypes

By carefully investigating a disease, evidence for genetic variation for resistance to a wide variety of diseases has been observed. Evidence for genetic variation for resistance to a number of different diseases is available for small ruminants and this includes resistance to scrapie, footrot, mastitis, paratuberculosis (Johne's disease), dermatophilosis, salmonellosis, cowdriosis (heartwater), trypanosomosis, GI nematodes, liver fluke (*Fasciola hepatica* and *Fasciola gigantica*) and flystrike[2,5]. The reliability of the evidence for resistance to these diseases is quite variable and ranges from well-designed experimental studies to strong anecdotal evidence.

Genetic resistance may be found with discrete or continuous phenotypic variation and inherited by simple Mendelian mechanism or polygenic inheritance. The measure of disease resistance such as morbidity, immune response, survival period, loss of performance are governed by polygenic inheritance. However the genetic control of disease resistance has been observed to be controlled by very few pair of genes. The example of single gene inheritance of disease resistance occurs in chicken and pigs. In small ruminants scrapie is controlled by major gene. Before it is possible to embark on any study on genetics of disease resistance it is essential that indicator traits of host response to infection should be identified. It is very important to identify phenotypic expression of disease and to identify indicator traits, which should be variable, able to measure and heritable.

Genetic Variation in Resistance to Diseases in Small Ruminants

The knowledge of extent of genetic variation to a disease is an essential requirement for inclusion of disease resistance in breeding programme. The possible source of genetic variation between breed, strains, families, and individuals should be analysed and confounding effects of environment should be partitioned. As a first step requirement, it is essential to know how much disease resistance trait has been reported in number of diseases and it gives an idea about the feasibility of the trait to be included in breeding programme. The source of genetic variation in important diseases in farm animals will give an idea about the nature of variation and inheritance of the trait in the population. Therefore some of the important viral, bacterial and parasitic diseases are discussed briefly to have an idea regarding genetic variation in small ruminants in response to natural as well as experimental infection. The application of breeding for resistance to diseases should consider all the factors in relation to impact on production, control strategy and zoonotic potential.

The first indication of genetic variation for disease resistance often comes from the observation that some breeds or populations have the ability to tolerate endemic diseases or to survive an epidemic disease that more susceptible breeds cannot. The susceptibility of sheep to ovine lentivirus infection such as Maedi/visna has long been known to be subject to distinct genetic variation. Several sheep breeds have a high degree of resistance and within breed genetic resistance has also been observed. British goats were more susceptible to capripox challenge than north Yemen Black goat[14]. Similarly in an experiment to study susceptibility of British sheep breed to bluetongue virus infection revealed that Soya sheep were most susceptible and died before showing any clinical symptoms, whereas Dorset horn and Merino developed several clinical signs before death[15]. Davies[16] reported that the indigenous sheep and cattle breeds of Kenya were relatively resistant to rift valley fever as compared to exotic animals imported for breed improvement programme.

Infectious bacterial diseases continue to limit the productivity and health of sheep and goat worldwide. Genetic variation to many bacterial diseases has been described in studies between small ruminant breeds and within breeds and flocks. Genetic resistance to foot rot in Sheep has been well recognised and sources of genetic variation in resistance to foot rot have been studied in detail[17]. The heritability of resistance to footrot has been reported in specifically designed experiment in Merino Sheep, which was 0.14±0.04 on a single observation and 0.28±0.08 on the basis of multiple observations for natural exposure to footrot. Resistance to fleece rot and dermatophilosis is studied in different sheep breed and genetic variation between flocks within merino breed has been established[18]. Breed differences in susceptibility to Jhone's disease in goat have been reported in different countries[19,20,21,22].

Salmonellosis in sheep and Goat causes acute enteritis. No report is available on genetic differences between breeds of goats & sheep or any difference within flocks. However genetic difference to Salmonella in mice has been observed in an experiment, and identification of specific loci (Ity & LPs) associated with innate resistance to salmonella infection in mice provides a powerful option to screen for similar genetic variation in sheep & goat breeds[23].

Bacterial diseases that has been analysed for disease resistance include anthrax, Brucellosis, caseous lymphadenitis, entrotoxaemia, pasterulosis, salmonellosis and dermatophilosis in Sheep and Mycoplasmosis in goat. A Possible sources of genetic variation in resistance to bacterial infections in sheep and goat have been reported .In small ruminants almost all of the heritability estimates for disease resistance are from studies undertaken in the temperate developed world. A review of disease resistance studies in sheep in Australia and New Zealand[24] found an average heritability of 0.28 for the six diseases studied (i.e. resistance to GI nematodes, facial eczema, ryegrass staggers, fly strike, dermatophilosis and footrot) which is very similar to the heritability estimates for the production traits such as milk yield or body weight.

Evidence for genetic variation to parasites has been found both within and between host populations and for a number of important parasite species including nematode worm, coccidian, flies and ticks. Variation in resistance to gastro-intestinal nematodes between breeds and within breeds have been reported and reviewed by Kassai and Streter[25] and by Gray and colleague[5]. In goats, the resistance to H. Contortus has been reported by different researchers[26,27]. In the case of small ruminants in the tropics this is well illustrated by the resistance and/or tolerance of a number of sheep breeds to GI nematode infections including the Red Maasai in East Africa, the Djallonke in West Africa, the St Croix and Barbados Blackbelly in the Caribbean, the Javanese Thin Tail in Indonesia and the Garole in India[5,28]. The heritability estimates for resistance to GI nematodes (assessed by faecal egg counts) estimated from the relatively susceptible Romney and Merino breeds in New Zealand and Australia, respectively, average about 0.30 and range from 0.20 to 0.40[5,28]. Trypanotolerance in cattle is well documented, however it is well known that dwarf sheep and goats of West Africa can live under conditions of high tsetse challenge. Djallonke and Red-Massai sheep breed of Africa and dwarf goats of West Africa and East Africa are more resistant to trypanosomiasis[29].

Association of QTL and Major Genes

Most of the studies to identify Quantitative Trait Loci (QTL) for resistance to disease and ultimately the gene or genes underlying the QTL are being carried out in pigs and poultry[2]. However, experiments are being carried out at ILRI in Kenya to identify QTL for resistance to trypanosomes in N'dama cattle and for resistance to GI nematodes in Red Maasai sheep.

Scrapie is associated with major gene in sheep. The research to identify genetic markers for resistance to gastro-intestinal parasite infection in sheep has been carried out. There have been some successful reports of population association studies to identify genetic markers for endoparasite resistance in sheep using candidate gene approach (MHC DRB1 alleles with *Ostertagia circumcincta*). It is now generally accepted that a genomic scan and linkage analysis in segregating resource families is a more efficient approach for finding genetic marker for endoparasite resistance. Recently, Beh & colleague[30] have reported the first achievement of mapping quantitative trait loci (QTL) associated with resistance to helminthes (*Trichostrongylus colubriformis*) infection. In a second study, Merino sheep were artificially challenged with H. contortus and identified number of QTL[31]. Similarly two QTL was identified in response to H. contortus infection in the crosses of Thin Tail sheep of Indonesia and susceptible Merino sheep[32]. Davies & colleagues[33] identified QTL on chromosome 3 and 20 in response to parasite natural exposure in Scottish black face sheep. Similarly Crawford & colleagues[34] observed in his large study that a single QTL was associated with nematode resistance in sheep. The potential impact of QTL experiments will be to identify the gene(s) controlling resistance to endoparasites in sheep and then through functional genomics and proteonomics research identify the transcribed products of these genes and their functions. This in turn could lead to the formulation of new drugs or vaccines, which would not only benefit ruminant livestock production worldwide, but also possibly help control helminthosis in humans.

National Scenario

In our country, the research on breeding for disease resistance in livestock has been started in poultry under Pl-480 project at Indian Veterinary research Institute, Izatnagar In small ruminants, some reports of breed differences to disease susceptibility from experimental stations maintained by research institute and universities have been reported, however most of the studies are not aimed to assess genetic variation in breeds. Comparative disease susceptibility to sheep pox, blue tongue, Ovine adenovirus, FMD and haemonchosis has been reported[35,36,37,38,39]. Malpura and Chokla breeds were less susceptible to sheep pox in comparison to Sonadi and Nali[35]. Malpura, Marwari, Sonadi and Nali breeds were less susceptible to Blue tongue in comparison to exotic breeds[36]. Nellore, Mutton synthetic breeds of sheep were highly susceptible to FMD, whereas Malpura and Mandya were lowly susceptible to FMD[37,38]. Similarly in goats, breed differences in resistance to goat pox, FMD, Johne's disease, colibacilosis have been reported[22,40,41]. Results of analysis of data in natural infection revealed that Sirohi is less susceptible to goat pox in comparison to Kutchi and Marwari[40]. Barbari was less susceptible to FMD than Jamunapari and Jakhrana. Jamunapari goats are less susceptible to JD than Barbari goats[22]. Jamunapari kids were less susceptible to colibacilosis than Barbari kids[41]. Barbari goats were more susceptible to pneumonia than Jamunapari goats under similar system of management[42].

Genetic resistance to gastro intestinal nematodes (H. contortus) in goats has been initiated at CIRG. An investigation has been carried out in Jamunapari and Barbari kids to assess their resistance to natural infection with gastro-intestinal nematodes (GIN) and to establish indicator traits for such resistance in Indian goats. Age had a significant effect on FEC in both breeds, there being an increase in FEC from 3 months of age to 9 months of age for kids born in either season[43,44]. This is contrary what occurs in sheep. The correlation coefficient of FEC on both body weight and body weight gain was negative, and there was a loss of body weight in the individuals with a high FEC[44]. Resistance to naturally acquired GIN infections were studied in different physiological stages, namely pregnant, dry and lactating stages in semi-arid region of India. Jamunapari lactating goats had significantly higher FEC than that of dry and pregnant does but there was no peri-parturient rise in FEC in the resistant Barbari goats[45].

Conclusions

Recently livestock diseases in the tropical developing world have been ranked for their impact on poverty in terms of their direct effect on production and their indirect effects on international trade and marketing[46]. For sheep and goats the top ten ranked diseases (alphabetical order) were anthrax, ectoparasites, gastro-intestinal helminthes, haemonchosis, heartwater (*Cowdria ruminantium* infection), liver fluke (fasciolasis), neo-natal mortality, peste de petits ruminants (PPR), respiratory complexes (e.g. pneumonia, etc) and sheep and goat pox. Some diseases can by controlled by effective vaccination, adoption of effective management and some by incorporating genetic element with existing disease control.

Breeding for disease resistance experiment has been conducted in temperate regions. The breeding objective for small ruminants in tropical production systems will be different from those in temperate production systems. The genetic variation among indigenous sheep or goat breeds in the tropics is likely to be larger than that found among breeds of small ruminant in temperate regions. We must utilize the breed variation most efficiently to improve the performance of our stock and to check mortality as well as sub clinical infection. We have not yet evaluated our genetic resources for genetically based resistance in different agro-climatic zones and production system.

For each production system and each livestock species there is a need to document and prioritize the economic impact of different diseases[47]. After prioritizing diseases then an assessment is required for existing disease control strategies. When current disease control strategies are found to be inadequate and non-sustainable then breeding for disease resistance can be considered. Where there is already evidence of differences among breeds in resistance or tolerance to a particular disease then it is necessary to devise a strategy for continuing the work in this direction. Integrated disease control strategies including host genetic resistance will generally be more sustainable, cost effective, increase intermediate trade, enhance consumer acceptance, decrease spreading of disease to livestock and humans and will take care of animal welfare.

References

1. Muller, M and Brem, G. 1994. Transgenic strategies to increase disease resistance in livestock. Reproduction Fertility Devlopment 6, 605-613.

2. Bishop, S.C. Chesnais, J. and Stear, M.J. 2002. Breeding for disease resistance: Issues and opportunities. *Proceedings of the 7th World Congress on Genetics Applied to Livestock Production, Montpellier, France* 31: 597-604.

3. Owen, J.B. and Axford, R.F.E. 1991. *Breeding for Disease Resistance in Farm Animals.* CAB International, Wallingford, UK.

4. Axford, R.F.E., Bishop, S.C., Nicholas, F.W. and Owen, J.B. 2000. *Breeding for Disease Resistance in Farm Animals, 2nd edition.* CAB International, Wallingford, UK.

5. Gray, G.D., Woolaston, R. and Eaton, B.T. 1995. *Breeding for Resistance to Infectious Diseases of Small Ruminants.* Australian Centre for International Agricultural Research (ACIAR) Monograph No. 34, Canberra, Australia.

6. Nicholas, F.W. 1987. Veterinary Genetics, Clarendon Press, Oxford.

7. Waller, P.J 1990. Resistance in nematode parasites of livestock to the benzimidazole anthelmintics. Parasitology Today 6,127-129

8. Waller, P.J. 1997. Anthelmintic resistance. *Veterinary Parasitology* 72: 391-412.

9. Piper, L.R. and Barger, I.A 1988. Resistance to gastrointestinal strongyles: feasibility of breeding programme. In: Proceedings of the 3rd world congress of Sheep and Beef cattle breeding, Paris, 19-23 June, Vol1.pp593-611

10. Crittenden, L.B. 1975. Two levels of genetic resistance to lymphoid lukosis. Avian Disease. 19:281-292.

11. Pani, P.K. 1975. Genetic control of resistance of chick embryo cultures to RSV (RAV-0). Journal. of General virology, 27:163-172.

12. Gibbons, R.A., Sellwood, R, Burrows, M and Hunter, P.A 1977. Inheritance of resistance to neonatal E.Coli diarrhoea in the pig: Examination of the genetic system. Theoretical and Applied Genetics, 51,65-70.

13. Sellwood, R. 1980. The interaction of the K88 antigen with procine intestinal epithelial cell brush borders. Biochimica and Biophysica Acta, 632, 326-335.

14. Kitching, R.P., Hammond, J.M. and Taylor, W.P. 1987. A single vaccine for the control of capripox infection in sheep and goats. Research in Veteinary Science 42, 53-60.

15. Jeggo, M.J., Corteyn, A.H., Taylor, W.P., Davidson, W.L. and Gorman, B.M. 1986. Virulence of bluetogue virus for British sheep. Research in Veterinary Science 42, 24-28.

16. Davies, F.G. 1981. Rift Valley fever in Kenya. Office International des Epizooties, Technical Series No. 1 Paris, France.

17. Egerton, J.R. and Raadsma, H.W. 1991. Breeding sheep for resistance to foot rot. In:Owen, J.B. and Axford, R.E.F. ed., Breeding for disease resistance in farm animals. C.A.B. International, Wallingford, U.K. pp. 347-370.

18. Raadsma, H.W., Egerton, J.R. and Nicholas, F.W. 1991. Investigations in genetic variation for resistance to footrot. In: Gray, G.D. and Woolaston, R.R. (eds) Breeding for Disease Resistance in sheep. Proceedings of a Workshop, UNE, Armidale, Nov. 1990, pp. 41-50.

19. Srivastava, C.P. & More, B.K.1987. Livestock Adviser, 12:33-37.

20. Kumar, R.; Prasad, M.C. and Paliwal, O.P. 1988. Indian vety. Journal, 26:582-584.

21. Singh, N. ; Kala. S.N.; Vihan, V.S. and Singh, S.V. 1990. Indian Journal of Animal Science, 60.1163-1165.

22. Molina, A., Delgado, J.V., Madueno, R., Morera, L. and Llanes, D.,1991. Paratubercuosi incidence in different goat breeds; Canary Island breed, a case of genetic resistance. In: Owen, J.B. and Axford, R.F.E. ed., Breeding for diseases resistance in farm animals, Proceedings of an International symposium, Bangor,, Wales, Sept. 1990. Wallingford, U.K., CAB International, pp, 479-480.

23. Briles, D.E(ed) 1986. Genetic control of the susceptibility to bacterial infection. Contemporary Topics in Microbiology and Immunology, vol.124,

24. Morris, C.A. 1998. Genetics of disease resistance in Bos Taurus cattle. Animal Genetics Research Information 23,1-11.

25. Kassai, T and Streeter, T. 1992. Genetic aspects of the host resistance to helminthic infection. Research and rReviews in parasitology,53,67-74.

26. Gruner, L.1991. Breeding for helminth resistance in sheep and goats In: Owen, J.B. and Axford, R.F.E. ed., Breeding for diseases resistance in farm animals, Proceedings of an International symposium, Bangor, Wales, Wallingford, U.K., CAB International, pp, 479-480.

27. Wakelin, D. 1992. Genetic variation in resistance to parasitic infection: experimental approaches and practical applications. Research in Veterinary science, 53, 139-147.

28. Baker, R.L. 1998. A review of genetic resistance to gastrointestinal nematode parasites in sheep and goats in the tropics and evidence for resistance in some sheep and goat breeds in sub-humid coastal Kenya. *Animal Genetics Resources Information Bulletin* 24: 13-30.

29. Griffin, L. and Allonby, E.W. 1979. Trypanotolerance in breed of sheep and goats with an experimental infection Trypanosoma congolense. Veterinary Parasitology, 5,97-105.

30. Beh KJ, Hulme DJ, Callaghan MJ, Leish Z, Lenane I, Windon RG, Maddox JF 2002. A genome scan for quantitative trait loci affecting resistance to Trichostrongylus colubriformis in sheep. *Animal Genetics* 2002, 33(2): 97-106.

31. Marshall K, van der Werf JHJ, Maddox JF, Graser HU, Zhang Y, Walkden-Brown SW, Kahn L: 2004. A genome scan for quantitative trait loci for resistance to the gastrointestinal parasite Haemonchus contortus in sheep. *Proceedings Association for the Advancement of Animal Breeding and Genetics*, 16:115 -1118.

32. Raadsma HW, Margawati ET, Piedrafita D, Estuningsih E, Widjajanti S, Beriajaya, Subandriyo, Thomson P, Spithill TS 2002. Towards molecular genetic characterization of high resistance to internal parasites in Indonesian Thin Tail sheep. In *Proceedings of the 7th World Congress on Genetics Applied to Livestock Production: 2002;Montpellier; France* INRA, 2002:13-19.

33. Davies G, Stear MJ, Benothman M, Abuagob O, Kerr A, Mitchell S,Bishop SC: 2006 Quantitative trait loci associated with parasitic infection in Scottish blackface sheep. *Heredity*, 96(3): 252-258.

34. Crawford, Allan M, Korena A Paterson, Ken G Dodds, Cristina Diez Tascon, Penny A Williamson, Meredith Roberts Thomson, Stewart A Bisset, Anne E Beattie1, Gordon J Greer1, Richard S Green, Roger Wheeler, Richard J Shaw, Kevin Knowler and John C McEwan 2006. Discovery of quantitative trait loci for resistance to parasitic nematode infection in sheep: I. Analysis of outcross pedigrees *BMC Genomics* 2006, 7:178 doi: 10.1186/1471-2164-7-178

35. Sharma, M.M., Uppal. P.K.; Lankar, P.S. and Mathur, P.B. 1986. Indian J. of Animal Science-es 56, 1183-1186.

36. Dubey, S., Rao, T.R.M.: Srivastava, C.P.; lanker, P.S. and Maru, A. 1988. Indian J. of Animal Sciences. 58: 68-70.

37. Khurana. K.L.: Maru, A. ; Bhagwan, P.S.K.; Dubey, S. and Srivastava, C.P. 1991. In: Recent advances in the control of disease of crossbred and companion animals, Bombay,

38. Rao, T.M. ; Charyulu, E.K.; Prasad, R.D.D.; Srivivasulu, D. And Munirathnam, D. 1993. Indian J. of Animal Health, 32, 77-78.

39. Yadav, C.L.; Grawal, H.S. and Banerjee, D.P. 1993. International J. for Parasitology, 23: 819-821.

40. Nagpal, A.K.; Vihan, V.S.; Singh, D.; Singh, S.V. and Singh, N. 1990. Indian J. of Animal Sciences 60: 544-546.

41. Vihan, V.S. and Singh, N. 1988. Indian J. of Animal Sciences, 58: 233-236.

42. Rana, R.,A. Kumar, VK gupta and VS vihan 2002. Moratality pattern in caprine pneumonia in semi-intensive system of management under semi-arid zone- a retrospective study. Xth International Congress of Asian-australian Association of animal Production, New –Delhi.

43. Rout, P.K., Sharma, D.K., Saxena, V.K., Singh, S.K and Roy, R. 2002. Resistance of Jamunapari and Barbari goat breeds to natural infection with gastro-intestinal nematodes, Indian Journal of Animal sciences, 72, 770-773.

44. K.K.Chauhan, P.K.Rout, Singh, P.K., Mandal, A., Singh, S.K and Roy, R 2003a. Genetic resistance of Barbari and Jamunapari kids to natural infection with gastro-intestinal nematodes. Tropical Animal Health and Production. 35(5), 397-408.

45. K.K. Chauhan, P.K. Rout, P.K.Singh, A. Mandal, R.Roy and S.K.Singh 2003b. Susceptibility to natural gastro-intestinal nematode infection in different physiological stages in Jamunapari and Barbari goats in the semi-arid tropics. Small Ruminant Research, 50,219-223.

46. Perry, B.D., Randolph, T.F., McDermott, J.J., Sones, K.R. and Thornton, P. K. 2002. *Investing in Animal Health Research to Alleviate Poverty.* International Livestock Research Institute, Nairobi, Kenya.

47. Perry P.D. and Randolph, T.F. 1999. Improving the assessment of the economic impact of parasitic diseases and their control in production animals. *Veterinary Parasitology* 84:145-168.

-ΔΔΔ-

25

Alternative Therapies for the Treatment of Animal Diseases in India

ASHOK KUMAR AND V S VIHAN

Division of Animal Health, Central Institute for Research on Goats, Makhdoom, Farah, Mathura (U.P.) - 281 122

In veterinary practice, mainly modern therapy is practiced in our country; however, other alternative approaches are also gaining momentum. Alternative medicine is the best recognized term for a group of treatment or therapeutic system that lie outside mainstream or conventional medicine. More recently, the term complementary and integrated medicines have come in to use to indicate that they are used with in addition to conventional medicine. It is also describe as holistic medicine, the simplest definition comprised of any system that diagnose and treat a diseases in the context of the whole patient that is in relation to the patient general state of health, the presence of other diseases and the environmental influences, both physical and mental. Holistic system does not address a diseases entity alone[19].

Traditional medicine system that developed from specific culture includes Chinese medicine and ayurvedic medicine. Our current conventional medical system has been called western medicine, modern medicine, scientific medicine and biomedicine[16]. Integration of the different medical system may appear challenging because of the widely varying metaphors and paradigms by which practitioners view health and diseases. Recently on workable synthesis of various systems has been proposed[12], a new system for categorizing the various holistic and conventional therapies.

1. *Biochemical therapies:* Drugs, herbs, nutritional supplements.
2. *Life style therapies:* Nutrition, exercise, environmental therapies, and Mind body regulation.
3. *Bio mechanical therapies:* Massage, spinal manipulation and surgery.
4. *Bioenergetic therapies:* acupuncture, homeopathy.

In 1980, saw a interest in both scientific and traditional medicine and ethanopharmacology. Traditional medicine is now considered a potential source of research in drug development.

Veterinary Medicine like all profession is rapidly changing. Veterinary Medicine is developing and expanding in to a new horizon. Our horizon is complementary and alternative medicine. The integration of these therapies in mainstream medicine is inevitable. Now Veterinarians have the opportunity to enhance animal health care and benefit all life on earth. Besides , Conventional therapy , There are other alternative therapies are practiced in animals, are as follows.

1. Ayurveda

Ayurveda originated in India long back in pre-vedic period. *Rigveda* and *Atharva-veda* (5000 years B.C.), the earliest documented ancient Indian knowledge have references on health and diseases. Ayurved texts like *Charak Samhita* and *Sushruta Samhita* were documented about 1000 years B.C. The term *Ayurveda* means 'Science of Life'. It deals elaborately with measures for healthful living during the entire span of life and its various phases. Besides, dealing with principles for maintenance of health, it has also developed a wide range of therapeutic measures to combat illness. These principles of positive health and therapeutic measures relate to physical, mental, social and spiritual welfare of human beings. Thus *Ayurveda* becomes one of the oldest systems of health care dealing with both the preventive and curative aspects of life in a most comprehensive way and presents a close similarity to the WHO's concept of health propounded in the modern era.

In human being ayurveda have , eight disciplines are generally called "Ashtanga Ayurveda" and are ; Internal Medicine (*Kaya Chikitsa*) ,Paediatrics(*Kaumar Bhritya*),Psychiatry (*Bhoot Vidya*), Otorhinolaryngology and Ophthalmology(*Shalakya*) ,Surgery (*Shalya*),

Toxicology (*Agad Tantra*), Geriatrics (*Rasayana*), Eugenics and aphrodisiacs(*Vajikarana*). Compendia on these subjects like *Charak Samihta, Sushruta Samhita* etc. were written by the ancient scholars during B.C. period. These were used for teaching of Ayurveda in the ancient universities of *Takshashila* and *Nalanda*. Basic theme of ayurveda treatment consists of restoring the balance of disturbed body-mind matrix through regulating diet, correcting life-routine and behaviour, administration of drugs and resorting to preventive Panchkarma and *Rasayana* therapy.

There are two streams of practices; one is folk stream, comprising mostly the oral traditions practiced by the rural villages. The carriers of these traditions are millions of housewives, thousands of traditional birth attendants, bone setters, village practitioners skilled in acupressure, eye treatments, treatment of snake bites and the traditional village physicians/herbal healers, the vaidyas' or the tribal physicians. This stream of inherited traditions is together known as Local Health Traditions (LHT). At the second level of traditional health care system are the scientific or classical systems of medicine. This comprises of the codified and organized medicinal wisdom with sophisticated theoretical foundations and philosophical explanations expressed in classical texts like 'Charka Samhita', 'Sushruta samhita', 'Bhela samhita', and hundreds of other treatises including some in the regional languages covering treaties of all branches of medicine and surgery. Systems like Ayurveda, Siddha, Unani, Amchi and Tibetan, etc. are practiced.

Indian medicine is not reflected merely in the treatment of human beings but other important dimensions like veterinary medicine are addressed in detail through these systems. This represents a whole new spectrum of knowledge and opportunity. This area has not been exploited at all. The definition of Ayurvedic medicines under the Drugs & Cosmetics Act includes veterinary medicines. There are several authentic books of Ayurveda dealing with veterinary medicines such as Nakul Samhita, Palkapya Shastra, Go Ayurveda, Hasti Ayurveda, Baj Nama etc. Central Government will encourage two institutions in the 10th Plan for introducing courses and undertaking documentation of the classical books in various languages. Homoeopathy also has effective treatment for care of animals and a similar approach would be followed.

In veterinary practice, ayurvedic drugs are used by utilizing academic knowledge of basics of physiology , biochemistry and diseases manifestation and as per indication given on the products. Among ayurvedic drugs, medicinal plants continue to be an important therapeutic content in it. In many eastern cultures such as those of India, China and the Arab/Persian world this experience was systematically recorded and incorporated into regular system of medicine that refined and developed and became a part of the Materia Medica of these countries. Ayurveda is much superior among organized ancient systems of medicine. From history we learn that in the ancient times India was known as a place of rich natural resources, knowledge, wisdom and scholarship.

In Veterinary practice, there are many ayurvedic products are available, mainly in the form of powder , syrup, boluses derived from plants . In India, Natural Remedies Private Ltd Bangalore, Indian Herb, Saharanpur UP, Cattle Remedies, UP etc are some of the herbal pharmaceuticals. These products are not in the form of standardized drug. There is a need to standardize these drugs with modern instrumentation and techniques, so it can compete international Market.

2. *Traditional Veterinary Medicine (Ethnoveterinary Medicine)*

Historically , Right from the Mahabharatha times, veterinary medicine was practised in India and documented in Sanskrit scriptures and literature in vernacular languages. The information in Sanskrit scriptures like the 'Asva Vaidyaka by Nakula', 'Palakpya', 'Garuda Purana', 'Asvayur Veda Sarasindu by Vysampayana', and 'Asvayur Veda Sara Sindhu by Malldeva' is well classified and documented according to the principles of the humoral theory of the Ayurvedic system. A few vernacular scriptures like 'Sahadeva Pasu Vaidya Sastramu' (Telugu) and 'Mattu Vaidya Bodhini' (Tamil) are dealing with traditional veterinary medicine. Apart from the above, hundreds of references on treating animals with the locally available plant resources can be traced in vernacular articles which do not have the authenticity and authorship of the literature. Furthermore, numerous undocumented veterinary practices are in vogue in rural areas of India. Although Ayurvedic veterinary treatments are well documented in various Sanskrit scriptures, language barriers prevent their effective use.

Today we find a renewed interest in traditional medicine. During the past decade there have been an ever-increasing demand especially from developed countries for more and more drugs from plant sources.

This revival of interest in plant-derived drugs is mainly due to the current widespread belief that 'green medicine' is safe and more dependable than the costly synthetic drug many of which have adverse side effects. This resurgence of interest in the plant based drugs have necessitated an increased demand of medicinal plants leading to over-exploitation, unsustainable harvesting and finally to the virtual decimation of several valuable plant species in the wild. Moreover, the habitat degradation due to increased human activities (human settlements, agriculture and other developmental programmes, illegal trade in rare and endangered medicinal plants, and loss of regeneration potential of the degraded forests have further accelerated the current rate of extinction of plants particularly the medicinal plants.

In remote areas , livestock owners practiced traditional treatments because of no side effects ,Low costs and lack of modern veterinary facilities.

The diseases and conditions commonly treated with EVM included mastitis, fever, bloat, diarrhoea, and foot-and-mouth disease. Especially with husbandry-related problems such as mastitis, farmers believed that no external expertise was necessary but that they could manage treating these ailments with EVM. But the level of effectiveness of the treatment depended on the stage and severity of the disease. Nearly all farmers and traditional healers regarded fever as a disease. They were satisfied with their treatment for fever.

Neem (*Azadirachta indica*) and tamarind (*Tamarindus indica*) were the most popular treatment plants. Both were commonly available in every village. All parts (bark, leaf, root, fruit etc.) were utilised in the preparations. Other plants commonly used included the spices: pepper (*Piper nigrum*), ginger (*Zingiber officinale*), turmeric (*Curcuma longa*), and garlic (*Allium sativum*). In contrast to other reports, the farmers did not seem to use any of the following plants for animal treatment: thulasi (*Ocimum sanctum*), adalodakam (*Adhatoda*), kurumthotti (*Sida rhombifolia*), thippali (*Piper longum*), papaya (*Carica papaya*), and areca nut (*Areca catechu*).

In general, in India, the commercialisation of herbal medicines may be connected with the long tradition of Ayurvedic medicine in this country. Ayurvedic scripts contain information on many plants and are often cited as an indication for the efficacy of these plants. Still, commercialized herbal medicine may be expensive for smallholders, especially when compared to self-made drugs[2]. And if commercial herbal drugs are exported to other countries, they may there become nearly as expensive as other imported allopathic drugs. Unfortunately, there is little information on the economics of commercialized herbal drugs versus ethnoveterinary remedies prepared from scratch at the field level.

In India, ethnomedico-botanical surveys and inventories receive little attention and detailed information and documentation on the uses of medicinal plants in indigenous communities are lacking. This disinterest in the existing knowledge and neglect by authorities and public slowly result in the loss of this knowledge. It is high time that effective measures and programmes are adopted to save this information before it is lost forever. Ethnomedico-botanical surveys in each district are needed. Comparing and crosschecking the results will provide us with information about herbal medicines presently in use. Furthermore, concerted efforts are needed towards the usage of plant resources, linking conservation strongly with utilization.

Validation of Indigenous Technical Knowledge (ITK)

To prepare a package of simple, effective traditional veterinary treatments for rural India and require to validate treatments on an empirical basis by keeping quantitative and qualitative records of cases and their treatments. Documenting and validating ethnoveterinary practices is a first step towards preserving and promoting them

ICAR New Delhi, under a Net work project (NATP) on collection and documentation of Indigenous technical knowledge in the field of treatment of animals and subsequently their scientific validation to find out potential ITKs for their use by farmers and drug development by pharmaceutical Industry. As a result, Inventory have been prepared and published for scientist of the country.

Strategies and approaches for exploiting the existing potentials of ethnoveterinary medicine: It mainly composed of (i) Establishment of a central apex body for the formulation of policies and programmes and identification of educational and training needs of the system to popularise ethnoveterinary medicine in India, (ii) Setting up regional work stations for the collection and documentation of information on prevailing ethnoveterinary practices and imparting education and training for the people who are interested in ethnoveterinary medicine, (iii)Standardization of ethnoveterinary techniques and the dissemination of standardized techniques into areas with similar agro-climatic conditions through a well-designed extension system, (iv)Identification of critical states in which the ethnoveterinary system would have to play a lead role with appropriate infrastructure facilities and related incentives, (v) Translation of information in vernacular languages into Hindi and English to have a fruitful debate in national and international forums with the objective to refine the techniques and also to ensure the geographical spread of such techniques, where feasible,(vi)Establishment of improved databases on livestock population, veterinary institutions of the government, co-operative and private sectors, practicing veterinarians, and ethnoveterinarians and (vii) Quantification of the veterinary infrastructure along the lines of the analytical approach.

3. *Medicinal Plant/Herbal Therapy*

Interest in medicinal plants as a re-emerging health aid has been fuelled by the rising costs of prescription drugs in the maintenance of personal health and well being and the bio-prospecting of new plant-derived drugs[23]. Based on current research and financial investments, medicinal plants will, seemingly, continue to play an important role as health aid. Medicinal plants, since times immemorial, have been used in virtually all cultures as a source of medicine. The widespread use of herbal remedies and healthcare preparations, as those described in ancient texts such as the Vedas and the Bible, and obtained from commonly used traditional herbs and medicinal plants, has been traced to the occurrence of natural products with medicinal properties. The use of traditional medicine and medicinal plants inmost developing countries, as a normative basis for the maintenance of good health, has been widely observed (UNESCO, 1996). Furthermore, an increasing reliance on the use of medicinal plants in the industrialized societies has been traced to the extraction and development of several drugs and chemotherapeutics from these plants as well as from traditionally used rural herbal remedies (UNESCO, 1998). Moreover, in these societies, herbal remedies have become more popular in the treatment of minor ailments, and also on account of the increasing costs of personal health maintenance. Indeed, the market and public demand has been so great that there is a great risk that many medicinal plants today, face either extinction or loss of genetic diversity. As defined by WHO, health is a state of complete physical, mental, and social well being and not merely the absence of disease or infirmity.

There are certain examples for extensive use of plants such as *Phytolacca dodecandra* extracts , commonly known used as an effective molluscicide to control schistosomiasis[15]. Other notable examples are *Catharanthus roseus*, which yields anti-tumour agents such as vinblastine and vinvristine; and *Ricinus communis*, which yields the laxative—castor oil. The Maltese islands constitute an apt example where medicinal plants are widely used in every day life as part of folk medicinal remedies[14].

Encompassing concepts and methods for the protection and restoration of health, traditional medicine has served as a fount of alternative medicine, new pharmaceuticals, and healthcare products. The best known examples of traditional medicine, differing in concept and protocol, are well-developed systems such as acupuncture and ayurvedic medicine that have been widely used to conserve human health in China and India in heman beings, however, little used in veterinary practice, though homeopathy is gaining momentum. Developed countries, in recent times, are turning to the use of traditional medicinal systems that involve the use of herbal drugs and remedies.

Traditionally in animals practice, farmers and pastoralists in several countries use medicinal plants in the maintenance and conservation of the healthcare of livestock. Intestinal disorders in cows, in Mexico, are treated with herbal extracts of *Polakowskia tacacco*. Dietary supplements such as vitamin A in poultry feeds in Uganda are supplied through enrichments of amaranth (*Amaranthus sp.*). It is estimated that medicinal plants, for several centuries, have been widely used as a primary source of prevention and control of livestock diseases. In fact, interest of such use in the veterinary sector has resulted primarily from the increasing cost of livestock maintenance and the introduction of new technology in the production of veterinary medicines and vaccines[24]. Recent and renewed interest in medicinal plants coupled to developments in information technology has fuelled an explosion in the range and content of electronic information concerning medicinal plants as a re-emergent health aid. It was recently reviewed diverse sources of such information in traditional abstracting services as well as in a variety of online electronic databases[7]. As a result of such developments, access to indigenous peoples and cultures concerning medicinal plants are greatly facilitated. Furthermore, the active participation of such natural custodians and practitioners of valuable knowledge is guaranteed in the generation of research focusing on screening programmes dealing with the isolation of bioactive principles and the development of new drugs.

The ethano botanical use of planrts are described in books and updated version of 80,000 plants are entered[10] are available on internet (http:// www.ars-grin.gov/~ ngrlsb). Many herbalists prefer using whole herbal extracts as opposed to the isolated active constituents of the herb[17]. These herbalist believe that by using the isolated active compounds, with some of other constituents that may act as synergistic are lost , there by decreases the efficacy . It was noted that evidence indicates that crude extract have greater efficacy than dose that active constituents. Some other believes that it is the best use the purified active ingredients. Many herbs contain ingredient toxins and other principle may decrease that activity and produce side effect[21].

In veterinary practice, mainly drugs are ayurevedic formulation/ feed supplement, however, plant based, but not marketed as herbal drug.

4. *Homeopathy*

Although in recent years we have witnessed a renaissance of interest in homeopathy in both human and veterinary practice. The reliability of its main principles is the 'simile' and the 'dilution/ dynamization' of medicines, has still to be demonstrated on the experimental ground and few studies have been conducted to understand the underlying mechanism(s). On the other hand, inspection of the literature and experiments carried out by the authors of this review show that the principle of similarity—brought back to its biological meaning, i.e. the inversion of effect of the same or similar compounds—can be found to be operative in various experimental and reproducible phenomena[14,6,20]. The concept of 'similia', according to which there is a similitude between the symptoms evoked by the medicine and the symptoms of natural disease; in animal research, this modality has been pursued utilizing both the single medicines and the complexformulations, even if the latter does not follow the classical Hahnemann's rules.(ii) 'Isopathy', according to which the same substance that causes the disease can be used in low doses or high dilutions to treat the disease; this concept is analog to the hormesis effect that was discussed in a previous article[5]; when the preparation is from pathological tissues or microbial products the term 'nosode therapy' is also used.(iii) 'Iso-endopathy', where therapeutic effects are obtained from highly diluted endogenous molecules (hormones, inflammatory mediators).

Application in veterinary treatment

Disease can affect an animal to a greater or lesser depth and therefore so can the so-called remedies of homoeopathy since they are also disease-producing agents when administered to a healthy body. These different levels of prescribing can be classified as: Pathological, local, organ-specific, institutional, regulatory, detoxifying, historical; specific and preventive.

Some of the medicines are used in treatment of different condition ,may be given orally, by injection, topically or orally via the drinking water.) -

Acetonaemia (Lycopodium, nux vomica), Arthritis (Bryonia, caulophyllum, phytolacca, rhustox, ruta graveolens) , Bloat (Carbo veg., lycopodium, nux Vomica) , Calf diphtheria (Mercurius cy.),Constipation (Nux vomica, sepia),Convulsions (Belladonna, nux vomica)Dystokia (Caulophyllum, pulsatilla), Fever (Aconitum, belladonna,pyrogenium),Foot-Rot(Hepar sulplllris) ,Haemorrhage (Aconitum, arnica, Phosphorus), Infertility (Apis mell., aurum, pulsatilla, sepia), Injury(Aconitum, arnica, calendula, hypericum, rhus tox.,rutagrav., symphytum Hepar sulphuris) Joint ill (Hepar sulphuris),Mastitis (Aconitum, belladonna, bryonia, carbo veg., hepar sulphuris, mercurius sol.,phytolaeca, .silicea, urtica), Metritis (Caulophyllum, pulsitilla,pyrogenium, sabina, sepia) ,Navel ill (Hepar sulphuris, silicea), Papillomatosis (Thuja) Placenta (retained) (Caulophyllum, sabina, sepia), Pneumonia Aconitum, antimonium tart.,bryonia, lycopodiumn,phosphorus, pyrogenium[8].

5. Veterinary Acupuncture

This Chinese approach to disease is very holistic, and emotional, hereditary and environmental factors are considered important factors in disease patterns. The philosophy and aim of Traditional Chinese Medicine (TCM) is to restore equilibrium between physical, emotional and spiritual factors, thus restoring and maintaining health. Treatment involves using needles in specific acupuncture points (often in combination with herbal therapy) to achieve this balance, by addressing imbalances in Yin and Yang and the flow of Qi and Blood, etc. It based on following principle[3].

1. Immune inflammatory reaction due to physical act of traumatizing tissue as it stimulated causing a generalized response.

2. Peripheral neural stimulation at the acupuncture point, modulating central neural regulation and causing more specific response.

3. Psychological response.

   It can be performed by many ways[1].

   a. Acupressure- transdermal pressure therpy

   b. Temperature Variation (Moxibustion) by heat and cold therapy

   c. Ultrasound,- Sona puncture, 10- 30 Sec per point require by a probe

   d. Aquapuncture – Injection of solution in to acupuncture point by hypodermic needle

   e. Electro-acupuncture : electric devise used to stimulate point (Laser puncture on similar way)

Acupuncture is used predominantly in the treatment of musculoskeletal disorders and, in particular, chronic pain states in animals. The effect of acupuncture is largely segmental, that is nerves, muscles and acupuncture points are needled in the same spinal segmentas the affected or painful area. "Pain gating", as well as descending pain inhibition (via various brain and spinal cord pathways), are involved. Acupuncture also stimulates the release of pain relieving chemicals in the brain and spinal cord (endorphins, seratonin, noradrenalin, etc.) which produce more generalised or extra segmental analgesia. These effects, combined with local needling of painful trigger points in taut muscle bands, result in exceptional relief of pain.

Acupuncture anesthesia was first developed in humans in 1958, and first applied to horses and donkeys in 1969. Analgesia was first used outside of China on humans in 1972 in Austria, where the first surgery for a tonsillectomy in the Western world was performed.

The National Association for Veterinary Acupuncture was formed in 1973; the International Veterinary Acupuncture Society was founded in 1974. By 1975, symposiums were being conducted by the Chinese Academy of Medicine and held at the University of Cincinnati, University of Georgia, Purdue University, University of California

School of Veterinary Medicine, and others. Since those early days in the United States, veterinary acupuncture has become increasingly popular, with more than 500 certified doctors of veterinary medicine as practitioners in North America alone. Its clinical use ranges from equine to feline to canine and avian. The success rate borders on the astonishing for a variety of conditions both known and unknown to the human patient. Most licensed acupuncturists' state licenses deal only with the treatment of the human and do not include animal applications.

In veterinary practice, It is used in small animals treatment , particularly in treatment of paresis , paralysis , painful condition , ataxia, and epilepsy etc. In a caesarean section of a ewe was done under aquapuncture anesthesia[13]. Veterinary acupuncture is practiced globally and has reached to level of acceptance by some section of the veterinary establishment of country.

6. *Physical Therapy*

It involves physical measure in the treatment of diseases. It includes electrical stimulation, ionotophoresis, ultrasound, Therapeutic heat/ cold, compression, stretching, therapeutics laser (LLLT- Low level laser therapy). Physical therapy has certain limitations. It has the potential for doing harm[9,11].

7. *Massage Therpay*

It is used to increase circulation, release scar tissue, balance muscle function and relax. Technique mainly adopted are trigger point massage (Myotherpy), Acupressure massage, support massage. Craniosacral therapy. Ice massage, tapolment, Deep cross - fiber friction etc[18, 22,25].

In all countries, including India, the interest is growing tremendously to use all approaches judiciously to treat the animals. Besides, allopathic (Modern medicine), ayurvedic, Homeopathy, Herbal medicine, Traditional medicine are now commonly practices, However, other ways of treatments like acupuncture is not practiced in our country, might be because of poor academic knowledge of acupuncture and not a part of syllabus of veterinary degree. Homeopathy is in practice in country, but need more rational attention considering it suitability in organic farming of animals.

References

1. Altman, 1994 *Techniques and Instrumentation,* In Schoen A E, Veterinary Acupuncture, : ancient and Modern St loius Mosby

2. Anjara, J. 1996 *Ethnoveterinary Pharmacology in India: Past, Present and Future.* In: *Ethnoveterinary Research and Development.* eds. McCorkle, C.M., Mathias, E. and chillhorn van veen, T.W., Intermediate Technology Publications, London, UK, pgs. 137 – 147.

3. Beasler DE and Kroening, R J 1976 *Three essential factors in effective acupuncture therapy* (Fide : Schoen, AM and Wynn S G (1997) *Complementary and alternative veterinary Medicine* . Mosby, St loius)

4. Bellavite P, Andrioli G, Lussignoli S, Signorini A, Ortolani R, Conforti A 1997 A scientific reappraisal of the 'principle of similarity'. *Med Hypotheses* 49:203–12.

5. Bellavite P, Conforti A, Pontarollo F, Ortolani R. Immunology and homeopathy. 2006 Cells of the immune system and inflammation. *Evid Based Complement Alternat Med* ,3:13–24.

6. Bellavite P, Signorini A 2002 *The Emerging Science of Homeopathy.* Berkeley (CA): North Atlantic.

7. Bhat, K K S 1997. Medicinal and plant information databases. In: *Medicinal Plants for Forests Conservation and Health Care.* eds. Bodeker, G. and Vantomne, P.,FAO, Non-Wood Forest Products Series No. 11, FAO, Rome, pgs. 158.

8. Days, C E I, 1994) Bovine Medicine ed AH Andrews, Mosby publication

9. Downers A 1978 Physical therapy for animals , Springfield III Effect of therapeutic temperature on tendon extensibility. *Arch Phys Med Rehab.* 51:481

10. Duke Wain, 1981 (Medicinal plants of the world) are available on internet (http://www.ars-grin.gov/~ ngrlsb

11. Gentzkow, G D 1993 Electrical stimulation to heal dermal wound . *J. dermatol surgery Oncol,* 19:753

12. Kemper, 1996 Kemper KJ Separation or Synthesis: a holistic approach to therapeutics pediatrics review 17:279

13. Lakshmipati O R and others 1984. Caesarean section in an ewe was done under acupuncture anesthesia, 12:161.

14. Lanfranco G.1992. Popular Use of Medicinal Plants in the Maltese Islands, *Insula,* 1: 34 – 35.

15. Lemma, A 1991. The Potentials and Challenges of Endod, the Ethiopian Soapberry Plant for Control of Schistosomiasis. In: *Science in Africa: Achievements and Prospects,* American Association for the Advancement of Sciences (AAAS), Washington, D.C., USA.

16. Micozzi, M 1996 editor Fundamental of complementary and alternative medicine, New york , Churchills , Living stone.

17. Mills S 1991 *The essentials of book of herbal Medicine,* Arkana London.

18. Porter M 1990 *Equine sport therapy* Wildomar calef veterinary data.

19. Schoen, AM and Wynn S G 1997.*Complementary and alternative veterinary Medicine* . Mosby, St loius.

20. Teixeira MZ 2005 Paradoxical strategy for treating chronic diseases':a therapeutic model used in homeopathy for more than two centuries.*Homeopathy* 94:265–6.

21. Tyler , V E , Brady L R and Robbers , E 1988 Pharmacognosy ed Philadelphia , Lea fabiger.

22. Walker J 1984 Deep transverse friction in ligament healing. *J ortho sports physical therapy* 6:89.

23. Walker, A F 1996. Of Hearts and Herbs, *Biologist* 43:177 - 180.

24. World Bank 1997. *Medicinal Plants – Rescuing a Global Heritage.* eds. Lambert, J., Srivastava, J. and Vietmeyer, N., Technical Paper No. 355, pgs. 61.

25. Zidonis, N, Soderberg M and Pedersons , S 1991 *Equine acupressure* , Panercolo, Equine pressure).

26

Nutritional Disorders Associated with Macro and Micro Elements in Goats

N P SINGH, V S VIHAN AND ASHOK KUMAR

Central Institute for Research on Goats
Makhdoom, Farah, Mathura (UP) - 281 122

In goats , nutritional requirements are for various types of production such as maintenance, growth, fattening, milk production for raising kids, maximum milk production , and mohair or cashmere production. Role and contribution of goats and sheep in developing country has been prominent. They have occupied a place of honour in there long association with man over the millennia. India ranks first with regard to its goat population of 123 million in the world .The intensification of animal agriculture have created complex animal health and production problems. This poses an even increasing emphasis on production diseases in food producing animals. Intensification of sheep and goat industry is likely to result in an increase in the number of high producing animals. The intensive method of management and feeding may result in a stress and animal becoming vulnerable to diseases. In order to meet the high milk yield and body weight gain, a commensurate balanced nutritional ration is to fed the high potential animals. This lead to graver risk of disease that go far beyond the specific disease conditions in terms of economic gain expected[10]. Goat has tremendous capacity for digestion of fibrous and woody substances like grasses, leaves, small twigs and is a very efficient feed converter with a far larger forage capacity that cows or sheep (25-40% of body weight, as compared to 12.5-15% for cows,

12.5-20% for sheep), which are strictly grazers. Goats are have a higher digestive efficiency for cellulose than cattle, buffaloes and sheep and their inquisitive feeding behavior is particularly significant in areas where the vegetation is coarse such as in many semi arid and arid areas[1]. Nutritional disorders in goats are common in several parts of world including India though well-defined deficiencies are not common in goats. Countries providing informations on animal diseases, 80% reports on nutritional disease of moderate or high incidence and trace element deficiency or toxicities are involved in more than half of those whose cause were unidentified. The soil- plant- animal interaction in relation to the incidence of trace element deficiencies in livestock are being examined.

The trace element are involved as component parts of many tissues and one or more enzyme activities and their deficiency leads to a wide variety of pathological consequences and metabolic defects[26].

A. Mineral and vitamin deficiency

1. Copper and Cobalt

All these trace element have their main function to synthesize hemoglobin and prevent nutritional anemia. From dietary point of view, copper and cobalt are deficient in feed in India and their requirements are not met unless supplemented. Milk is deficient in iron and copper; therefore, its supplementation is important from neonatal feeding point of view. There are problem areas where feeding are not able to provide adequate quantities of these element to provide adequate quantities of these element to plants and inturn bring sub clinical and clinical syndrome[19] . Excess in molybdenum in certain areas may bring about indirect copper deficiency in lambs and kids and they suffer from sway back, a paralytic syndrome and enzootic ataxia are related to defective myelination. Molybdenum interferes metabolically in the copper intake by liver cells and this may disturb the primary intracellular metabolism of ceruloplasmin and other Cu protein compounds[18]. Low-level copper in lush green pasture compared with conserved forage and inhibitory effects on absorption of molybdenum and sulfur antagonism from iron ingested in soil[33]. In adult goats, Abortion and still birth, anemia, frequent bone fracture, poor appetite , weight loss, decreased production are the symptoms of copper deficiency related to decreased cytochrome oxidase activity. Deficiency symptoms occur when dietary copper is less than 7 ppm and Mo is normal, there fore a suitable ration formulation is 10 ppm[16], and it is generally recommended that to keep the Cu:Mo ratio above 2:1 and below 10:1[8]. Enzootic anemia due to copper deficiency has been reported in young goat kids. Kids may be affected at birth or the clinical signs may be delayed until the animals are several weeks of age. Cerebellar hyperplasia is a frequent finding in goats. Adult goats are not as susceptible to copper toxicity as sheep[32]. Younger kids are more susceptible to copper toxicity because pre ruminant kids absorb copper more efficiently than adult goats. Cobalt is a component of Vita min B_{12} and in the absence of cobalt rumen microflora can not synthesize the essential vitamin and also required as a coenzyme for transformation of propionic acid to succinic acid. The disease is caused by deficiency of cobalt in the diet and characterized by anorexia, poor production, weight loss, anemia and, diarrhea may be noted and increased susceptibility to strongyls[34]. White liver disease, (Hepatic dystrophy) with low Vitamin B_{12} has been reported in young goats in New Zealand and responded to Vitamin B_{12} and cobalt therapy[5]. Similar to sheep, 0.1 ppm of cobalt in the diet is necessary in goats. Primary cobalt deficiency occurs only on soil deficient in cobalt. In such areas, salt should be fortified with cobalt chloride or cobalt sulphate at the rate of 12 g/ 100 kg salt and alternatively to dietary supplementation, top dressing of deficient pasture with cobalt salt may be instituted.

2. Iodine

Iodine is as essential in caprine diet as it is in human. Iodine deficiency has been reported from hill arrears in India such as Himalaya region. , Jabalpur and even Delhi region etc, where goiter cases have been reported. The major clinical manifestation is neonatal mortality, with alopecia and visible and palpable enlargement of the thyroid glands. The iodine deficiency may be primarily due to deficient intake or secondarily by a high intake of Calcium. The diet high in linseed meal (20% of ration) given to pregnant goats may also result in a high incidence of goitrous in kids. Soyabeen products are also considered to be goitrogenic. Iodized salt is a good preventive treatment against goiter, its requirement is 0.1ppm and it should be given with caution because high level of iodized salt in diet may also cause goiter, if given in excess. Potassium Iodide and sodium iodide are good iodine supplement for livestock. Certain crops which are high in levels of goiterogens would need supplementation of extra iodine as a preventive measure against goiter.

3. Iron

Unless goats are anemic, iron deficiency in caprine not a problem in foraging goats. Stomach worm (*Heamonchus*), suckling lice and blood loss are the most common cause of anemia in goats. An increased proportion of *Allium* (onion type) plants in the diet can also cause iron deficiency because n- propyl disulphide in these plants induces hemolytic anemia. 30 mg/ kg dry matter (ppm) in the ration is the standard ruminant recommendation. Young kids on all milk diet routinely become deficient and could be treated with iron dextran (150 mg/kid at 2-3 week interval) if diet modification is not possible[7].

4. Sulfur

Sulfur is the part of sulfur containing amino acid such as methionine and cystine. Dietary supplementation of Cashmere goat with dl-methionine increased hair yield much more than it increased cashmere yield[4]. Deficiency sign are poor growth rate, salivation, lacrimation and alopecia with marked deficiency. Consider offering sulfur in the form of a sulfur-salt block free choice.

5. Magnesium

Magnesium has received far less study than have calcium and phosphorus, although the metabolism of these mineral is interrelated[34].It is required for many enzyme system, necessary for energy metabolism. The physiological levels for goats are 2.8-3.6 mg/ dl. Deficiency of magnesium may lead to anorexia or hyperexciatibility. Goats deficient in Mg have lowered urine production and milk production, which seems to a compensatory mechanism in deficient state. Magnesium tetany has been recorded in livestock in India. Though "Grass stagger" problem did not seem to be prevalent. Magnesium requirement is 0.35% of the total DM intake. Grasses growing rapidly in cool, wet weather or after heavy fertilization are often low in magnesium, need special attention in feeding strategy. Whole goat milk also occasionally supplies inadequate magnesium, and kids might develop tetany, if not supplemented[14].

6. Zinc

Zinc content of the body is only approximately 3 mg% because it is not stored. Certain soils have been declared deficient in Zinc because of intensive cultivation and no attention has been paid towards its supplementation. It causes deficiency disease in plants and animals named "Khera" disease in paddy and "parakeratosis" in livestock. Sub-optimal levels may lead to poor growth, testicular size and spermatogenesis reduced. There is a marked delay in wound healing. Apart from its function as a co- factor in activation of many enzymes, it is helping in mobilization vitamin A from liver, which is the storage organ for Vitamin A after absorption[34]. It is also a factor in preventing Vitamin A resistant night blindness[2]. Zinc is considered essential for the optimum rate of microbial fermentation and is deficient in milk which makes it all the more important as a dietary essential for neonates.

Normal diets in certain deficient areas including India do not contain more than 25 ppm Zinc, whereas the requirement is 70 ppm[9]. Further more, its requirement will very depending upon the Ca level and phytic acid in the diets. The oral administration of Zinc at the rate of 250 mg zinc sulfate daily for 4 Weeks resulted in a clinical cure of zinc deficiency in goats in 12-14 weeks.

7. Fluorine and Molybdenum

Both of these elements are dietary essentials and have physiological function in the animal's metabolism. At higher levels, they are toxic. Up till now molybdenum toxicity has not been recorded in India where as fluorine toxicity is quite widespread in certain states such as Andhra Pradesh, U.P. and Rajasthan involving all species of animals including human beings. Fluoride is one of the potent anions which is considered essential trace element, but with a narrow margin of safety. It is established that where fluorine occurs in abnormal amount in soil or water, a chronic endemic fluorosis, Characterized by mottling, excessive wear of teeth and osteoporosis is likely to exist[35]. Some investigations are of opinion that of al pollutant which affects farm animals, fluorine has caused most severe and wide spread damage[3]. Dental lesions in young ruminants which occur on the intake of 10-14 ppm DM of hay contaminated. The osseous changes fluorine has been described as osteosclerosis, exostosis, hyperstoses, osteomalacia and rickets. Fluoride might act as a possible mutagen in the ova of animals taking excess of it with feed and water[15]. Goats are also susceptible to high fluoride levels causing dental lesions and its accumulation in bones and teeth, while the levels in soft tissues and milk remain unaltered and the reproductive performance of adult and the growth of kids were unaffected[21,22].

8. Manganese

Deficiency of manganese in goats consisted of deformed forelimb, reluctances to walk and reduced fertility and abortion, and reduced growth rate in growing kids.

9. Sodium, Potassium and chlorine

All these three elements occur in soft tissues and body fluids to maintain osmotic pressure, acid base equilibrium, controlling the passage of nutrients into cells and water metabolism. Sodium and potassium are not stored in the body; therefore, their daily supply in the feed is essential. Subclinical syndromes to effect growth and milk production and even infertility use to occur due to their deficiency. Both are required 0.2% each of total DM intake. Sodium which is commonly supplemented as NaCl is required at a level of 0.5% of the total DM. Goats out on forage usually get all the potassium they need. Emaciation and muscle weakness are signs of severe potassium deficiency.

10. Calcium, phosphorus and vitamin D

A dietary deficiency or disturbance in the metabolism of calcium, phosphorus or vitamin D, including imbalance of calcium phosphorus ratio, is the principal cause of the osteodystrophies. A severe imbalance of calcium and phosphorus (relatively too much phosphorus from mostly grain diet) can cause osteodystrophia fibrosa in goats, against the normal Ca: P ratio of 2.2:1, which clinically characterized by soft enlargement of lower jaw, commonly encountered when owners increases the grain in the diet with the thought of speeding the younger's goats growth. The normal level of calcium and phosphorus for goats is approximately 9.0-11.6 mg/dl and 4.2-9.9 mg/dl respectively. Hypocalcaemia in goats is occurring most commonly at the time of parturition[41]. It is characterized by low levels of blood calcium, general muscular weakness, circulatory collapse and depression of consciousness, low energy and low protein feed given to goats leads to decreased retention of calcium and phosphorus[6]. The maternal demand for calcium and phosphorus; and their absorption increases during pregnancy and lactation as a result of additional requirement for the fetus and for milk[40]. Hypocalcaemia as one of the possible cause of non - dilation of cervix in particular in parturient goats[24]. A primary deficiency due to the lack of calcium in the diet seldom occurs, although a secondary deficiency due to marginal calcium intake aggravated by high phosphorus intake is common. Phosphorus deficiency is usually primary under field conditions but may be exacerbated by deficiency of vitamin D and possibly by an excess of calcium. Vitamin D deficiency alone in young growing kids, particularly raised in dark conditions develop rickets, rachitic rosary on the ribs, enlarged joints and bowed limbs etc. Growth rate and body condition are poor. Adult deficient in vitamin D develop' osteomalcia or osteoporosis.

11. Vitamin B

There are two instances where B vitamin deficiencies have primary clinical importance. One with cobalt deficiency and other associated with Vitamin B1 (Thiamin) .Thiamine is made by the normal bacteria in the rumen. Kids or does on high carbohydrate may have an upset in normal flora. A change in bacterial type may cause a deficiency of thiamine or production of an enzyme which inhibit thiamine activity, resulting polioencephalomalacia (softening and necrosis of grey matter of the brain). Overdosing of amprolium in the treatment of coccidiosis and exposure to high level of sulfur in diet may result same clinical situation. Such cases are responded to thiamine injection.

B. Deficiency / imbalance of carbohydrate, protein and energy

Intensive method of management and feeding may result in the environmental stress on the metabolic health of the animals rendering them more prone to the metabolic disorders. It is not surprising that production diseases/metabolic diseases are becoming increasingly serious problems with intensive husbandry/practices. Amongst domestic farm animal, the metabolic diseases achieve the greatest importance in pregnant and lactating sheep and goat. Metabolic diseases such as pregnancy toxemia and hypomagnesaemia, hypocalcaemia are attributable to an imbalance between the rates of input of dietary nutrients and the output of production. The metabolic disorders achieve their greatest importance in pregnant and lactating goats. The effect of pregnancy is important in the she goats particularly those carrying more than one kids. The bulk of economic losses are due to the loss of production and failure to return to full production after recovery in production disease in farm animals including sheep and goats.

In small ruminants, most of these production related disorder are mainly caused by dietary irregularities viz. inadequate feed, irregular feeding, sudden change in feed, spoiled food, change in rumen microbial population indigestible substances and inadequate nutrients etc. Because of the difference in the soil type, climate condition, livestock husbandry practices and difference in the feed and fodder among counties, the incidence and gravity of disease is also likely to vary country to country. Peculiar problem of sheep and goat industry in India, like hot climate, course roughages, irregular availability of fodder, imbalance of nutrient causing sub clinical starvation lead to the metabolic diseases, particularly in pregnant and lactating animal. Hence in most cases, the disease picture is so complicated that the diagnosis becomes difficult.

Pregnancy toxemia is thought to result from disruption of the dam glucose homeostasis mechanism in response to increased nutrient demand of the rapidly developing fetal placental unit. Commercial production system are comprised of a variety of nutrient metabolic, gastric ,physiological ,environmental, economic, health and managemental factors that singularly as a group influence the clinical expression of pregnancy toxemia[27].Pregnancy toxemia in goats is primarily due to impaired metabolism of carbohydrate and volatile fatty acids characterized by hypoglycemia and hyperketonemia. There is negative nutritional balance and dietary deficiency of net energy exacerbated by the increased demand for energy in the later part of pregnancy. The most important etiological factor in pregnancy toxemia is decline in the plain of nutrition during their last two months of pregnancy particularly in goats carrying more than kids[29,38,39]. The onset of clinical sign was always preceded by hypoglycemia and hyperketonemia. The value of measured glucose and ketone body concentration in the blood of ewes and goats and found average glucose concentration of 57.6 mg% in pregnant and 48.2 mg% in lactating goats. About 10% and 2% of blood sugar values below the hypoglycemia limit (30 mg %) in lactating and pregnant goats respectively[40]. In goats, the syndrome is of nervous form. There is apparent blindness, declination to move, salivation, grinding of teeth, and tonic -clonic convulsion. The only rational treatment in pregnancy toxemia is to relive the need for glucose formation from tissues and allow ketone body utilization by simply administration of glucose replacement therapy. Therefore, management practice, adequate concentrate, best quality roughage and avoidance of predisposing factors are the key points to prevent the occurrence of pregnancy toxemia/ lactation ketosis in ewes and does in farming system.

Supplementation of feed with Monensin (20 to25mg) or Nicotinic acid (0.8 gram) daily in sheep between 3 to5 days before lambing and 35 days after words prevented hypoglycaemia and ketonuria[12]. The benefit of Monensin feeding include less mobilization of body fat as evidenced by reduced level of NEFA, ketones and increased glucose levels [20].

Pregnant doelings need extra energy not only for their pregnancy but also to continue their growth sufficiently. Shortages of energy, especially under range condition are known to cause not only stunted growth but also abortion in goats. This may occur mostly between 90-110 days of gestation when under nutrition is especially critical to normal fetal development. So called Stress abortion is triggered by low maternal abortifaecient estrogen level and premature expulsion of a live foetus. Thus abortion can be prevented by proper sufficient nutrition, which needs attention in goat husbandry.

C. Nutritional causes of reproductive losses

There are no nutrients specifically required for reproduction which would not be needed for other physiological functions[37]. Reproduction requires most of the same nutrients that are essential for maintenance, growth and milk secretion. Nutritional factors that cause reproductive failures or reduced efficiency will also have detrimental effects on other physiological functions. The most important nutritional factors influencing reproduction are Energy, Protein, Phosphorus, vitamin D, Vitamin A, Selenium, vitamin E, Salt and trace elements. Fertility is impaired in drought seasons, by poor pastures, change to lower quality feedstuffs, high production,where an increase in energy supply is often followed by improved rates of ovulation and conception. Energy supplies have a marked effect on age of puberty and thus on age of first kidding. Pregnant doelings need extra energy not only for their pregnancy but also to continue their growth rate sufficiently. Shortages of energy, especially under range conditions, are known to cause not only stunted growth but also abortion in goats. So called stress abortion is triggered by low maternal blood glucose levels which initiates hyperactivity of the fetal adrenal gland resulting in elevated abortifacient estrogen level and the premature expulsion of a live or fresh fetus. The relationship of protein to reproduction is similar to that of energy and the two nutrients interact to a large extent. Even when energy supplies are adequate, a shortage of protein will impair fertility, cause delayed onset of puberty, lengthen anestrus of goats and result in weak expression of estrus if it occurs. Phosphorus deficiency is more likely than a calcium deficiency in grazing goats because of phosphorus deficient forages. Adequate phosphorus supplementation for high producing dairy goats is more critical. A level of 0.4P in the total ration is recommended. Wide Ca:P ratios and to phosphorus deficiencies; such as low first service conception rates and silent heats. Vitamin D has also been implicated through its effect on phosphorus utilization. Vitamin D supplementation is advised for young, poorly growing kids and goats in confinement and exposed to little sunlight. Vitamin A Deficiencies ,its carotene precursors or interference in their conversion all are implicated in reproductive problems in goats, although more studies exist on cattle and sheep, and species differences have been noted. Vitamin A is essential for normal spermatogenesis in quantity and quality. It is also essential in combating various respiratory and gastro-intestinal diseases. As a result, dead or weak kids may be born; even abortions or retained placenta may occur. Newborn kids may have low vitamin A liver reserves and suffer high mortality. Commercial supplementation of vitamin A is relatively inexpensive, as is that of vitamin D or E, which all three are usually provided in commercial feeds .Retained placenta can be a selenium and vitamin E responsive disease when not caused by mechanical or pathogenic factors. Specific vitamin E roles in improving goat reproductive efficiency have been alleged for some time but reliable evidence is difficult to obtain. Salt is a convenient carrier also for the trace elements needed by goats for normal reproduction such as zinc, manganese, iodine, cobalt, iron, copper and sulfur. Zinc and manganese in particular, affect spermatogenesis, libido and oogenesis when deficient.

D. Metabolic/feeding disorders

1. Subabool feeding

Sole feeding of subabool caused anemia in goats[25] . In addition, varying degree of degenerative changes occur in liver, kidney, brain and ovaries. Changes in adrenal were indicative of stress. Thyroid exhibited goitrous changes. Reproduction problems (Abortion, non-conception) were correlated with hypothyroidisim which was caused by the goitrogenic effect of subabool.

2. Urea toxicity

As similar to cattle, urea can be fed to the goats as NPN source which converted in to microbial protein in rumen, if adequate energy sources are available[13]. Urea toxicity can occur in over mixing of urea in feed or improper mixing and / or consumed at the level of 30-50 g/ 100 kg bwt in a single meal. Excess urea is degraded in to ammonia and absorbed systemically and further converted in to urea, and cause interference in to citrate cycle[17]. The main clinical signs consisted of muscle and skin tremors, salivation, frequent urination and defecation, incoordination, bloat and death. Urea toxicity can be best managed by administration of acetic acid (Vinegar) @ 200-500 ml by stomach tube, which reduces the rumen pH and retard absorption of urea.

3. Ruminal acidosis

Most commonly, acidosis develops due to an imbalance in the feeding program or because an animal, usually a growing kid and adult goats, is gorging itself on grain (grain overload). Acidosis is develops particularly in kids with ill developed rumen is not yet fully functional, where lowered gastric pH creates an ideal environment for the proliferation of harmful bacteria, such as *Clostridium perfringens,* resulting in potentially fatal enterotoxaemia. In adult goats which are being pushed for higher production often fed large amounts of concentrates resulting in a lowered rumen pH and preponderance of gram-positive bacteria and production of lactic acid and subsequently systemic acidosis. Only sub acute cases respond to standard therapy. Treatment depend upon the severity of stage included the administration of antacids, yeast and chlortetracycline, and the intravenous infusion of isotonic sodium chloride and 5 per cent sodium bicarbonate solutions. The critical level amount of grain that may induce rumen acidosis is variable. Goats are highly susceptible to acute ruminal acidosis induced by sorghum at dose rate of 50 g/kg B.wt.[23] , whereas, wheat at a dose rate of 100 g/kg B.wt. is required to produce the classical signs in goats[36].

In preventive strategy, good feeding management is the key to reduce the mortality rate due to rumen acidosis. Prophylactic approach have been attempted and demonstrated that lactic acidosis can be reduced by immunization against *Strptococcus bovis,* and that live Sb-5 vaccine is effective in invoking mucosal as well as systemic antibody responses in sheep[11]. Similarly it was suggested that the risk of lactic acidosis can be reduced by immunization against S. bovis, and that the immunization primed I/M is more effective than the immunization primed[31].

In general, nutritional disorders can only be prevented by offering balanced nutritional ration to minimize production losses.

References

1. Acharya , R M and Singh, NP 2004. The role of goats in conservation of ecology and livelihood security. *A.P. Veterinarian*, April : 26-33.
2. Arora, S P; Hatfield, E E: Hinds, F C and Garriegus, U S 1968. *Journal of Anim. Sci.* 27: 1765-66.
3. Aschbaker, PW 1973. *J. Air Pollution Assoc.* 23: 267-72.
4. Ash, AJ and Norton, BW 1987 *J. Agric. Sci. Camb.*, 109: 197-199.
5. Black, H; Hutton, J B; Sutherland and James, MP 1988 .*N.Z. Vet. J.*, 15-17
6. Braithwaite, G D 1976 *J. Dairy Res.* 43: 501
7. Bretzlaff, K., Haenalien, G., and Huston, E. 1991: Common nutritional problems. Feeding the sick goats. *In*, Large Animal Clinical Nutrition. Edited by J.M. Naylor and S.L. Ralston. St. Louis, Mosby-year-book, Inc., 351-355.
8. Buck, W B 1986 Copper- Molybdenum. In. current Veterinary therapy Food animal practice 2 edited by J L Howard, Philadelphia, W B Saunders Co, 437-439.
9. Chhabra, Aruna (Mrs); Arora, S P and Jaikishan 1980 *Journal of Anim. Sci.* 50: 879- 81.
10. Dirksen, G. 1981 Acidosis. In: Phillipson, A.T. (Ed.), Proceedings of the Third International Symposium on the Physiology of Digestion and Metabolism in the Ruminant. Oriet Press, Newcastle, England, pp. 612 -625.
11. Gill HS, Shu Q, Leng RA 2000. Immunization with *Streptococcus bovis* protects against lactic acidosis in sheep . *Vaccine.* 18(23):2541-8.
12. Goranov, K H, Gabrzovski, P Zhelyazkov and Popgeogieva, M 1984 *Veterinarnomeditsinski narki* 21:43-50.
13. Harmeyer, J and Marten, H 1980 J *Dairy Sci.,* 63:1707-1728.
14. Hines, T G; Jacobson, N L; Beitz, D C Littledeke, E T 1986 *J Dairy Sci.,* 8: 2868-2876.
15. Jagiello, G and Lin, J S 1974 *Archives of Environmental Health* 29: 230-35
16. Lamand, M and Oligoelements 1978 *Alimentation des ruminants.* Edit by R. Jarrige. Versailes, France, INRA pub, 143-159.
17. Lloyd, W E 1986 Urea and other Non protein nitrogen sources. In. current Veterinary therapy Food animal practice 2 edited by J L Howard, Philadelphia, W B Saunders Co, 354-356.
18. Mariciless, N A; Ammerman, C B; Valcecchi, R M; Dunavant, B G and Davis, G K 1969 *Journal of Nutrition* 99: 177-63.
19. Maynard, L A and Loosli, J K 1969 *Animal Nutrition*, 6th Ed. McGrew Hill Book Co. Inc Newyork.
20. McGuffey RK, Richardson LF and Wilkinson J.D. 2001 *Journal of Dairy Science* Vol. 84, No. Elect. Supplement pp E 194: E201.
21. Milhaud, G.; Zundenl, E and Crombet, M 1980b *Recueil de Medicine Veterinaire.* 156 : 37-47

22. Milhaud, G: Mathieu, E.; Chary, J F and Parodi, A L 1980 a *Recueil de Medicine Veterinaire*. 156: 211-18.

23. Mohamed Nour M.S., Abusamra, M.T and Hago B.E.D 1998 Experimentally induced lactic acidosis in Nubian goats: Clinical, biochemical and pathological investigations. *Small Ruminant Research* 31 :7- 17.

24. Popoff, M 1978 *Bulletin Mensuel de la societe Veterinaire Pretique de france* 62: 183-96.

25. Prasad, M C and Sastry, M S 198) National symposium on Pathology and Biotechnology in the diagnosis of diseases of livestock and poultry, IVRI, Izatnagar.

26. Radostits, O.M., Blood DC and CC Gay 1994 *Veterinary Medicine*, 8th edit ELSB, BT, London.

27. Rook J.S. 2000 *Vet. Clin. North. Am.* Equine part 7 (16) 393-397.

28. Saraswat, R C and Arora, S P 1972 *Indian J Anim. Sci.* 42: 358-62.

29. Sauvant, D and Morand-Fahr, P 1979 *Vet. Bull.* 49: 1569.

30. Schmidt, H. 1941 *Am. J. Vet. Res.*, 2: 373-389.

31. Shu Q, Gill HS, Leng RA, Rowe JB. 2000 Immunization with a *Streptococcus bovis* vaccine administered by different routes against lactic acidosis in sheep. *Vet J.* 59(3):262-9.

32. Soli, N E and Nafstad, I 1978 *Acta.Vet. Scand.,* 19: 561-568.

33. Suttle, N F1986 *Vet. Rec.* 119: 148-152.

34. Smith MC and Sherman DM 1984 Goat Medicine, Ist Ed, Lippincott, Williams and wilkins, Philadelphia.

35. Swarup, D 1989 Proceeding National Symposium on emerging diseases of livestock due to environmental pollution. IVRI, Izatnagr.

36. Tanwar, R.K., Mathur, P.D. 1983 Studies on experimental rumen acidosis in goats. *Ind. Vet. J.* 60, 499- 500.

37. Van Horn; U, Gainesville G F and W Haenlein U (2002 Nutritional causes of reproductive losses). Genetics and Reproduction Text 1 Pennsylvania State U., University Park.

38. Venkataraman, R. and Rajamani, S. (1975 *Indian Vet). J.* 52:69.

39. Vihan, V S 1983 Studies on metabolic derangement with special reference to pregnancy toxemia in sheep and goat. Ph.D. Thesis, CSAUVAT, Kanpur.

40. Vihan, V S and Rai, P 1983 *Indian Vet. Med. J: 3*: 1-8.

41. Vihan, V S and Rai, P 1985 *Indian Vet. J.* 62:573-576.

–▲▲▲–

Index

E

F

G

H

I

–ΔΔΔ–